ISBN 978-1-5280-1240-9
PIBN 10909294

1 MONTH OF
FREE
READING

at

www.ForgottenBooks.com

By purchasing this book you are eligible for one month membership to ForgottenBooks.com, giving you unlimited access to our entire collection of over 1,000,000 titles via our web site and mobile apps.

To claim your free month visit:

www.forgottenbooks.com/free909294

English
Français
Deutsche
Italiano
Español
Português

www.forgottenbooks.com

Mythology Photography **Fiction**
Fishing Christianity **Art** Cooking
Essays Buddhism Freemasonry
Medicine **Biology** Music **Ancient**
Egypt Evolution Carpentry Physics
Dance Geology **Mathematics** Fitness
Shakespeare **Folklore** Yoga Marketing
Confidence Immortality Biographies
Poetry **Psychology** Witchcraft
Electronics Chemistry History **Law**
Accounting **Philosophy** Anthropology
Alchemy Drama Quantum Mechanics
Atheism Sexual Health **Ancient History**
Entrepreneurship Languages Sport
Paleontology Needlework Islam
Metaphysics Investment Archaeology
Parenting Statistics Criminology
Motivational

THE AGRICULTURAL BOARD

The Agricultural Board, a part of the Division of Biology and Agriculture of the National Academy of Sciences—National Research Council, studies and reports on scientific aspects of agriculture in relation to the national economy. It was established in 1944 upon joint recommendation of the Association of Land–Grant Colleges and Universities and the Academy—Research Council's Division of Biology and Agriculture.

The Board has four primary functions: (1) to mobilize scientific talent from government, industry, and universities to survey the broad problems of agriculture and establish priorities for study of these problems; (2) to evaluate present policies and practices in agriculture in the light of current knowledge; (3) to determine trends in current research and select neglected areas most likely to yield profitable long-range results; and (4) to disseminate knowledge and expedite the application of research findings to technological practice, governmental policies, and socio-economic affairs.

Financial support for the meetings and publications of the Board is provided primarily by the Agricultural Research Institute, an organization composed of representatives of industry, trade organizations, academic institutions, and governmental agencies concerned with agriculture. Members of the Agricultural Board and of its committees serve without compensation beyond their actual expenses. Funds for the work of the Agricultural Board are received and administered by the Academy—Research Council.

The National Research Council was established by the National Academy of Sciences in 1916, at the request of President Wilson, to enable scientists generally to associate their efforts with those of the limited membership of the Academy in service to the nation and to science at home and abroad. Members of the National Research Council receive their appointments from the President of the Academy.

Receiving funds from both public and private sources, by contribution, grant, or contract, the Academy and its Research Council thus work to stimulate research and its applications, to survey the broad possibilities of science, to promote effective utilization of the scientific and technical resources of the country, to serve the Government, and to further the general interests of science.

Symposium on

MUTATION

and

PLANT BREEDING

Sponsored by the

COMMITTEE ON PLANT BREEDING AND GENETICS
of the
AGRICULTURAL BOARD

Cornell University, Ithaca, N. Y.
November 28 to December 2, 1960

Publication 891
NATIONAL ACADEMY SCIENCES—NATIONAL RESEARCH COUNCIL
Washington, D. C.
1961

Library of Congress
Catalog Card Number 61–60045

Foreword

THIS Symposium developed from the deliberations of the Committee on Plant Breeding and Genetics which had been asked by the Agricultural Board of the Division of Biology and Agriculture, National Academy of Sciences—National Research Council, to make a realistic evaluation of the present status and future prospects of the use of induced mutations in the breeding of improved varieties of plants. A comprehensive symposium in this broad area of research had not been held for several years. In the meantime many laboratories were active in both the theoretical and applied aspects of research in induced mutations. The deliberations of the Committee were directed primarily toward the research involving mutations produced by radiation. The scope of the program was broadened necessarily to include mutations regardless of origin, but with considerable emphasis remaining on those produced by radiation. The results of research in both the practical and theoretical investigations in genetics and breeding were emphasized. From the start of the Committee's deliberations it was felt that the results of the Symposium should be of immediate, as well as long-time value, to geneticists and breeders, and should bring their research into closer juxtaposition.

The meetings were held from the 28th of November through the 2nd of December, 1960. Afternoon and evening work sessions met during each of the first 4 days. Approximately 160 participants attended by invitation of the sponsoring Committee. The program consisted of 15 invitational papers, 5 formal discussions, and a formal résumé. In addition, 34 volunteer contributions were given during the work sessions. These are not to be published by the Committee, but brief abstracts were multilithed and distributed to those in attendance.

The sponsoring Committee gratefully acknowledges the financial support received from the United States Atomic Energy Commission, the National Institute of Health, the National Science Foundation, and the Agricultural Research Service of the United States Department of Agriculture. The encouragement of the Agricultural Board and the Agricultural Research Institute was helpful to the Committee in bringing the Symposium to fruition.

I wish to acknowledge the help of all members of the Committee on Plant Breeding and Genetics, and especially that of H. F. Robinson and G. F. Sprague who contributed greatly to the final development of the program.

I especially wish to record my own as well as the Committee's appreciation and thanks to all of the formal participants whose papers are presented here; to the discussants, M. M. Rhoades, B. S. Strauss, W. M. Myers, I. J. Johnson, and H. H. Kramer; to the Editor, James D. Luckett; and, finally, to S. G. Stephens who undertook the major task of developing the résumé, for their contributions to the Symposium.

R. P. Murphy, Chairman

Committee on Plant Breeding and Genetics

R. A. Brink	H. F. Robinson
W. M. Myers	W. R. Singleton
F. L. Patterson	G. F. Sprague

J. D. Luckett, *Editor*

R. P. Murphy, *Chairman*

Session I

The Nature and Characteristics of Mutations

R. P. MURPHY, *Chairman*
Cornell University,
Ithaca, N. Y.

The Nature of Mutations
in Terms of Gene and Chromosome Changes

JOHN R. LAUGHNAN[1]

University of Illinois, Urbana, Illinois

FOR THE BREEDER, variability of the biological entity, regardless of its source, is essential. But considered from an operational standpoint, his success in the selection of improved strains has depended to only an insignificant degree, or not at all, on an understanding of the ultimate basis for that variability. The fact that geneticists have so far been unable to provide a convincing demonstration of intragenic mutation means that, for the breeder, the question of whether, in a particular instance, variability is attributable to intragenic or extragenic causes must for the time being remain academic. Mainly then, his approach to the problem has been to deal with the potential variability available in his source materials and to develop operational techniques aimed at manipulating this variability to the advantage of his program. Thus, the main impact which genetics has so far had on breeding programs has been to provide general techniques based on the principles of transmission genetics and, in my opinion, it is our lack of understanding of the physiological actions and interactions of genes, including their capacities to change, which is mainly responsible for differences of opinion and earnest debate over the breeding method to invoke in a particular instance.

Nor is it always clear that a certain apparent degree of success in an improvement program is a function of the breeding technique employed. For one thing, superiority is often relative, difficult to assess, and sometimes embarrassingly short-lived. Then, too, nature is on the side of the breeder, and while he is therefore reasonably confident that he will finish with no worse than he started, he must, by the same token, ofttimes concede that his efforts may have been quite passive, alongside those of nature, in achieving a proved measure of success.

---

[1]Currently Guggenheim Fellow on leave, Biology Division, California Institute of Technology, Pasadena, California. Work reported here was aided by a grant from the National Science Foundation.

3

That there has been some success in the development of artificial methods to increase variability cannot be denied, and perhaps the introduction of various mutagenic agents is the outstanding example of this. However, I think it is safe to say that these agents have not had a significant impact on the development of improved techniques of selection, but are now viewed by the breeder primarily as a means of enhancing variability, with the added hope that an occasional spectacular, useful variant will appear. Rather, it should be anticipated that the use of these agents, along with other techniques available to the geneticist and plant breeder, in the interests of a better understanding of the nature and action of the genetic elements, stands to add far more to the development of improved breeding techniques than their direct use to enhance, through increased variability, the prospects of success in a conventional selection program.

But the traits of particular concern to the plant breeder are most often quantitative and are usually controlled by numbers of genes whose individual genetic analysis appears particularly unrewarding. Unless we hold with the idea that genes governing quantitative traits are unique in their actions, a premise that I consider indefensible, the reasonable alternative is to deal individually with genes having so-called qualitative effects.

For some time it has been apparent that extragenic events make up a considerable portion of the occurrences we call mutations, and while we were at one time accustomed to think of variability in nature as due to the reshuffling of genetic entities whose differences were ascribable ultimately to qualitative changes that we preferred to think of as gene mutations, a great deal of painstaking work has led to increased emphasis on extragenic events as more immediate sources of variability in a population. As increasing numbers of mutations are resolved as extragenic events, the classical gene mutation appears more and more to be an elusive phenomenon and some, perhaps, would even doubt its validity as a biological concept.

If, in fact, the propagation of coded genetic information from mother to daughter cells, from parent to offspring, is a far more accurate process than had earlier been considered, and we suppose that the intramolecular alteration we think of as gene mutation may range in frequency for individual loci from 10^{-6} or 10^{-7} to lower levels, it is obvious that, where technical difficulties place a limit on the size of population that may be analized effectively, our studies of the mutation phenomenon have unwittingly selected against the gene mutational event. We should be prepared to concede then that however important may be the ultimate qualitative genetic change, the rarity of its occurrence in natural and even in experimentally manipulated laboratory populations may make it relatively impregnable to attack compared with other types of changes which make a greater immediate contribution to variability.

I should like to consider here, in some detail, genetic analyses of certain derivatives of the A^b complexes in maize which, on the basis of conventional criteria, appear not to be extragenic in origin and which, for this reason, are of particular interest in connection with the problem of gene mutation.

The various forms of the A_1 gene in maize control the synthesis of varying amounts of anthocyanin pigment in the aleurone layer of the endosperm and in certain vegetative tissues. Under appropriate conditions they are also in control of the type of pigment deposited in cells of the pericarp. The phenotypes of aleurone and plant may range from deep purple, as in the presence of A, the type allele, through intermediate levels, to colorless aleurone and brown plant characteristic of individuals that are homozygous for the recessive a allele. In pericarp tissue the type allele A produces a red pigment and is dominant to recessive a, associated with brown pigmentation. The several alleles that go under the general designation A^b, and their mutant derivatives, are unique in that they have divergent effects which will not allow their placement in consistent linear array with sister alleles. Thus, A^b, which was first described in a stock from Ecuador (5)[2] has a brown pericarp phenotype that is dominant to A, yet is weaker than the latter in its effect on plant pigmentation (6).

Following the original finding (22) that A^b mutates to an intermediate allele designated A^d (dilute), with a frequency of about 5×10^{-4}, detailed analyses (7, 8) employing suitable marker genes have established that A^b consists of two closely linked, but separable, elements, both concerned with anthocyanin pigmentation.

This conclusion is based on an analysis of marker constitutions of A^d strands derived from A^b/a heterozygotes that carried various

---

[2]See References, page 25.

combinations of markers.

In summary, these experiments indicated that of 161 independently occurring changes to A^d, 148 were recombinants for the distal marker gene, while 13 were nonrecombinants. Thus, the member elements of A^b consist of (a) the left-most A^d element (hereafter designated alpha or α) which has intermediate effects on aleurone and plant pigmentation and carries the dominant brown pericarp effect of the parental complex, and (b) an adjacent element on the right (hereafter designated beta or β) associated with purple plant and aleurone. Because in regard to aleurone and plant phenotypes the beta element is an isoallele of its parental A^b complex, it is technically more difficult to isolate by crossing over than the alpha component; however, there have been 10 independent isolations of the beta element (13) and, as might be predicted from the strand constitutions of alpha isolations from the same complex, each of these was a recombinant for the proximal marker and each was shown to have a red pericarp effect. It may be concluded that the beta member of the A^b complex is similar to the wild type A allele, and that the alpha element, which retains the dominance of its brown pericarp effect in spite of its intermediate plant and aleurone phenotypes, is the basis for the anomalous phenotypic behavior of A^b. The genetic distance between alpha and beta is about 0.05 of a unit and the sequence of these elements in the long arm of chromosome 3 is centromere:alpha: beta.

More recent studies (9, 11) indicate that most of the alpha cases from A^b homozygotes also occur in association with crossing over. This observation, along with evidence on the derivatives from certain special compounds involving A^b, leads to the conclusion that alpha and beta, or the segments in which they reside, are members of an adjacent duplication in which the genetic materials are ordered in the same sequence. Thus A^b, like bar in Drosophila, is a tandem, serial duplication whose members may engage in oblique synapsis.

While these experiments indicate a strong association between the occurrence of the alpha derivative and crossing over in the A^b segment, it is apparent, from the data cited above, that about 8 per cent of the alpha exceptions occur without an associated exchange in this region. The possibility was earlier considered (7) that these nonrecombinant alpha cases may represent double exchanges, one crossover occurring between alpha and beta thus isolating the former, the second occurring between beta and the distal marker to reconstitute the parental combination. This explanation was not taken seriously, however, since it was found (8) that $lg\ A^b\ et/a\ sh$ heterozygotes in which sh (shrunken endosperm–2), the inside distal marker, is located 0.25 unit from A (14), gave rise to 43 alpha isolations, 4 of which were nonrecombinants for the sh marker. It was discarded altogether on finding that alpha derivatives are obtained from hemizygous A^b/a–X1 plants in which the X-ray induced deficiency a–X1 is substituted for recessive a. Since the deficient segment here includes the A locus as well as the Sh locus, crossing over between homologues can not be considered as a basis for these alpha occurrences. It should be noted that these experiments did not afford the opportunity to decide between several possible mechanisms to explain the anomalous nonrecombinant alpha derivatives, although gene mutation of the beta element of the A^b complex was one of the possibilities suggested (8).

A detailed analysis of the factors influencing the occurrence of the noncrossover alpha derivatives and, in particular, a test of the hypothesis that they occur as a result of gene mutation of beta are seriously hampered by the relatively low frequency of their occurrence (ca. 5×10^{-5}) among A^b gametes of A^b/a individuals. For this reason we have shifted the emphasis in these studies to other A^b complexes whose yield of noncrossover alpha exceptions is greater than that of the original A^b complex discussed above.

We have carried out intensive analyses of two A^b "alleles" of Peruvian extraction (the designation A^b–P will be employed hereafter in general references to these) which share with the original A^b of Ecuador the determination of purple plant and aleurone and a dominant brown pericarp. While it has been established (10) that these A^b–P forms are likewise complexes consisting of separable alpha and beta elements, they differ from the original A^b complex of Ecuador origin in several ways:

(a) The alpha element of the A^b–P complex is weaker in its effect on plant and aleurone pigmentation than that of A^b.

(b) The sequence of alpha and beta elements in A^b–P is the reverse of that in A^b. Using the C notation for the centromere, the order of the members of the A^b complex is $C : \alpha : \beta$,

whereas the sequence of members in the A^b–P complex is
C : β : α. The evidence suggests that one complex is not simply a gross inversion of the other, but rather that the individual members of the duplication have exchanged position
while retaining the same serial order of the duplication as
a whole.

(c) There is a striking difference in the frequency of occurrence
of the noncrossover alpha derivatives from the Ecuador A^b
complex (ca. 5×10^{-5}) as compared with that from the A^b–P
source (ca. 5×10^{-4}), whereas the crossover alpha derivatives
occur with about the same frequency from both complexes.

In attempting to test the hypothesis that the anomalous nonrecombinant alpha derivatives are ascribable to gene mutation, the
reduced phenotype of the alpha from A^b–P and its enhanced frequency of occurrence are decided advantages. In what follows we
have brought together the available evidence from several years' study
in this laboratory bearing on the question of the gene mutation origin
of the nonrecombinant alpha derivative.

The markers employed in the experiments to be discussed are
given in Figure 1 which provides a map of a portion of the long arm
of chromosome 3, including the A locus with which we are concerned.

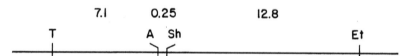

FIGURE 1.—*Map of a portion of the long arm of chromosome 3 showing position of A locus and of the marker loci employed in these studies.
Centromere is located to the left of T.*

The symbol T refers to Translocation 2–3d (1) which has been used
extensively in heterozygous condition to define a segment proximal
to the A locus. Whenever it is used to designate chromosome constitutions, it refers specifically to an interchanged 2^3 chromosome carry the
A locus. Similarly, in translocation heterozygotes the symbol N refers
to a normal (noninterchanged) chromosome 3. Since all individuals
to be tested were crossed with homozygous normal testers, the presence or absence of the interchanged chromosome in exceptional
individuals among the offspring was easily determined by the pres-

ence or absence of aborted pollen, the former of which is typically associated with plants that are heterozygous for the translocation. This classification was confirmed by scoring the ears of exceptional individuals for the aborted condition.

The sh_2 (shrunken–2) factor which produces a striking collapse of the mature endosperm is distal to A and, whenever possible, because of its close proximity to the latter, has been used as a marker in preference to et (etched endosperm, virescent seedling).

Table 1 summarizes tests of over a million gametes from A^b–P/a heterozygotes marked with T and sh. The A^b–P alleles designated Lima and Cusco are extractions from two different plants of Peruvian origin and have been treated as separate entities throughout this presentation. Referring to the alpha derivatives in Table 1, we note

TABLE 1.—CONSTITUTIONS OF ALPHA-BEARING STRANDS FROM A^b–P/a INDIVIDUALS MARKED WITH T AND sh.

Source	A^b–P gametes tested	Number and distribution of alpha-bearing strands among offspring*			
		nco–1	nco–2	co–1	co–2
N A^b–Lima/T a sh	30,390	20 N α Sh	0 T α sh	4 T α Sh	0 N α sh
N A^b–Lima sh/T a	53,190	24	0	12	0
T A^b–Lima/N a sh	324,360	126	0	31	0
Totals	407,940	170	0	47	0
N A^b–Cusco/T a sh	38,410	14	0	3	0
N A^b–Cusco sh/T a	56,950	13	0	22	0
T A^b–Cusco/N a sh	1,750	1	0	1	0
T A^b–Cusco sh/N a	18,760	6	0	7	0
Totals	115,870	34	0	33	0

*As illustrated by the strand constitutions provided in the first row of this table, here and in similar tables to follow, nco–1 refers to a strand carrying the nonrecombinant parental markers of the parental A^b–P chromosome; nco–2 refers to a strand carrying the parental markers of its homologue; co–1 refers to a recombinant strand that carries the distal marker of the parental A^b–P chromosome and the proximal marker of its homologue; and co–2 refers to a recombinant strand that carries the proximal marker of the parental A^b–P chromosome and the distal marker of its homologue.

that only two of the four possible marker combinations are represented among these strands. One of these, the recombinant type designated co–1, carries the shrunken marker of the parental A^b–P chromosome and the T marker of the homologue. Since this recom-

binant type constitutes over 20 per cent of the alpha strands from the
A^b–Lima parent and almost half of those from the A^b–Cusco heterozygote, it is apparent that the crossover event is not independent of the
event that isolates alpha from the complex. If these events were independent, the frequency of recombinant alpha-bearing strands, based
on the genetic length of the T–sh segment, should be barely in excess
of 7 per cent of the total alpha exceptions. This evidence, combined
with the observation that the complementary recombinant type (co–2)
is not represented among the alpha strands, confirms the original
conclusion (10) that the order of elements in A^b–Lima and A^b–Cusco
is centromere:beta:alpha; accordingly, isolation of the alpha component from the parental A^b–P complex by a crossover between the
beta and alpha elements gives rise to recombination for the proximal
marker, as illustrated:

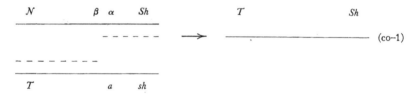

The alpha-bearing strand which is of particular interest, and
which is predominant among those presented in Table 1 (nco–1),
carries the nonrecombinant markers of the parental A^b–P chromosome. As we shall see, the alpha strand of nco–1 type is found among
the progeny of various A^b–P compounds. Its regular occurrence and
the fact that it is not associated with recombination for marker loci
suggest that it may be the result of gene mutation of the closely linked
beta element.

The possibility that multiple exchanges within a short segment
or a copy-choice phenomenon might account for the anomalous nco–1
alpha derivatives should be considered. A detailed treatment of these
possibilities will be given elsewhere, but it may be noted here that
although the nco–1 alpha strands are expected on either or both of
these schemes, they also call for the occurrence of alpha strands of
nco–2 and co–2 constitutions, both of which are conspicuous by their
absence in the data reported here.

Tables 2 and 3 present the constitutions of alpha strands among offspring from A^b–P$/A$ and A^b–P$/A^b$–Ec (original Ecuador source) heterozygotes. Again we note that the nco–1 alpha strands constitute a significant proportion of the total alpha occurrences and the prominence of co–1 and the absence of co–2 alpha strands are further evidence in support of the aforementioned order of components in the A^b–P complexes. The absence of nco–2 derivatives (Table 2) from A^b–P$/A$ heterozygotes is expected since there is

TABLE 2.—CONSTITUTIONS OF ALPHA-BEARING STRANDS FROM A^b-P$/A$ INDIVIDUALS MARKED WITH T AND sh OR WITH T AND et.

Source	A^b–P* gametes tested	Number and distribution of alpha-bearing strands among offspring			
		nco–1	nco–2	co–1	co–2
N A^b–Lima sh/T A	5,460	2	0	0	0
T A^b–Lima$/N$ A sh	16,325	4	0	1	0
N A^b–Cusco sh/T A	22,600	.	0	2	0
N A^b–Lima$/T$ A et	4,000	3	0	0	0
N A^b–Lima et/T A	550	2	0	0	0
T A^b–Lima$/N$ A et	585	1	0	1	0
N A^b–Cusco$/T$ A et	2,075	2	0	0	0
Totals	51,595	23	0	4	0

*Here and in the following tables population is expressed in terms of A^b–P gametes tested, though, of course, an equivalent number of gametes from the homologue were also scored.

no evidence to indicate that the type A allele is a complex, let alone that it carries an alpha component. And the rare occurrence of nco–2 alpha derivatives from the A^b–P$/A^b$–Ec heterozygotes (Table 3) is anticipated, since it is established independently that the A^b of Ecuador extraction carries an alpha component that is infrequently isolated as a nonrecombinant.

One further point may be mentioned here. The occurrence of crossover alpha derivatives from the A^b–P$/A^b$–Ec heterozygote indicates that these complexes have at least one homologous member in common. Evidence on crossover derivatives from A^b–P homozygotes, omitted here because it does not bear on the order of the components, indicates that the members of this complex engage in oblique synapsis from which it may be concluded that A^b–P, like A^b, is a duplication

TABLE 3.—CONSTITUTIONS OF ALPHA-BEARING STRANDS FROM A^b–P/A^b (ECUADOR)
INDIVIDUALS MARKED WITH T AND sh. '

Source	A^b–P gametes tested	Number and distribution of alpha-bearing strands among offspring			
		nco–1	nco–2	co–1	co–2
N A^b–Lima/T A^b sh	98,730	54	2	101	0
T A^b–Lima sh/N A^b	45,660	14	0	29	0
N A^b–Cusco/T A^b sh	42,075	15	0	52	0
Totals	186,465	83	2	182	0

whose members have retained synaptic homology.

Mention was made at an earlier point of the possibility that the nonrecombinant alpha cases might represent double exchanges in which one crossover occurs between the alpha and beta elements and another within the marked segment such as to reconstitute a parental marker combination. Critical evidence is provided on this point by an analysis of A^b–P hemizygotes in which the homologue is deficient for a segment including the A locus. If both the recombinant (co–1) and nonrecombinant (nco–1) alpha derivatives from A^b–P/a parents (Table 1) are dependent on a crossover between the beta and alpha elements, it is expected that A^b–P/Df a–X individuals in which the opportunity for synapsis of the A^b–P complex and hence for exchanges between homologues in that region is removed, would yield no alpha offspring.

The data summarized in Table 4 indicate that alpha occurrences are common among the offspring of such hemizygotes. The deficiencies a–X1 and a–X3 are of X-ray origin (23) and are known to be deficient for a segment including the A locus and extending to the right beyond the Sh locus. That the deficiency extends to the left beyond beta in the homologue is apparent from the absence of N a Sh recombinants (co–1) among the alpha strands from T-marked hemizygotes. (See first four rows of Table 4.) Their complete absence is somewhat surprising since they might be expected as a result of coincidental exchange in the $T - \beta$ segment. However, this coincidental event may be rarer than anticipated either as a result of an interfering effect of the event that gives rise to alpha, or because the

TABLE 4.—CONSTITUTIONS OF ALPHA-BEARING STRANDS FROM A^b–P/Df a–X HEMIZYGOTES. *

Source	A^b–P gametes tested	Distribution of alpha-bearing strands among offspring
T A^b–Lima Sh/N a–X1	132,640	93 T α Sh
T A^b–Cusco Sh/N a–X1	400	1 T α Sh
T A^b–Lima Sh/N a–X3	20,580	12 T α Sh
T A^b–Cusco sh/N a–X3	2,350	3 T α sh
No marking:		
N A^b–Lima Sh/N a–X1	128,280	83 N α Sh
N A^b–Cusco Sh/N a–X1	86,355	44 N α Sh
N A^b–Lima Sh/N a–X3	4,555	4 N α Sh
N A^b–Cusco Sh/N a–X3	2,500	1 N α Sh

*The deficient segments in both Df a–X1 and Df a–X3 include the A locus and the Sh locus. See text for further details.

deficiency may extend well to the left of the A locus. In any case, the regular occurrence of alpha strands from these deficiency heterozygotes indicates clearly that the phenomenon leading to the nonrecombinant alpha occurrence can not be attributed to any mechanism requiring the direct participation of the homologue.

One obvious explanation for the occurrence of the nonrecombinant alpha derivatives that does not conflict with the evidence so far presented would attribute them to mutation of the beta element. Since the beta element of the complex has a purple effect, it masks the pale or intermediate phenotype associated with the adjacent alpha component. A qualitative change in the beta element of the complex, rendering it ineffective in the production of pigment, would allow the expression of the adjacent alpha element under circumstances which would not involve the participation of the homologue and which would thus be independent of recombination for marker loci. However, since the crossover alpha derivatives (co–1) which have lost the beta element in the exchange event are phenotypically indistinguishable from the nonrecombinant alpha individuals (nco–1), the restriction is imposed that the presumed mutation of the beta element be to the null level, hereafter referred to as beta$_0$ or β_0.

The gene mutation hypothesis proposed here equates the occurrence of the nonrecombinant alpha with a change in the A^b–P complex from $\beta : \alpha$ to $\beta_0 : \alpha$, where the former has the typical purple effect, the latter the typical alpha (pale) effect, and the hypothetical

β_0 (mutated element) has a null effect which, so far as the aleurone phenotype is concerned, is equivalent to the colorless phenotype associated with recessive *a*. The validity of this hypothesis may be tested objectively by isolating the active beta element from the complex through a crossover event and subjecting it to appropriate mutational analysis to determine whether it has the predicted capacity to mutate to the null level, and if so, whether the frequency of this event corresponds to the frequency of occurrence of the nonrecombinant alpha derivatives from the $\beta\alpha$ complex.

In order to provide a situation that permits the simultaneous comparison of mutation rates of the isolated beta element and the beta element in the beta:alpha complex, and that would eliminate the effect of modifiers, crosses were made to produce marked heterozygotes having the constitution $T \ \beta\alpha \ Sh/N \ \beta \ sh$, in which the isolated beta element is the same as that in the beta:alpha complex of the homologue. The steps involved in producing the desired heterozygotes are shown diagrammatically in Figure 2, which emphasizes the

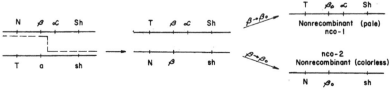

A.   B.   C.

FIGURE 2.—*Diagrammatic presentation of the steps involved in testing the hypothesis of gene mutatoin of beta to beta$_0$. A, An exchange between beta and alpha isolates beta on N β sh strand. B, Appropriate crosses are made to produce marked F_1 individuals of the type shown here, carrying the isolated beta in one chromosome and the beta:alpha complex from which it was isolated in the homologue. C, On the hypothesis, gene mutation of beta to beta$_0$ in the F_1 parent should produce type 1 nonrecombinant strands (pale phenotype) and type 2 nonrecombinant strands (colorless phenotype), with equal frequencies.*

common origin of the two beta elements in the tested individual and also indicates the strand constitutions of exceptional offspring expected on the hypothesis that mutation of the beta element is responsible for the nonrecombinant alpha derivative in question. If this hypothesis is valid, the frequency of nonrecombinant N and sh, colorless derivatives, should equal the frequency of nonrecombinant T and Sh, alpha derivatives.

Since a number of independent isolations of beta derivatives were employed in these experiments, it will be helpful before proceeding to the results to consider the criteria that were employed in establishing that an isolated beta was really dealt with. In all cases prospective beta isolates were tested for pericarp phenotype and only those which had lost the dominant brown effect and hence the alpha component in the crossover event were used in the present experiments. Incidentally, we have never obtained an alpha derivative from a crossover isolate whose pericarp test indicated a loss of the brown phenotype. A more serious problem is posed by the fact that a considerable number of beta isolates (lacking the alpha component) are new complexes of the $\beta : a$ type in which the a element from the homologue has taken the place of alpha in the complex. Detailed analyses of this phenomenon will be published elsewhere and it will suffice here to point out that this substitution is expected if a is paired to the right (see Figure 2A) of the crossover which takes place between beta and alpha. The $\beta : a$ crossover derivative has a pericarp phenotype indistinguishable from that of β, and since it is established independently that homozygotes of the $\beta : a$ complex yield both crossover and noncrossover a derivatives much as the $\beta : \alpha$ complex yields crossover and noncrossover alpha derivatives, the unwitting use of $\beta : a$ in the present experiment would lead to the isolation of noncrossover a derivatives which are not attributable to mutation of the beta component. Consequently, we have adopted the criterion that any beta isolate that has yielded crossover a derivatives among its progeny is a $\beta : a$ complex and is not legitimately included in the data to be presented. At the same time it is apparent that this runs the risk of mistakenly including data from those $\beta : a$ complexes whose progenies, speaking statistically, were too small to yield even a single crossover a case, even though they might produce occasional noncrossover a cases, which in that event would be mistakenly taken to represent mutation of beta to $beta_0$.

One particular beta extraction (designated isolate #1) has had extensive testing. The absence of the alpha component in isolate #1 has been independently established by an analysis of over 50,000 gametes from marked heterozygotes carrying beta and recessive a, among which no alpha exceptions were found. That this particular beta isolate does not carry an adjacent a element is suggested by the absence of a derivatives among over 15,000 offspring from marked compounds carrying the beta isolate in one chromosome and the original A^b complex of Ecuador extraction in the other.

The results of progeny analyses of heterozygotes carrying beta isolate #1 in one chromosome and the beta:alpha complex of A^b–Lima, from which it was extracted, in the other, are presented in Table 5. We note that from T and sh marked heterozygotes of this

TABLE 5.—SUMMARY OF EXCEPTIONAL CASES FROM HETEROZYGOTES INVOLVING BETA
ISOLATE NO. 1 AND THE BETA:ALPHA COMPLEX OF A^b–LIMA FROM
WHICH IT WAS EXTRACTED.

Source	A^b–P gametes tested	Distribution of exceptional cases among the offspring			
		nco–1	nco–2	co–1	co–2
T and sh marked:					
Sibs:					
$T\ A^b$–Lima/$N\ \beta$–Lima sh	32,030	17 $T\ \alpha\ Sh$	0	5 $N\ \alpha\ Sh$	0
$T\ A^b$–Lima/$N\ a\ sh$	30,685	17 $T\ \alpha\ Sh$	0	9 $N\ \alpha\ Sh$	0
No proximal marker:					
A^b–Lima/β–Lima sh	37,905	34 $\alpha\ Sh$; no colorless derivatives			

type five co–1 and 17 nco–1 alpha derivatives were obtained. The five recombinants are expected since they represent crossover isolations of the alpha element of the parental A^b–Lima and carry the recombinant markers predicted from the knowledge that alpha is the distal element in this complex. Our special interest in these data concerns the 17 nonrecombinant alpha cases since, according to the mutation hypothesis, they have the constitution $\beta_0 : \alpha$ and are presumed to have resulted from the mutation of beta to beta$_0$ in the A^b–Lima complex. However, since the isolated beta element in the homologue is a replica of that carried in the A^b–Lima complex, an equivalent number of mutations of the former to the β_0 level is expected. These would be scored phenotypically as a (colorless) derivatives, and since they should carry the N and sh parental marker combination, they would be expected to fall in the nco–2 registry of Table 5. Contrary to this expectation, not a single *a* derivative was found.

Data presented in the second row of Table 5, dealing with the number and distribution of alpha cases from $T A^b$–Lima$/N a sh$ sibs, indicate a close agreement with regard to the frequencies of nco–1 alpha strands from the two sources and may be taken as evidence that the frequency of this type of derivative from the sib $\beta : \alpha/\beta$ heterozygote is not subject to any peculiarity having to do with its special genotype.

In the third row of Table 5 are presented data from additional heterozygotes, involving beta isolate #1, which are without proximal marking. While it is thus not possible to distinguish between nco–1 and co–1 alpha derivatives, the data of Table 1 lead us to expect that at least half of these are noncrossovers; yet we note here the occurrence of 34 alpha cases and again the absence of *a* derivatives. Thus, the data from beta isolate #1 pose a striking contradiction to the hypothesis that would attribute the noncrossover alpha to gene mutation of the adjacent beta element of the complex.

Analyses similar to the foregoing have been carried out with nine other independently isolated beta derivatives from A^b–Lima and with eight such isolates from the A^b–Cusco complex. A summary of derivatives obtained from T and sh marked heterozygotes involving these beta isolates and A^b–Lima is given in Table 6. With the exception of isolate #162–1, the data from individual beta isolates are too meagre to justify inferences concerning the mutation hypothesis but, taken as a whole, the occurrence of 41 nco–1 alpha cases as compared with two nco–2 *a* cases is in excellent agreement with the results obtained in the more extensive tests involving beta isolate #1. We conclude, therefore, that gene mutation of the beta element can not be the mechanism responsible for the vast majority of noncrossover alpha derivatives; nor is it certain that the two cases of nco–2 *a* derivatives shown in Table 6 are attributable to that phenomenon since some of these beta isolates may actually represent $\beta : a$ complexes, of the type discussed above, that were inadvertently included here because, in the relatively small populations of tested gametes, they gave no crossover *a* derivatives to identify them as such. While further analyses of these questionable beta derivatives should resolve this question, their possible removal as nonvalid cases would only serve to emphasize the strong evidence, from the data as they stand,

TABLE 6.—SUMMARY OF EXCEPTIONAL CASES FROM COMPOUNDS INVOLVING 17 INDEPENDENTLY ISOLATED BETA ELEMENTS AND THE A^b–LIMA COMPLEX.

Source	Beta isolate	A^b–P gametes tested	Distribution of alpha and colorless derivatives among the offspring			
			nco–1	nco–2	co–1	co–2
T A^b–Lima/N β–Lima sh	151–1	4,580	1 T α Sh	0	0	0
"	151–2	5,945	1 "	0	0	0
"	153–1	4,470	1 "	0	1 N α Sh	0
	161–3	2,650	0	0	1 "	0
	162–1	26,475	14 "	0	0	⁀0
	165–2	4,670	1 "	0	1 "	0
	165–3	4,620	2 "	0	0	0
	174–3	4,135	0	0	0	0
"	183–2	8,945	1 "	0	3 "	0
Totals		66,490	21	0	6	0
T A^b–Lima/N β–Cusco sh	155–5	2,530	2 T α Sh	0	1 N α Sh	0
"	155–9	8,585	2 "	0	1 "	0
"	155–12	7,745	4 "	0	1 "	0
	156–3	8,675	1 "	0	3 "	0
	168–1	5,160	2 "	1 N a sh	1 "	0
	178–1	8,330	7 "	0	1 "	0
	190–2	6,295	2 "	1 "	1 "	0
"	280–1	1,815	0	0	0	0
Totals		49,135	20	2	9	0
T A^b–Lima/N a sh*		88,890	31 T α Sh	0	4 N α Sh	0

*These individuals occurred as sibs of the heterozygotes listed above. Data presented in this row represent totals for all such sibs.

against the gene mutation hypothesis.

It is appropriate to consider at this point certain other experiments designed to test the mutation hypothesis for the origin of the noncrossover alpha cases. These studies, like those reported above, are based on the argument that if gene mutation of the beta element in the complex is responsible for the occurrence of the nonrecombinant alpha, the latter, represented as $\beta_0 : \alpha$ (see Figure 2), should carry a null beta form which is susceptible to isolation by crossing over in advanced generation tests of these derivatives. These studies (20), carried out in this laboratory, not only fail to support the gene mutation hypothesis, but provide strong empirical support for the supposition that, if the event in question is not due to a qualitative change in beta, it must be assigned to its removal altogether from the complex.

One more line of evidence, less decisive perhaps than those presented above, may be brought to focus on the question of gene mutation of the beta element, and in this case the original A^b complex of Ecuador origin is involved. As pointed out earlier, the elements of this complex are ordered in reverse ($\alpha : \beta$) of those of the A^b–P complex. Moreover, the alpha component of the former is isolated as a nonrecombinant with a ten-fold lower frequency compared with the corresponding event from the A^b–P complex. Among the crossover alpha isolations from the A^b/a heterozygote, an occasional one is found (8) that is mutable (α–m) when the Dt gene is present in the genome. Kernels that carry this mutable alpha form are pale in phenotype, with dots (clusters of purple cells) distributed throughout the aleurone. Since Dt is known to condition the mutation of recessive a to A (18, 19) and since the mutability of the α–m derivative is also controlled by Dt, it is inferred that α–m itself represents a complex of the type $\alpha : a$, in which, as a result of a crossover in the A^b/a heterozygote, the mutable a allele has been traded for, and taken the position of, beta in the alpha:beta complex. The mutability of this new complex is then considered to reside not in alpha itself but in the now adjacent a element. This argument is supported by the crossover extraction of the mutable a element from the $\alpha : a$ complex in experiments which also confirm that the order of the alpha and a elements in this complex, as expected, is centromere : alpha : a.

Taking advantage of the influence of Dt on the mutation of a, we have obtained from $\alpha : a$ homozygotes a full-colored revertant whose phenotype is indistinguishable from A^b. There is good reason to conclude that this revertant is a synthetic A^b of the type $\alpha : A$, in which A, originating from a as a result of mutation *in situ*, is substituted for the original beta of the complex (Figure 3). This is supported by tests which reveal that the revertant no longer yields a derivatives and that the mutant form, in heterozygotes with recessive a, yields alpha derivatives whose occurrence is associated with recombination for the distal marker.

The substitution of A for beta in A^b affords the opportunity to compare the frequencies of crossover and noncrossover alpha cases from the original and the synthetic complexes. This experiment

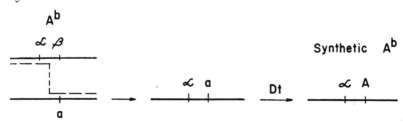

FIGURE 3.—*The steps involved in obtaining the synthetic A^b. First step involves crossover isolation of the α:a complex. The second step involves Dt-induced mutation of a in this complex giving α:A, the synthetic A^b in which A replaces β.*

would seem to be particularly appropriate, since, according to the mutation hypothesis which attributes the noncrossover alpha occurrences to mutation of the beta element, there is no reason to expect that the synthetic A^b, which carries A instead of beta, should yield alpha derivatives at the same rate that A^b does. The data presented in the first two rows of Table 7 indicate that the original and the

TABLE 7.—YIELD OF ALPHA DERIVATIVES FROM A^b (ALPHA:BETA) AND FROM ITS SYNTHETIC COUNTERPART (ALPHA:A) IN HETEROZYGOTES WITH RECESSIVE a.

Source	A^b gametes tested	Alpha derivatives	
		nco	co
Sibs:			
A^b sh/a Sh..............	9,180	0	11 α Sh
A^b–synthetic/a	12,340	Total of 15 α cases*	
A^b–synthetic Sh/a sh.	36,700	3 α Sh	51 α sh

*These could not be scored for noncrossover or crossover origin since the parent lacked distal marking.

synthetic complexes yield alpha derivatives with similar frequencies. Moreover, it is apparent that the noncrossover alpha is a rare occurrence among the progeny of the synthetic A^b (Table 7, third row) just as it is among the progeny of the original A^b. The 3 noncrossover cases among a total of 54 alpha derivatives reported here for the synthetic complex is in good agreement with the 10 noncrossovers among a total of 127 alpha cases reported for the original A^b (8). From these results, indicating that the substitution of A for beta in the Ecuador complex has little if any effect on the frequency of noncrossover alpha derivatives among the progeny, there is again no support for the gene-mutation hypothesis.

On the basis of the evidence presented above, the contention that gene mutation of the beta element is responsible for the occurrence of nonrecombinant alpha derivatives is indefensible. This should come as no surprise, however, since it fits a pattern that has by now become commonplace. The *A* alleles are among the most intensively studied in maize and, of the many "mutations" recorded at this locus, the noncrossover alpha derivatives analyzed here are among the select group that had survived previous tests designed to identify extragenic changes. It is difficult to avoid the conclusion that, if the currently favored hypothesis of gene mutation (25, 26) based on the alteration of structure at the molecular level is valid, we should not expect to encounter the phenomenon in maize in experiments which at best can deal statistically with events at the level of 10^{-5}. It is increasingly apparent that the classical gene mutation eludes the investigator, not so much because it defies characterization, but because it is of such infrequent occurrence that it is swamped by events that are extragenic. We can only conclude that nature is conservative in its display of gene mutations and that the predominant and immediate contributions to variability in the population are functions of various extragenic events.

In regard to the case discussed here we may summarize as follows:

(a) The noncrossover event leading to the expression of alpha involves a loss of the beta (purple) phenotypic expression. This follows since the crossover and noncrossover alpha derivatives are identical in phenotype.

(b) The noncrossover alpha derivative originates through a physical loss, not gene mutation, of the adjacent beta element.

(c) The experiments dealing with heterozygous deficiencies indicate that the occurrence of the noncrossover alpha derivative (loss of beta) does not involve the participation of the homologue.

(d) Loss of the beta element of the complex is conditioned by the presence of the adjacent alpha element since it has been shown that the isolated beta element "mutates" rarely, if at all.

Mention should be made of a phenomenon at the R locus in maize which is a rather close parallel of that involving A^b. It is now apparent (3, 4) that loss of the plant color effect of the (P) element in the (P) (S) Cornell complex, whether or not it is associated with recombination of marker genes, is attributable to physical loss rather than to gene mutation of the (P) element in the complex. Spontaneous deficiency of the (P) element or, alternatively, loss of this element in response to some type of activator, are two of the mechanisms proposed (4) to account for the event in question. However, neither of these explanations satisfactorily accounts for the noncrossover alpha derivatives from the beta:alpha complex since it has been shown that the isolated beta element, removed from its association with alpha, is surprisingly stable.

Thus, an acceptable model on which to account for the loss of beta from the complex must satisfy the evidence that the event does not require the participation of the homologue and that it is uniquely dependent on the *cis* association with alpha. Moreover, the event must be highly restricted as to time of its occurrence since the vast majority of the derivatives occur as single kernels on ears of tested individuals, thus giving strong indication that meiosis is involved. Several years ago we suggested (12) a mechanism which accommodated the evidence available at that time and which appears not to be in conflict with that available now. It is based on the finding (9, 11) that the A^b complexes are tandem, serial duplications whose members, at meiosis, engage in oblique synapsis with their complements in the homologue; and it assumes only that the adjacent members of the duplication have the alternative of pairing with each other, intrachromosomally, in a double-loop configuration at meiosis which will be referred to as A.A. (auto-association). As illustrated in Figure 4, rare exchanges between the beta and alpha segments comprising the double loop would result in the loss and/or gain of one complete member of the duplication, depending on the strands involved in the event. Occasional losses of the beta-carrying segment by this mechanism would yield apparent noncrossover alpha derivatives which, because of their origin through "deficiency" of beta, would be indistinguishable in phenotype from the crossover alpha derivatives. Moreover, since the event is dependent on a pairing phenomenon ordinarily restricted to meiosis, the observed individual occurrences of the alpha derivatives

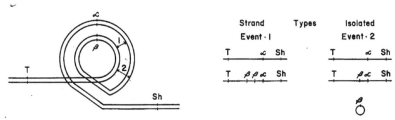

FIGURE 4.—*Diagrammatic representation of proposed mechanism of auto-association. At the left, adjacent, homologous beta and alpha segments of a single chromosome are shown paired with each other at meiosis to form the double loop. The results of exchange event 1 (involving sister chromatids) and of exchange event 2 (involving a single chromatid) are shown at the right. Both events yield a "noncrossover" alpha strand, but the first event produces a complementary strand carrying two beta members plus an alpha, while the second event yields a parental-type strand carrying beta and alpha members plus an acentric ring representing the beta member.*

are explained. The observation that loss of beta is conditioned by the presence of alpha in the parental complex is also predicted since, on the model proposed here, the removal of the beta element is dependent on the adjacent alpha-carrying segment to provide the double loop. Finally, since the A.A. hypothesis involves a strictly intrachromosomal event, the occurrence of noncrossover alpha cases among the progeny of plants heterozygous for the a–X deficiencies is also anticipated.

In fact, assuming that the A.A. type of pairing at meiosis is in competition with conventional interhomologue pairing, the frequency of the former, and therefore of noncrossover alpha occurrences, should be enhanced in the deficiency heterozygote, in which the opportunity for interhomologue pairing of members of the complex is removed. Results of experiments testing this prediction will be published elsewhere, but it may be mentioned here that studies of this type dealing with over 400 alpha cases and with tests of over a million gametes indicate a significant enhancement of the alpha rate in the hemizygote.

The possibility that sister strand crossing over might account for the origin of the noncrossover alpha derivatives was treated in an earlier publication (8) but was not considered a likely explanation because (a) there was evidence against the occurrence of the phenomenon in Drosophila (15, 16), and (b) to explain the occurrence of noncrossover alpha cases, the sister strand exchanges would have to be unequal. More recently, evidence has been presented (21) in support of the occurrence of sister strand crossing over in maize, but it is apparent that the question of the existence of the phenomenon is still debated among geneticists. In any case, if it is argued that crossing over is a function of the reduplication process or that it occurs at the time of reduplication of strands, it is difficult to visualize how old and new member elements of the A^b complex are, as required, in precise juxtaposition at the time of and immediately following replication, yet at the same time are in oblique association to facilitate the required unequal sister event. Moreover, according to the sister strand hypothesis, there is no apparent basis on which to explain the enhancing effect of the deficient homologue on the frequency of occurrence of the noncrossover alpha derivative.

We are now searching for independent cytological evidence bearing on the validity of the A.A. scheme proposed here and have chosen the bar locus in Drosophila for this purpose because it too represents a tandem, serial duplication and moreover, presents the opportunity for cytological analysis. To be sure, we cannot expect to observe the double loop in meiotic tissues of either Drosophila or maize, nor would such an observation be decisive. But we note (Figure 4) that the A.A. mechanism calls for the removal (or addition) of one complete member of the duplication, no more, no less. Thus, in the case of the bar duplication, the hypothesis calls for the occurrence from bar of both noncrossover normal and noncrossover double-bar individuals, corresponding to the loss and gain, respectively, of one complete member of the duplication. Since the duplicated segment in bar is clearly defined as carrying the seven bands of the 16A region of the salivary map, cytological analysis of noncrossover derivatives should constitute a critical test of the A.A. hypothesis.

Studies carried on in this laboratory over the past year (17) indicate that noncrossover wild-type reversions from bar do occur, and with a sufficiently high frequency to permit their statistical analysis. Thus far, in addition to the expected crossover cases, reversions not associated with recombination for marker genes have been obtained from bar individuals heterozygous for the deficiency B–263–20, from homozygous bar individuals heterozygous for the ClB chromosome, and from bar homozygotes carrying two normal X chromosomes. In addition, double bar has given rise to nonrecombinant bar offspring, and bar in turn, has produced noncrossover double-bar types. These derivatives are consistent with expectations on the A.A. hypothesis though their occurrence is not of itself decisive. It is of interest to note that Sturtevant, in his classical work on bar (24), recorded several cases of changes at the locus which were not associated with crossing over, though he considered their status as valid cases somewhat doubtful. Also, Braver (2), in the course of an experiment designed to test the effect of a nearby inversion on unequal crossing over at the bar locus, has obtained (personal communication) a nonrecombinant wild-type revertant from a homozygous-bar parent.

Thus, it appears that the bar duplication has its share of anomalous derivatives, and cytological analysis of those now in hand may be expected to shed light on a phenomenon which, though it is not attributable to gene mutation, is certainly a subtle substitute. On the basis of available evidence it may be anticipated that the phenomenon dealt with here is significant not only for the contribution it makes to variability through regressive changes, but for the progressive evolution of gene systems as well.

References

1. Anderson, E. G., and Brink, R. A. 1940. Translocations in maize involving chromosome 3. *Genetics, 25: 299–309.*

2. Braver, G. 1960. The influence of an adjacent inversion breakpoint on unequal crossing over in the Bar region of *Drosophila melanogaster. Records Genetics Soc. Amer., 29: 59.*

3. Emmerling, M. H. 1956. Unequal crossing over between elements of the *R* complex in *Zea mays. Genetics, 41: 641.*

4. ————. 1958. An anaylsis of intragenic and extragenic mutations of the plant color component of the *Rr* gene in *Zea mays. Cold Spr. Harb. Symp. Quant. Biol., 23: 393–407.*

5. Emerson, R. A., and Anderson, E. G. 1932. The *A* series of allelomorphs in relation to pigmentation in maize. *Genetics, 17: 503–509.*

6. Laughnan, J. R. 1948. The action of allelic forms of the gene *A* in maize: I. Studies on variability, dosage, and dominance relations. The divergent character of the series. *Genetics, 33: 488–517.*

7. ————. 1949. The action of allelic forms of the gene *A* in maize:

II. The relation of crossing over to mutation of A^b. *Proc. Nat. Acad. Sci.*, **35**: *167–178*.

8. ————. 1952. The action of allelic forms of the gene A in maize: IV. On the compound nature of A^b and the occurrence and action of its A^d derivatives. *Genetics*, **37**: *375–395*.

9. ————. 1952. The A^b components as members of a duplication in maize. *Genetics*, **37**: *598*.

10. ————. 1955. Structural and functional aspects of the A^b complexes in maize: I. Evidence for structural and functional variability among complexes of different geographic origin. *Proc. Nat. Acad. Sci.*, **41**: *78–84*.

11. ————. 1955. Structural and functional bases for the action of the A alleles in maize. *Amer. Nat.*, **89**: *91–104*.

12. ————. 1955. Intrachromosomal association between members of an adjacent serial duplication as a possible basis for the presumed gene mutations from A^b complexes. *Genetics*, **40**: *580*.

13. ————. 1956. The beta member of A^b complexes. *Maize Genetics Coop. News Letter*, **30**: *68*.

14. Mains, E. B. 1949. Heritable characters in maize: Linkage of a factor for shrunken endosperm with the a_1 factor for aleurone color. *Jour. Hered.*, **40**: *21–24*.

15. Morgan, L. V. 1933. A closed X chromosome in *Drosophila melanogaster*. *Genetics*, **18**: *250–283*.

16. Muller, H. J., and Weinstein, A. 1933. Evidence against the occurrence of crossing over between sister chromatids. *Amer. Nat.*, **67**: *64–65*.

17. Peterson, H. M., and Laughnan, J. R. 1960. Noncrossover derivatives from serial duplications. *Maize Genetics Coop. News Letter*, **34**: *44–45*.

18. Rhoades, M. M. 1938. Effect of the Dt gene on the mutability of the a_1 allele in maize. *Genetics*, **23**: *377–397*.

19. ————. 1941. The genetic control of mutability in maize. *Cold Spr. Harb. Symp. Quant. Biol.*, **9**: *138–144*.

20. Sarma, M. S. 1959. Studies on the origin of the noncrossover derivatives from the A^b complexes in maize. *Ph. D. thesis, Department of Botany, University of Illinois.*

21. Schwartz, D. 1953. Evidence for sister-strand crossing over in maize. *Genetics*, **38**: *251–260*.

22. Stadler, L. J. 1941. The comparison of ultra-violet and X-ray effects on mutation. *Cold Spr. Harb. Symp. Quant. Biol.*, **9**: *168–177*.

23. ———— and Roman, H. 1948. The effect of X-rays upon mutation of the gene A in maize. *Genetics*, **33**: *273–303*.

24. Sturtevant, A. H. 1925. The effects of unequal crossing over at the Bar locus in Drosophila. *Genetics*, 10: *117–147.*

25. Watson, J. D., and Crick, F. H. C. 1953. Genetical implications of the structure of desoxyribose nucleic acids. *Nature*, 171: *964.*

26. ————. 1953. The structure of DNA. *Cold Spr. Harb. Symp. Quant. Biol.*, 18: *123–131.*

Comments

STEPHENS: Strains of corn from Ecuador and Peru have the alpha and beta elements arranged in different serial order. Do you have any simple mechanism which would derive one from the other?

LAUHGNAN: A scheme to derive one from the other was produced in connection with the original report (PNAS 41: 78–84) on the changed order of elements in the Peruvian complexes. It is based on the finding that adjacent alpha and beta segments represent tandem serial duplications whose members retain homology and may thus be obliquely synapsed in meiosis. Given the opportunity for crossovers within these members, and the occurrence of alpha-carrying chromosomes in the population, there are various ways of deriving the beta:alpha order from the alpha:beta order, and *vice versa*. For example, with exchanges in successive generations, we may go from alpha:beta to alpha:beta: beta to alpha:beta:alpha to beta:alpha.

PETERSON: Can the physical loss of beta be considered in the light of transposition events (simple drop out) associated with mutable gene phenomena?

LAUGHNAN: It may be assumed that a Ds–like element resides at the beta locus and conditions an infrequent loss of the latter. Against this is the evidence indicating that the isolated beta element, removed from its association with alpha, is remarkably stable. Moreover, somatic alpha occurrences, expected as the rule on the insertion-transposition hypothesis, are all but absent, the vast majority of the alpha cases originating in connection with meiosis. And finally, progeny analysis of the alpha occurrences, a routine procedure in our studies, gives no hint of lowered gametophytic or sporophytic viability which would accompany the grosser aberrations predicted to occur at least occasionally on the transposition hypothesis.

GROBMAN: Examination of a large number of archeological ears of corn in Peru has not disclosed the presence of phenotypes with lower pigmentation intensity than those corresponding to the A^b or a^p levels (brown and brown-red pericarp, respectively), during at least the first 2,000 years of a 2,900–year period represented by those ears. It is believed that A^b, a^p, or both represent the wild condition alleles at the A_1 locus in maize. The stability of A^b phenotype seems to have been relatively high during a long period in the evolution of corn, suggesting stability of A^b ($\alpha\beta$) allele complex, until such time as tripsacoid races of corn and lower level expression of A^b down to a became prevalent.

This apparent evolutionary stepping up of the changes at A, which we find coincides with the appearance of other conspicuous mutational events in other genes, suggests a direct or indirect effect of Tripsacum genetic material incorporated into corn.

In the light of the information he has accumulated, would Doctor Laughnan care to comment on the possibility that the mutational events at A^b were accelerated and are now being influenced by extragenic elements originating from some kind of Zea-Tripsacum interaction?

LAUGHNAN: Experiments conducted by R. A. Emerson and E. G. Anderson reported in 1932 established that A^b and a^p, in the presence of the P factor, govern the synthesis of brown pigment in the pericarp of maize. However, since a number of other loci are known to affect pigmentation of the pericarp, even in the absence of P, and since a variety of intergrading phenotypes is possible with respect to both quality and pattern of pigmentation, there is no way, other than by crossing and progeny test, to establish with certainty whether a particular pericarp phenotype is governed by the P locus. Under these circumstances, I consider it unsafe to make inferences about the A-locus constitution on the basis of pigmentation in archeological ears of corn, and even more hazardous to argue a shift in frequencies of alleles on this criterion.

So far as I can see, there is nothing about the A^b events which can be taken to support or contradict the mentioned Zea-Tripsacum interaction. I suppose some better understanding of this would come from genetic analysis of the presumed A elements of Andean Tripsacum.

AUERBACH: I wonder whether you think it is safe to generalize from maize to other organisms, in particular to animals. You said in your opening remarks that in maize intergenic changes are so much more frequent than gene mutations that, in over-all estimates, the *bona fide* mutations are quite swamped. The opposite is true for Drosophila, where intergenic changes are much less frequent than mutations. It seems that plant chromosomes are much more easily broken than animal ones, at least those of Drosophila germ cells.

If one assumes that most of the presumed Drosophila mutations are minute rearrangements, one would have to assume a very high frequency of rearrangements of undetectably small size. As far as I know, the size distribution of detectable rearrangements, in particular small deficiencies, does not lead us to believe that this is so.

LAUGHNAN: I prefer not to generalize from maize to other organisms, including Drosophila, concerning the relative frequencies of gene mutations and extragenic events. But I should like to emphasize that, as with the alpha occurrences (beta losses) reported here, the change may be subtle to the point that only intensive genetic analyses of the changed form, or of the circumstances surrounding its origin, would be expected to reveal its extragenic nature; and upon this point I should not hesitate to generalize.

It is clearly not established that most of the presumed Drosophila mutations are minute rearrangements, nor is it evident that, at the other extreme, they represent molecular alterations in the genetic material. But it seems to me, that to favor the latter interpretation on the basis of size distribution of detectable rearrangements, is to make the unwarranted assumption that opportunities for extragenic events are restricted to changes whose origin is similar to those we observe in gross rearrangements.

Comparison of Spontaneous and Induced Mutations

K. C. ATWOOD

University of Illinois, Urbana, Ill.

Paper presented, but no manuscript available.

Mutation, Selection, and Population Fitness

C. C. LI

Graduate School of Public Health,
University of Pittsburgh, Pittsburgh, Pa.

THE PROBLEMS of mutation and selection in natural populations are varied and complicated. Probably each case merits a separate study, and it is doubtful if any attempt of abstract generalization is warranted. This situation, however, is in principle no different from that in physical sciences. A physical phenomenon may also be very complicated and yet physicists start out from very simple models and gradually build more and more realistic models as experimental evidence accumulates. It is in this sense that biologists may formulate mutation and selection models, granting and remembering that they are only first approximations of natural situations. Simple models do help us to understand certain biological phenomena and may serve as starting points for further research.

Before proceeding, it may be well to define the scope of our discussion. The disciplines of population genetics, ecology, and evolution are interlocked in such a way that it is impossible to disentangle one from the others. In a very general way, however, ecology deals with population growth in size and the joint distribution of more than one group of organisms under a variety of environmental conditions. Evolution is concerned with long-term changes whatever the causes may be. In the present discussion, I will limit myself to the comparatively short-term changes in the genetic composition of a population in the face of recurring mutation and persistent selection.

Probably all genes mutate. The history of mutation must be as old as the gene itself. Mutation is one of the most fundamental properties of a gene. It is also probable that each allele produces a different, however slight, effect on the organism in one respect or another which may or may not be detectable by our present technique. As the title implies, the present discussion is further limited to mutations which affect the reproductive capacity of the organism.

30

I. Definition of Selection and Fitness

When a common word, in lieu of a new one, is drafted into science and used as a technical term, it is likely to introduce semantic difficulties if it is not precisely defined at the outset. Selection and fitness are such common words that have been drafted into biology and, regretfully, have caused a great deal of misunderstanding and unnecessary arguments, because different biologists take them to mean different things. The definition of these terms, given below, is for the purpose of the present discussion. This does not prevent other scientists defining them in some other way to mean something else, just so long as they are used consistently in the sense they are defined and not in any other sense.

Consider two types, A and B, of organism (e.g. A = red wheat, and B = white wheat) that grow in the same environment and there is no intermixture. Suppose that in the initial period there are equal proportions of A and B, and that for every 100 living offspring produced by the A type in the next generation, type B produces only 80. The relative numbers of the two types of organism in the entire population will be as follows:

Generation	Type A	Type B		Generation	Type A	Type B
0	100	100		0	100	100
1	100	80		1	125	100
2	100	64	òr	2	156.25	100
.	.	.		.	.	.
.	.	.		.	.	.

It makes no difference which type is taken as the "standard"; the ratio of the relative numbers is the same in the two systems of presentation, *viz.*, 100:64 = 156.25:100. In other words, the initial 50%:50% distribution becomes, after two generations, approximately 61%:39%. In a situation like this we say there is selection, meaning that there is differential reproduction, or differential contribution to the next generation, between the two types. Further, we say that type A is favored by selection, or that selection is against type B. Alternatively, we may say that type A has a greater fitness than B, or B has a lower fitness than A. All of these statements, inspite of the different wording, are equivalent and mean exactly the same thing. Selection and fitness are two words describing the same phenomenon of differential reproduction within a given population under a given set of environmental conditions.

Several observations may be made with respect to the definition of selection given above. First, if a population consists of only one type of organism (either all A or all B), there will be no selection to speak of and the statement that A or B is fit or unfit has no meaning. Second, selection or fitness is not something that can be determined by merely examining the organism itself as is the case in morphology, anatomy, and to a large extent, systematics. Rather, it is a description of the result of reproductive performance of one type relative to other types of the same population. The retrospective definition of fitness is quite analogous to that employed by Chinese historians:

> "The victorious is called a king;
> The defeated, a bandit."

This leads to the third observation, *viz.*, that selection and fitness are an overall verdict applying directly to the fact rather than the causes of differential reproduction. Thus Wright (15)[1] says: "Selection is a wastebasket category that includes...such diverse phenomena as differential viability at any stage, dispersal beyond the range of interbreeding, differential maturity, differences in mating tendencies, fecundity, and duration of reproductive capacity." In brief, selection may operate through various mechanisms at any stage of the life cycle of the organism. Finally, a word of warning may well be injected here; in using the terms selection and fitness, we must get rid of the connotations of these words in common usage. Selective fitness has no necessary connection with physical appearance, vegetative growth, market value, or social desirability.

The definition of selection and fitness given for the two types of plants is equally applicable if we identify Type A and Type B as two alleles of a locus. When one allele is reproduced, multiplied, represented, or transmitted proportionally more frequently than another allele so that the relative frequency of the two alleles in the next generation changes, we say that there is selection and that the allele whose relative frequency has been increased has a greater fitness. The same definition may be extended to genotypes.

II. Mutation-selection Balance

Very broadly speaking, there are two kinds of equilibrium in

---

[1]See References, page 46.

natural populations. One is that the balanced condition is maintained by the opposing forces between selection and mutation. The other is that the equilibrium is the result of conflicting selection effects alone and that mutation plays a very little role. In this communication, only one or two examples of each kind will be given to illustrate the properties of equilibrium. A minimum bibliography is cited from which further discussions and other references may be found (1, 8, 10, 16).

A. Gametic Selection

Consider the two alleles A and a which reproduce or transmit to the next generation in the ratio $1:w$, where w is called the (relative) fitness of gene a. If w is smaller than unity, we may write $w = 1 - s$, where s is called the selection coefficient against gene a. Suppose that the initial frequencies of A and a in a population are p and q. Then, after one generation of selection, their frequency ratio will be $p:q(1-s)$; that is, new $p = p/(1-sq)$ and new $q = q(1-s)/(1-sq)$. The decrease in frequency of a per generation is

$$\Delta q = \text{new } q - \text{old } q = \frac{q(1-s)}{1-sq} - q = \frac{-spq}{1-sq}$$

Gene a will be eliminated from the population if the selection continues unopposed. Now, suppose that the rate (or probability) of mutation from A to a is u per generation, where u is a small number of the order of 10^{-5} or 10^{-6}. A mutation may occur at any time and at any stage of the life cycle; but for the sake of algebraic simplicity, let us assume that mutations occur after the operation of selection. Then, among the $p/(1-sq)$ A genes, a small fraction u of them will mutate to a per generation. Selection tends to decrease q, but mutation tends to increase it. These two forces will balance each other so that there will be no change in gene frequency when

$$\frac{spq}{1-sq} = \frac{pu}{1-sq}$$

That is, $sq = u$, or $q = u/s$

This is the simplest type of equilibrium between mutation and selection. The equilibrium is *stable*. That means, if q in any generation is greater than u/s, it will decrease until the value of u/s is reached. If q is smaller than u/s, it will increase until the value of

u/s is reached. As long as the "rules of nature" *(u and s)* hold, the gene frequency will stabilize at the value u/s. Disturbances through artificial means can only increase or decrease the value of q temporarily. When the artificial agent is withdrawn and the population is left alone, the gene frequency will gradually restore to u/s. As long as there is a finite probability of mutation from A to a, there is really no permanent way of getting rid of gene a from a population for any appreciable length of time. The condition $q = u/s$ must be accepted as a fact of nature, and by all definitions of normality, it must be regarded as the normal condition.

B. Genotypic Proportions

Preparing for the discussion of genotypic selection, let us say a few words about the genotypic proportions in a population. Consider a locus with two alleles A and a, and let p and q be their respective frequencies in the population $(p + q = 1)$. The proportion of the three genotypes AA, Aa, aa (Table 1) depends upon the mating system being practiced in the population. Continued close inbreeding will result in complete homozygosis, and random mating yields the binomial proportions. The more general situation, especially in plants, is probably intermediate between the two extreme cases. The inbreeding coefficient F is formally defined as the coefficient of correlation between uniting gametes (Wright, 1922).

Consider a plant population in which a fraction θ is self-fertilized and the remaining fraction $(1 - \theta)$ is open-crossed (random mating) in each generation. It may be shown that after a few generations the inbreeding coefficient for such a population is

$$F = \frac{\theta}{2 - \theta}$$

For instance, if $\theta = 40$ per cent of the plants is self-pollinated in each generation, the inbreeding coefficient F will be $0.40/(2.00 - 0.40) = 0.25$. The genotypic proportions in terms of F are given in the fourth column of Table 1. Note that such a population may be regarded as consisting of two components, *viz.*, $(1 - F)$ random and F inbred.

Mating system determines how often the genes A and a are associated into the three types of pairs (genotypes) but does not change the frequency of the genes. Thus, in each of the middle three

TABLE 1.—GENOTYPIC PROPORTIONS IN A POPULATION.

Geno-type	Random mating	Continued inbreeding	Mating with inbreeding coefficient F		General notation
AA	p^2	p	$(1-F)p^2 + F p = p^2 + F pq$		D
Aa	$2pq$	0	$(1-F)2pq = 2pq - 2Fpq$		H
aa	q^2	q	$(1-F)q^2 + F q = q^2 + Fpq$		R
Total	1	1	$(1-F) + F = 1 - 0$		1

columns of Table 1, freq$(A) = p$ and freq$(a) = q$, although the genotypic proportions for the three mating systems are quite different. If the three genotypic proportions are denoted by D, H, R, whatever the mating system, the gene frequencies can always be calculated according to the relation

$$p = D + \tfrac{1}{2}H, \qquad q = \tfrac{1}{2}H + R$$

where $D + H + R = 1$. Should $D + H + R = W$ for some reason, the obvious modification is $p = (D + \tfrac{1}{2}H)/W$, and $q = (\tfrac{1}{2}H + R)/W$, so that $p + q = 1$.

C. Genotypic Selection

Let w_1, w_2, w_3 denote the selective fitness values of the three genotypes whose frequencies in the population are D, H, R, as shown in Table 2. The direct meaning of the w's is the probability with which the genotypes survive to reproduce; that is, we assume for the sake of simplicity that selection operates prior to reproduction. Then it is the selected group (that survives to reproduce) that determines the gene frequency of the next generation. The general procedure of calculating the new gene frequency after selection is given in Table 2, and the selection effect on gene frequency is $\Delta q = q' - q$ per generation. In this simple model the w's are assumed to be constants and, as far as the selection effect on q is concerned, only their relative magnitude is relevant. Any other set of three fitness values proportional to $w_1:w_2:w_3$ will yield the same value of Δq. Hence, for practical calculation, one of the three w's can always be taken as unity.

To illustrate the balance between selection and mutation, I will mention only two examples: (a) When selection is against a dominant mutant gene (Table 2A) and (b) when selection is against the recessive genotype (Table 2B). In the former case, the frequency of the dominant deleterious gene is so low that we may assume homozygous

TABLE 2.—GENERAL PROCEDURE OF CALCULATING SELECTION EFFECT.

Geno-type	Frequency f	Fitness w	Selected $f w$	New gene frequency
AA	D	w_1	$D\,w_1$	
				$(Dw_1 + \frac{1}{2}Hw_2) / \bar{w} = p'$
Aa	H	w_2	$H\,w_2$	
				$(Rw_3 + \frac{1}{2}Hw_2) / \bar{w} = q'$
aa	R	w_3	$R\,w_3$	
Total	1		$\bar{w}$	1
		A. Selection against dominant mutant gene A'		
$A\,A$	D	1	D	
				$[D+ \frac{1}{2}H(1-s)] / (1-Hs) = p'$
$A\,A'$	H	$1-s$	$H(1-s)$	
				$\frac{1}{2}H(1-s) / (1-Hs) = q'$
$A'A'$	0	0	0	
Total	1		$\bar{w} = 1 - Hs$	1
		B. Selection against the recessive genotype		
AA	D	1	D	
				$(D + \frac{1}{2}H) / (1 - Rs) = p'$
Aa	H		H	
				$(R - Rs + \frac{1}{2}H)/(1 - Rs) = q'$
aa	R	$1-s$	$R(1-s)$	
Total	1		$\bar{w} = 1 - Rs$	1

dominants are nonexistent in the population, so that $D + H = 1$ and $q = \frac{1}{2}H$ is the frequency of the dominant mutant gene. The change in gene frequency due to selection is $\Delta q = q' - q = -s\,q (1-H)/(1-Hs) \doteq -s\,q\,(1-H)$. Again, if u is the mutation rate from the type gene A to dominant mutant A', there will be $up' \doteq u (1-H + \frac{1}{2}H\,(1+s))$ new mutations. Equating the loss due to selection and the gain due to new mutation, we obtain the approximate equilibrium condition that could be written in various forms as follows:

$$sq = u, \qquad q = \frac{1}{2}H = u/s, \qquad H = 2u/s, \qquad u = \frac{1}{2}Hs.$$

When the mutant is dominant in its deleterious effect, the balanced condition $q = u/s$ is of the same form as that of gametic selection.

Table 2B gives the result of selection against the recessives. The change in freq(a) in one generation is

$$\Delta q = q' - q = \frac{q - Rs}{1 - Rs} - q = -s\,p\,R/\bar{w}$$

Equating this loss to the gain through mutation, *viz.*, $p \, u \, \bar{w}$, we obtain the equilibrium condition (in terms of the recessive proportion in population)

$$sR = u, \quad \text{or} \quad R = u/s.$$

In populations with inbreeding,

$$R = (1-F)q^2 + F\,q = u/s, \quad q = \frac{-F + \sqrt{F^2 + 4(1-F)u/s}}{2(1-F)}$$

In populations without inbreeding,

$$R = q^2 = u/s, \quad q = \sqrt{u/s}$$

The equilibrium value of q with inbreeding is always lower than that without inbreeding. But, with mutation rate and selection coefficient fixed, the recessive proportion in the population is the same, with or without inbreeding. The equilibrium condition $R = u/s$ is stable.

III. Selectional Balance

D. General Formula

The second broad category of equilibrium is that maintained by selectional forces alone, without introducing new mutations into the population each generation. Referring to the general procedure outlined in Table 2, we see that equilibrium will be achieved when the new gene frequencies $(p'$ and $q')$ after selection are equal to the old ones $(p$ and $q)$ before selection. Substituting $D = p^2 + Fpq$, etc. (Table 1) in the equation $q' = q$ and simplifying, we obtain the equilibrium condition (11)

$$q = \frac{(1 - F)\,(w_2 - w_1) + F(w_3 - w_1)}{(1 - F)\,[(w_2 - w_1) + (w_2 - w_3)]}$$

That is, when q assumes the value indicated above, the differential selective fitness of the genotypes does not change the gene frequencies. In other words, the population remains the same inspite of the selective operation. Note that when $w_1 = w_3$, the equilibrium condition is $q = \frac{1}{2}$, with or without inbreeding.

In order to make q a positive fraction in the expression above, the quantities $w_2 - w_1$ and $w_2 - w_3$ must be both positive $(w_2 > w_1$ and $w_2 > w_3)$ or both negative and the inbreeding coefficient F not too high (see below). In other words, the equilibrium is possible only when the fitness of the heterozygote, w_2, is greater or smaller than those of both homozygotes. One way of representing the former situation is as follows:

$$w_1 = 1-t, \left.\right\}$$
$$w_2 = 1, \left.\right\} \quad \begin{array}{l} w_2-w_1 = t, \\ \\ w_2-w_3 = s, \end{array} \left.\right\} \quad w_3-w_1 = -s+t$$
$$w_3 = 1-s, \left.\right\}$$

Substitution of these values in the general expression for equilibrium yields

$$q = \frac{t-F\,s}{(1-F)(s+t)}, \qquad p = \frac{s-F\,t}{(1-F)(s+t)}$$

where F must be smaller than t/s and s/t, one of them being a fraction. To facilitate the appreciation of this type of selectional balance, a numerical illustration is provided in Table 3. Even without the details of mathematics, it is quite obvious that this equilibrium is stable. Since Aa has the greatest fitness, neither gene A nor gene a can be eliminated from the population. Furthermore, if freq(A) or freq(a) is too low, the greater fitness of Aa tends to raise it to a certain level and *vice versa*. The greater fitness of Aa and the loss of AA and aa are balanced in such a way that the gene frequency remains unchanged.

It was mentioned previously that any set of three numbers proportional to $w_1:w_2:w_3$ will yield the same result. Since 80:100:70 = 100:125:87.5, the latter three numbers have been used in the lower portion of Table 3. The results are the same, of course.

On the other hand, if the heterozygote fitness value is lower than that of either homozygote, the equilibrium is unstable. This, again, should be obvious. The elimination of a certain number of Aa from the population implies equal loss of the number of A and a genes. This inflicts proportionally a greater loss of the gene whose frequency is already lower than that required for equilibrium, and hence its frequency will decrease further. The frequency of the more common gene will, in turn, increase. Consequently, selection against heterozygotes will lead to near elimination of one of the alleles (to be kept at a very low level by new mutations). Unstable equilibria are not expected to exist in nature.

TABLE 3.—SELECTIONAL BALANCE WHEN THE HETEROZYGOTE HAS A GREATER FITNESS VALUE THAN HOMOZYGOTES.*

Genotype	Frequency, f	w	fw	New gene frequency
AA	$0.4225 + 0.0455 = 0.4680$	0.80	0.3744	
				$0.5564/0.8560 = 0.65$
Aa	$0.4550 - 0.0910 = 0.3640$	1.00	0.3640	
				$0.2996/0.8560 = 0.35$
aa	$0.1225 + 0.0455 = 0.1680$	0.70	0.1176	
Total	$1.0000 + 0 \quad = 1.0000$		$\bar{w} = 0.8560$	1.00
	Another system of assigning fitness values			
AA	$p^2 + Fpq \quad = 0.4680$	1.000	0.4680	
				$0.6955/1.07 = 0.65$
Aa	$2pq - 2Fpq \quad = 0.3640$	1.250	0.4550	
				$0.3745/1.07 = 0.35$
aa	$q^2 + Fpq \quad = 0.1580$	0.875	0.1470	
Total	$1 + 0 \quad = 1.0000$		$\bar{w} = 1.0700$	1.00

*$F = 0.20$; $t = 0.20$; $s = 0.30$; $p = 0.65$; $q = 0.35$.

E. Variable Fitness

In all previous sections the value of the w's is assumed to be constant. This may be true for certain traits affecting reproduction and may not be so for others. It is quite possible that the fitness depends not only on the genotype itself, but also, possibly to a large extent, upon the genetic composition of the population as a whole. Some of the simplest examples of this nature are given in Table 4, in which the fitness value of a phenotype varies with the phenotypic frequency in the population. In part A it is assumed that selective advantage or disadvantage of one phenotype is proportional to the frequency of the other phenotype. Thus, in the first example, when recessives are rare in the population (R small and $D + H = 1 - R$ large), the recessive fitness becomes large, being $1 + s(1-R)$, and hence the recessive proportion will increase. The same is true with the dominants. This type of selection will lead to a stable equilibrium. Conversely, if as R becomes small, the recessive fitness also becomes lower, $1 - s(D + H)$, the equilibrium is unstable and one of the alleles will be reduced to near extinction. The fitness of a phenotype in part B varies with frequency of its own phenotype.

The equilibrium condition, when fitness varies with gene fre-

MUTATION AND PLANT BREEDING

TABLE 4.—SELECTIONAL BALANCE WHEN FITNESS VARIES WITH GENOTYPIC FREQUENCIES.

Type	Frequency, f	Advantage from encounter, w	Disadvantage from encounter, w
A. Advantage or disadvantage proportional to f of unlike phenotype			
Dominant	$D + H$	$1 + t R$	$1 - t R$
Recessive	R	$1 + s(D + H)$	$1 - s(D + H)$
Equilibrium		Stable	Unstable
B. Advantage or disadvantage proportional to f of own phenotype			
Dominant	$D + H$	$1 + t(H + D)$	$1 - t(H + D)$
Recessive	R	$1 + s R$	$1 - s R$
Equilibrium		Unstable	Stable

quency, may be found either through the general procedure outlined previously (Table 2), or by the method of Wright (15) and Lewontin (9). In such simple cases as listed in Table 4, however, an expeditious short-cut may be used. The phenotypic frequency and the selective fitness should be so balanced that the population remains unchanged. This condition obtains when the fitness values of the two phenotypes are equal, thus

for cases of Table 4A, $t R = s(1-R), R = \dfrac{s}{s+t}$

for cases of Table 4B, $s R = t(1-R), R = \dfrac{t}{s+t}$

where $R = q^2 + Fpq$ with inbreeding, and $R = q^2$ without inbreeding. The stability of these equilibrium values has been indicated in Table 4. The general principle which governs a large number of similar cases is

selective advantage when abundant $\longrightarrow$ unstable equilibrium
selective advantage when rare $\longrightarrow$ stable equilibrium

This principle is not only important in studying selection within a population, but it is also important in ecological studies which deal with the equilibrium between different populations.

F. Metrical Trait

Consider a metrical trait (say, size) whose measurement is x.

Probably in most cases, if not in all, natural selection favors not the extremely large or the extremely small, but some intermediate value. This value may then be called the optimum size with respect to fitness and be denoted by x_0. In any realistic situation in which the metrical trait is controlled by a number of loci, the optimum size x_0 is probably close to but does not necessarily coincide with the mean size $\bar{x}$ of the population.

One way of representing the fitness that decreases with large as well as with small sizes is to let

$$w = 1-c(x-x_0)^2$$

where c is a constant. That is, the selection coefficient against an organism of size x is proportional to the square of the deviation of x from optimum value. Suppose that the size of the three genotypes (AA, Aa, aa) are $x_1 = 11$, $x_2 = 10$, $x_3 = 1$, respectively, and that the optimum size is $x_0 = 8$. Then the fitness of the three genotypes will be (taking $c = 0.01$)

$$w_1 = 1 - c(11 - 8)^2 = 0.91$$
$$w_2 = 1 - c(10 - 8)^2 = 0.96$$
$$w_3 = 1 - c(\ 1 - 8)^2 = 0.51$$

Such a set of fitness values will lead to a stable equilibrium, as shown in section 4. If the optimum size is $x_0 = 4$, the three fitness values would be 0.51, 0.64, 0.91, leading to fixation of the small size.

For a metrical trait determined by a number of loci, the situation is very complicated and selection for an optimum size does not always lead to stable equilibrium. Wright (14) found no stable equilibrium when there is no dominance or complete dominance with respect to size at all loci.

Kojima (6, 7) recently showed that stable equilibrium points do exist when there is partial dominance or overdominance in size at all loci, and proposed a method of locating such stable points. In nature there must be a large number of metrical traits stabilized at certain optimum levels. It is unfortunate that the mathematics involved in this type of investigation should be so intricate that the most important type of selection in nature has been also the least understood. Kojima's finding must be hailed as a significant landmark in selection genetics.

In all of the previous examples there is only one equilibrium point for a population under a given selection scheme. In more complicated situations, however, such as that involving multiple alleles, multiple loci, local differential selection scheme, differential selection in sexes, mutations of all directions, etc., there may be more than one stable equilibrium value at which the population may be stabilized. The selection scheme may change from time to time and this fluctuation may also keep a population in equilibrium (2).

IV. Comparison of Population Fitness

The selection effect that arises from the differential reproduction or transmission of different alleles or genotypes within a population, though quite involved at times, is at least in principle easy to comprehend. It involves only the relative fitness of the genotypes concerned, and the problem of population growth, either in absolute size or in comparison with other similar populations, has not been taken into consideration. When we try to compare two or more populations, difficulties arise because there does not exist a unique criterion (or scale) by which different populations may be judged.

To illustrate the difficulty of comparing two populations, let us once more consider the mutation-selection equilibrium condition $Rs = u$, or $R = u/s$. The value of $\bar{w} = 1 - Rs$ at the bottom of Table 2B is the total of the selected individuals, but it may also be viewed as the *average* fitness of the three genotypes in the population (average fitness of the population in short), with the understanding that w_1 and w_2 are taken to be unity. Thus, Rs is the amount of recessive individuals to be eliminated by selection in each generation, although the same amount will be replaced through segregation of the old and new mutations so that there is no net change in the composition of the population. Mutation rate and selection intensity are parameters of nature, and have been taken as constants in our models; that is, for short range genetical purposes. (They may vary widely in the history of evolution.) Some hypothetical equilibrium populations are listed in Table 5. The average fitness at equilibrium is

$$\bar{w} = 1 - sR = 1 - s(u/s) = 1 - u,$$

independent of selection intensity. This is also true for several other cases (5).

In populations 1–3, mutation rate varies proportionally to selection intensity so that the recessive proportion R remains the same

TABLE 5.—EQUILIBRIUM CONDITION WHEN SELECTION IS AGAINST THE RECESSIVES.*

	Population 1	Population 2	Population 3
Parameters in nature	$u = 0.000,018$ $s = 0.02$	$u = 0.000,045$ $s = 0.05$	$u = 0.000,090$ $s = 0.10$
Recessive proportion	$R = 0.0009$	$R = 0.0009$	$R = 0.0009$
Average fitness	$\bar{w} = 0.999,982$	$\bar{w} = 0.999,955$	$\bar{w} = 0.999,910$
	Population 4	Population 5	Population 6
Parameters in nature	$u = 0.000,050$ $s = 0.02$	$u = 0.000,050$ $s = 0.05$	$u = 0.000,050$ $s = 0.10$
Recessive proportion	$R = 0.0025$	$R = 0.0010$	$R = 0.0005$
Average fitness	$\bar{w} = 0.999,950$	$\bar{w} = 0.999,950$	$\bar{w} = 0.999,950$

*$w_1 = 1 - t;\ w_2 = 1;\ w_3 = 1 - s.$

in all three populations. This means, the genetic composition of these three populations are the same at equilibrium. The average fitness of the populations, however, is decreasing from 1 to 3. In populations 4–6, the mutation rate and consequently the average fitness of the population is the same, but the recessive proportion R, and thus the recessive gene frequency q, decreases from 4 to 6.

How do we compare these populations? What criterion should be adopted? What is the value judgment involved here? In some discussions it is implied that a "good" population should have a low frequency of deleterious genes. This seems to suggest that the value of R should be used to compare different populations. If so, then populations 1, 2, 3 are considered to be equally good and 4 is worse than 5 which in turn is worse than 6. In other discussions the average fitness has been adopted as the criterion for judging populations on the theory that $\bar{w}$ shows the deviation of the actual population from an hypothetical one in which there is no deleterious mutation nor selection (all AA, $w_1 = \bar{w} = 1$). From this viewpoint, populations 4–6 are equally good, but 1 is better than 2 which in turn is better than 3.

The difference between these two criteria is essentially this: R is static in nature, describing the genetic composition of a population, while $sR = 1-\bar{w}$ is dynamic, indicating the amount of selectional turnover in each generation. As an analogy, we may consider two gamblers A and B. The former likes to play big stakes and wins or loses a large sum in each bet but comes out even at the end of the game; the latter plays cautiously and wins or loses a small sum in each bet but also comes out exactly even at the end of the game. Who, then, is a better gambler? When they do lose, A loses a total of $50 in two hands and B loses the same amount in ten hands, but both of them recover their losses before the end of the game so that there is no change in wealth distribution. Who shall we say is a better gambler?

Comparing two populations in selectional balance involves more or less the same difficulty. For the sake of simplicity, let us assume $F = 0$, so that the equilibrium condition is, according to section 4,

$$p = \frac{s}{s+t} \quad \text{and} \quad q = \frac{t}{s+t}$$

and the average fitness is

$$\bar{w} = 1 - tp^2 - sq^2 = 1 - \frac{st}{s+t}$$

In Table 6 are listed some examples of equilibrium populations of this type. It is seen that populations 1 and 2 have the same genetic composition but different average fitness, while populations 3 and 4 have different genetic composition but the same average fitness. Since in the case of selectional balance neither allele A nor allele a can be strictly considered deleterious, it may be easier to use $\bar{w}$ as an index which, as before, measures the amount of selection going on in the population, although the selection produces no change on the genetic composition of the population at equilibrium.

Finally, it should be pointed out that although $\bar{w}$, based on relative fitness of genotypes, measures the amount of selectional turnover per generation, it is not useful as an index for the *absolute* surviving ability of a population. If two populations, one with $\bar{w} = 0.94$ and one with $\bar{w} = 0.64$, are grown side by side in the same environmental conditions, there is no way to foretell which population will win out. The genetics of intrapopulation selection deal only with the genetic composition of a population and give no information as to inter-population competing abilities. The latter more properly belongs to the realm of ecology, to be studied by field naturalists and geneticists. In plants there are instances in which only heterozygotes survive,

TABLE 6.—EQUILIBRIUM CONDITION WHEN SELECTION FAVORS HETEROZYGOTES. *

	Population 1	Population 2
Parameters in nature	$w_1 = 0.90$ $\}\ t = 0.10$ $w_2 = 1.00$ $\}\ s = 0.15$ $w_3 = 0.85$	$w_1 = 0.40$ $\}\ t = 0.60$ $w_2 = 1.00$ $\}\ s = 0.90$ $w_3 = 0.10$
Gene frequency Genotype proportion	$p = 0.60,\quad q = 0.40$ 0.36, 0.48, 0.16	$p = 0.60,\quad q = 0.40$ 0.36, 0.48, 0.16
Average fitness	$\bar{w} = 1 - \dfrac{0.015}{0.250} = 1 - 0.06$	$\bar{w} = 1 - \dfrac{0.54}{1.50} = 1 - 0.36$

	Population 3	Population 4
Parameters in nature	$w_1 = 0.92$ $\}\ t = 0.08$ $w_2 = 1.00$ $\}\ s = 0.10$ $w_3 = 0.90$	$w_1 = 0.95$ $\}\ t = 0.05$ $w_2 = 1.00$ $\}\ s = 0.40$ $w_3 = 0.60$
Gene frequency Genotype proportion	$p = 5/9,\quad q = 4/9$ 25/81, 40/81, 16/81	$p = 8/9,\quad q = 1/9$ 64/81, 16/81, 1/81
Average fitness	$\bar{w} = 1 - \dfrac{0.008}{0.180} = 1 - 0.044$	$\bar{w} = 1 - \dfrac{0.02}{0.45} = 1 - 0.044$

*$w_1 = 1 - t; \; w_2 = 1; \; w_3 = 1 - s.$

e.g., Oenothera; that is, $w_1 = w_3 = 0$. The average fitness in such a case is very low, meaning that there is a great deal of selection going on, but the plant population may nevertheless thrive.

Summary and Conclusion

Two major types of genetic equilibrium have been discussed.

I. *The balance between recurrent mutation and persistent selection.* In this case, the gene frequency is always kept at a very low level, ranging from the order $\sqrt{u}$ to u, where u is the mutation rate to the gene under consideration. The small immediate harmful effect on the population as a whole must be viewed as the price the population has to pay for preserving the gene for possible further use in the course of evolution. Since mutations, the ultimate source of genetic variability, occur in nature, a population containing a small amount of mutants must be defined as the "normal" state of affairs.

II. *The balance maintained by selection scheme alone.* Mutation plays a very little role in determining an equilibrium value of this type, because selection intensity is so much greater than mutation rate. The gene frequencies involved in this type of equilibrium are usually intermediate in value and thus result in genetic polymorphism. Although greater fitness of heterozygotes is the simplest and the most familiar mechanism by which selectional equilibrium may be attained, it is not a necessary condition in more complicated situations. A vast number of combinations of various factors may lead to stable selectional balance. Probably most of the common genetic characteristics, especially quantitative traits, are controlled by selection rather than by mutation.

Only very simple examples have been chosen to illustrate the various points. The difficulty of comparing fitness of two different populations has been discussed. Populations with the same average fitness may have quite different selection schemes and different genetic compositions. More detailed discussions may be found else-where (3, 4, 12, 13, and 17).

References

1. Crow, J. F. 1948. Alternative hypotheses of hybrid vigor. *Genetics,* **33:** *477–487.*

2. Dempster, E. R. 1955. Maintenance of genetic heterogeneity. *Cold Spr. Harb. Symp. Quant. Biol.,* **20:** *25–32.*

3. Dobzhansky, Th. 1957. Genetic loads in natural populations. *Science,* **126:** *191-194.*

4. ————, Krimbas, C., and Krimbas, M. G. 1960. Genetics of natural populations, XXIX: Is the genetic load in *Drosophila pseudoobscura* a mutational or a balanced load? *Genetics,* **45:** *741–753.*

5. Haldane, J. B. S. 1937. The effect of variation on fitness. *Amer. Nat.,* **71:** *337–349.*

6. Kojima, K. I. 1959 A. Role of epistasis and overdominance in stability of equilibria with selection. *Proc. Nat. Acad. Sci.,* **45:** *984–989.*

7. ————. 1959. B. Stable equilibria for the optimum model. *Proc. Nat. Acad. Sci.,* **45:** *989-993.*

8. Levene, H. 1953. Genetic equilibrium when more than one ecological niche is available. *Amer. Nat.*, **87**: *331–333.*

9. Lewontin, R. C. 1958. A general method for investigating the equilibrium of gene frequency in a population. *Genetics*, **43**: *419–434.*

10. Li, C. C. 1955 A. Population Genetics. *Univ. Chicago Press.*

11. ————. 1955 B. The stability of an equilibrium and the average fitness of a population. *Amer. Nat.*, **89**: *281–295.*

12. Wallace, B. 1959 A. The role of heterozygosity in Drosophila populations. *Proc. 10th Intern. Congr. Genet.*, **1**: *408–419.*

13. ————. 1959. B. Studies of the relative fitness of experimental populations of *Drosophila melanogaster*. *Amer. Nat.*, **93**: *295–314.*

14. Wright, S. 1935. Evolution in populations in approximate equilibrium. *Jour. Genet.*, **30**: *257–266.*

15. ————. 1955. Classification of the factors of evolution. *Cold Spr. Harb. Symp. Quant. Biol.*, **20**: *16–24.*

16. ————. 1956. Modes of selection. *Amer. Nat.*, **90**: *5–24.*

17. ————. 1959. Physiological genetics, ecology of populations, and natural selection. *Perspectives in Biol. & Med.*, **3**: *107–151.*

Comments

AUERBACH: It seems to me that the examples discussed under I and II are not at all equivalent. In I, the equilibrium refers to a pair of alleles, in II to whole genomes, as in Oenothera, or whole chromosomes, as in Dobzhansky's Drosophila inversions. The balance in Oenothera is upheld by genes which are *not* allelic. This difference between equilibrium for alleles or for larger parts of the genome may not be important for the way a population achieves equilibrium in nature, but it is *the* important question in the assessment of genetical radiation damage. As far as I know, there are few examples for selectional equilibrium at the allele-level, the only level which has to be considered in this regard.

Discussion of Session I

M. M. RHOADES

Indiana University, Bloomington, Ind.

A^{T} a symposium concerned with mutation and plant breeding it would seem appropriate to consider briefly just what is meant by the term mutation and to describe some of the inherited changes in chromosomal organization which lead to modifications of the normal phenotype and hence are called mutations.

Many hold that mutations, either of spontaneous or induced origin, stem in large part from some kind of intragenic change at the molecular level, i.e., they are true or intragenic mutations. Inasmuch as the genetic information is determined by the combinations of the two pyrimidines and the two purines of DNA, it is believed that the substitution of one purine or of one pyrimidine for another, or of a purine for a pyrimidine and *vice versa,* would produce a new kind of genetic code which would lead in some cases to an altered phenotype. A change in gene action producing a mutant phenotype could also conceivably occur by an inversion in base order or by a deletion or duplication of one or more base pairs. That mutations of these kinds do in fact occur appears highly likely from the studies by Freese and Benzer on the mutagenic effect of base analogues and other chemical mutagens in phage. Some mutagens are thought to induce mutations in duplicating DNA and others in non-duplicating DNA. Comparable though less extensive results have been obtained with bacteria. Muller, Carlson, and Schalet suggest that a rotational substitution, whereby the two organic bases of a nucleotide pair become freed from the backbone and reversed in position upon reattachment, may be responsible for the whole body mutations in the non-replicating DNA of Drosophila sperm.

Granting that intragenic mutations occur in phage and bacteria there is virtually no convincing evidence for intragenic changes in mutation studies of higher forms. This does not mean that intragenic changes do not occur in these organisms, and indeed on theoretical grounds it is difficult to deny that they do arise, but only that there are a number of extragenic events which simulate gene mutation. In practice those mutations which cannot be ascribed to one or another extragenic mechanism are tentatively placed in the category of gene mutation. However, as Stadler points out, the mutations labeled as intragenic constitute a residual class at present unascribable to an extragenic event. As more sensitive and discriminating techniques have been developed the proportion of putative intragenic mutations becomes progressively smaller. The residual class remains suspect since no certain criteria exist which permit an unequivocal proof of true gene mutation. It has been suggested that the frequency of intragenic changes is so low that they escape detection by the investigator who, at least in working with higher plants, usually restricts his attention to those cases where the mutation rate is high enough to afford an adequate number of changes for further analysis. A true or intragenic mutation rate of 1×10^{-5} for a single locus, which may be typical for many genes, would require a vast amount of effort to obtain an adequate sample of mutations. An exception would be the S alleles for incompatibility where the screening technique for mutations is as efficient as those employed for microorganisms.

Different extragenic events which simulate gene mutation are as follows:

1. Minute deficiencies.

2. Position effects.

3. Recombination between different mutant sites within a cistron or functional unit which yields a crossover strand with no affected sites and one with two mutant sites.

4. The separation and isolation of the components of a complex locus by some kind of crossover mechanism. Mutations of this type have been described in Laughnan's paper at this symposium. Mechanisms 3 and 4 are associated with meiosis.

5. Restoration of gene action through the loss of an adjacent inhibitor. Conversely, normal gene action may be modified when an inhibitor is placed next to a normal allele. One of the best examples comes from McClintock's studies on the Ds–Ac mutator system in maize where an apparent mutation of C→c resulted from the inhibitory effect of Ds on the C allele when the two were brought into juxtaposition by transposition. Recovery of C action followed the loss of Ds. This occurs only when Ac is present in the nucleus.

I am uncertain where to place the phenomenon called "gene conversion" because of the diverse opinions of the causal mechanisms and because it may consist of a heterogeneous class of changes. However, there is no compelling reason to ascribe cases of gene conversion to intragenic changes since extragenic mechanisms are capable of producing the observed results. Whatever may be the cause of the intriguing cases of paramutation in maize described by Brink and his associates and by Coe, there is no evidence in these studies of intragenic modifications.

It should be emphasized that in none of the many carefully conducted and extensive mutational investigations in maize is there unassailable proof of true gene mutation. In all well-analyzed cases an extragenic mechanism has either been demonstrated to be responsible for the mutant change or else one appears to be highly probable. Insofar as the mutation spectrum is concerned, it would be surprising if a comparable situation does not exist in other higher plants. Mutational studies are of concern not only to the geneticist but also to the plant breeder who utilizes mutations as a source of variation and who needs to be aware of the diverse mechanisms which simulate true gene mutation.

Comments

KRAMER: Since the waxy gene in corn produces pollen grains whose starch stains brown rather than blue with iodine, Dr. O. E. Nelson, by looking at stained pollen from F_1 plants of crosses between wx stocks derived from independent reoccurrences of wx mutations, has been able to identify recombinants between different wx mutants by the rare occurrences of blue staining grains. A number of mutants have been placed in linear order within the waxy locus by this method. The large populations of pollen grains which can be screened, make this locus an ideal one and comparable to phage for genetic fine structure studies.

CASPAR: The latest chromosome theory, based on the information provided from electron microscope pictures of chromosomes and theoretical molecular biology, states that there are 32 to 64 strands of DNA in the chromosome of higher plants and animals. The theory holds that "gene" mutation is due to rearrangements in the base pair configurations of the DNA. A mutation is expressed when the chromosomes of a cell all carry the same mutated base pair configuration. It would seem that a chromosome would have to go through at least 16 reduplications before this could happen.

How, then, does the investigator, testing for induced mutations in the higher plants and animals, score, let alone find, mutations which are due to induced changes in the base pair configurations?

RHOADES: If chromosomes are multi-stranded it is, of course, difficult to understand how a mutational event consisting, for example, of a base substitution could occur simultaneously in all of the component strands and thus be immediately expressed. However, the genetic evidence for whole-body mutations occurring in gametes cannot be questioned. This has been interpreted as indicating that the chromosome of higher organisms consists of a single double helix of different DNA molecules which are linearly arranged and possibly connected by protein links. If the genetic data are deemed more reliable than the cytological studies, then one favors the latter concept of chromosome structure. This conflict has not been resolved.

CASPAR: With respect to the remark that corn does not respond in the same way as other organisms in radiation experiments, I should like to state that in some preliminary data of ours from an experiment with corn similar in design to the ones the Russells have done with mice in which the male gametes carrying dominant marker genes are radiated at different stages of gametogenesis and crossed on multiple recessive females, our results are quite similar. We find that the controls and material radiated after meiosis consist of losses of adjacent marker genes and most all the mutants also carry associated sterility effects, while the mutants from material radiated prior to meiosis consist of single gene losses and the mutants do not carry associated sterility effects.

Session II

Mutagenic Agents and Interpretation of Their Effects

W. R. SINGLETON, *Chairman*
University of Virginia,
Charlottesville, Va.

Types of Ionizing Radiation and Their Cytogenetic Effects[1]

ARNOLD H. SPARROW[2]

Brookhaven National Laboratory, Upton, New York

"PHYSICISTS now comprehend not only the structure of stars, the motion of our own and other galaxies, the curvature of space, the possible ways in which our universe has evolved, but also matter on a finer and finer scale: from familiar objects to molecules; then to the atoms of which the molecules are composed; the internal structure of the atom with its electrons orbiting around nuclei; the nucleus itself, made up of protons and neutrons and the mesons which bind them together; and lately even something of a picture of the inside of the proton itself, complex and containing yet other particles. We are peeling an onion layer by layer, each layer uncovering in a sense another universe; unexpected, complicated, and—as we understand more—strangely beautiful."

So begins a memorandum prepared for former President Eisenhower by a Special Advisory Panel on High Energy Accelerator Physics (120).[3] In a similar manner, geneticists are probing at the ultimate secrets of life or the fine structure of the chromosome, the gene, and of their nucleoprotein components. It is of considerable interest that the use by geneticists of the knowledge and tools of the atomic and nuclear physicists is making a significant contribution to our new knowledge of genetic fine structure and also, we hope, to the useful application of radiobiological techniques in plant breeding.

It is my assignment to try to outline the physical nature of the ionizing radiations of most interest to the geneticist, to explain something of their interaction with matter, i.e., the process of energy transfer from the radiation into the atoms or molecules, and to survey briefly some of the biological effects produced by the complex series

---

[1]Research carried out at Brookhaven National Laboratory, Upton, N. Y., under the auspices of the U. S. Atomic Energy Commission.

[2]The author gratefully acknowledges the many helpful suggestions regarding the manuscript offered by Doctors H. J. Evans and Rae P. Mericle and by Miss Virginia Pond and Mrs. Rhoda Sparrow.

[3]See References, page 105.

of changes initiated when ionizing radiations impinge upon living cells. Partly because of the space limitations and partly because of the large numbers of more detailed considerations of this topic already published elsewhere, this review will be brief and aimed primarily at readers looking for a fairly short and elementary survey of the subject. In the sections concerned with radiation effects, I shall emphasize chromosome breakage and derived phenomena rather than intragenic changes, partly because of my greater personal interest in this topic and partly because induced mutation in the more restricted sense is being covered in this Symposium by others who are better qualified to discuss it than I am.

I. Ionizing Radiations and Their Characteristics[4]

By definition, ionizing radiations have the ability to produce ionization (ion pairs) when they interact with matter. With the removal of each electron a positively charged atom or molecule is left. In addition to the process of ionization, energy transfer also occurs by a process known as excitation. The major effect of the ionizing radiations is considered to result from their ability to ionize and to rupture chemical bonds. In contrast to the ionizing radiations, ultraviolet, except for the very shortest wave lengths, does not have the capacity to ionize but transfers energy primarily through the process of excitation. The fact that ultraviolet radiation does produce genetic effects (both mutation and chromosome breakage) is adequate evidence that excitation itself can produce a biological effect.

Ionizing radiations include two different types: (a) electromagnetic radiations which include X and gamma rays, and (b) the so-called particulate radiations (alpha, beta, protons, deuterons, etc.). Neutrons are generally classed as ionizing radiations, but it should be pointed out that they do not ionize directly but indirectly through the nuclear reactions which occur following their absorption by atomic nuclei. After absorption they may emit α, proton, electron, or γ rays, or disintegrate.

Cosmic radiation is not of any serious interest to the plant breeder so far as artificially produced mutations are concerned but no doubt plays a part in the production of so-called spontaneous muta-

---

[4]For definitions of physical terms used, see Glossary of Terms in Nuclear Science (111).

tions. The cosmic rays are really a complex of radiations which are derived from interactions of very highly energetic particles. The less energetic particles interact with the atmospheric atoms and various reactions occur, including the production of mesons which in turn decay with the production of electrons, positrons, gamma rays, secondary pair production, and the associated showers or avalanches of low-energy electrons. The amount of cosmic radiation varies with the altitude and with the position on earth with respect to the magnetic poles (112).

With the rapid developments in space science, it is now possible to send seeds or spores into outer space and recover them. Since laboratory experiments using man-made radiations similar in nature to (but perhaps less energetic than) the particles from outer space are now under way, it will be possible to test the biological effects of both cosmic radiation and other ionizing radiations in the Van Allen belt. Reports of such experiments will appear shortly in the literature and should show significant effects if the predicted doses (up to 3×10^4 r per hour) really exist (112).

The characteristics of the various ionizing radiations are of considerable interest to physicists and some of their properties should be understood by radiobiologists. A partial list of the ionizing radiations and some of their more important properties are summarized in Table 1. This table may appear, at first sight, to be unnecessarily complicated. However, the minimum amount of information is given which, in my opinion, will allow the average biologist to use these radiations with some understanding of their practicality and the reason for differences in their relative biological effectiveness.

It should be clearly understood that ionizing radiation generally produces an effect proportional to the energy absorbed in the tissue in question. The fraction of energy which passes right through a cell or tissue produces no effect nor does the energy absorbed in the air or medium surrounding the object in question (except for very small structures). Thus, in general terms, it is obvious that a radiation must have sufficient energy to penetrate to the position where its effect is sought and not sufficiently penetrating that most of it passes right through. The penetration of the different radiations is thus a major factor in deciding which radiations should be used in many experiments, and it also has an important bearing on their relative hazards.

TABLE 1.—PROPERTIES OF RADIATION.*

Types of radiation	Source	Description	Energy†	Hazard	Necessary shielding	Penetration into tissue‡
X-rays	X-ray machine	Electro-magnetic radiation	Commonly 50—300 kv	Dangerous, penetrating	A few mm of lead except for very high-energy installations	A few mm to many cm
Gamma rays	Radioisotopes and nuclear reactions	Electro-magnetic radiation similar to X-rays	Up to several Mev	Dangerous, very penetrating	Requires very heavy shielding, e.g., inches of lead or several feet of concrete	Many cm
Neutrons (fast, slow, and thermal)	Nuclear reactors (piles) or accelerators	Uncharged particles slightly heavier than proton (hydrogen atom), not observable except through its interaction with nuclei in the material it traverses	From less than 1 electron volt to several Mev	Very hazardous	Thick shielding composed of light elements—such as concrete	Many cm
Beta particles, fast electrons or cathode rays	Radioactive isotopes or accelerators	Electrons (+ or −) ionize much less densely than alpha particles	Up to several Mev	May be dangerous	Thick sheet of cardboard	Up to several mm

Alpha particles	Radioisotopes	A helium-nucleus; ionizes very heavily	2—9 Mev	Very dangerous internally	Thin sheet of paper sufficient	Small fraction of a mm
Protons or deuterons§	Nuclear reactors or accelerators	Nucleus of hydrogen§	Up to several Bev (10⁹)	Very hazardous	Many cm of water or paraffin	Up to many cm

*After Sparrow and Konzak (167).
†Penetration is directly related to energy of the radiation and this is expressed in terms of electron volts, i.e., kv = kilovolt = 1000 volts; Mev = million electron volts; Bev = billion electron volts (10⁹).
‡Penetration is dependent upon many variables which cannot be considered in detail here. It is assumed here that penetration is into an ordinary plant tissue of average density.
§A proton is the nucleus of the common isotope of hydrogen and a deuteron is the nucleus of the heavy isotope of hydrogen.

The relative penetrations of several different radiations in various kinds of organic materials are indicated in Figure 1. Penetration into tissue depends upon density of the tissue, but is generally roughly comparable to that shown in Figure 1.

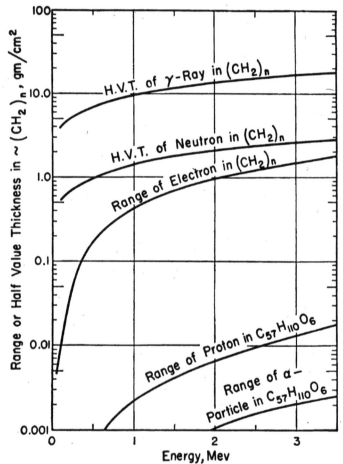

FIGURE 1.—*Range and half value thickness (H.V.T.) of various ionizing radiations or particles in $(CH_2)_n$. Range or H.V.T. would be roughly the same in plant tissues of similar density. After Sun (182).*

In addition to penetration one should know the pattern of distribution of absorbed energy in the irradiated object (131). The dose at a particular position below the surface is called the depth dose and is determined by a large number of factors, including such things as kind and energy of radiation (Figure 2), size of the area irradiated, chemical composition and density of the specimen or tissue, collima-

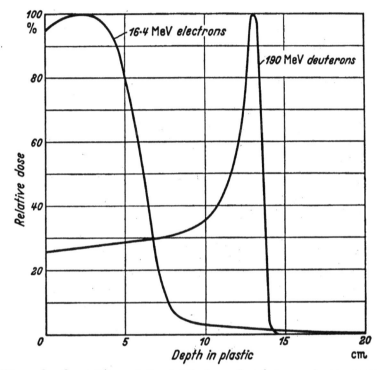

FIGURE 2.—*Comparison of the depth dose effect between deuterons and electrons. Curve taken from Tobias, Anger, and Lawrence (189). These curves were determined experimentally, using a plastic having an absorption coefficient similar to tissue.*

tion of the beam, etc. These factors are generally thoroughly discussed in books on radiological physics or radiation therapy (10, 51, 52, 76, 113, for example). These should be consulted for further detail.

Another important characteristic is the relative density of ionizations along the path of the ionizing particle. (See Figure 14.) As can be seen from Table 2, this varies widely for the different radiations, beginning at a minimum of approximately 8 ionizations per

MUTATION AND PLANT BREEDING

TABLE 2.—ION DENSITY PRODUCED BY DIFFERENT IONIZING RADIATIONS OR PARTICLES.*

Radiation	Mode of generation	Mean linear ion density †	Rate of transfer of energy kev/micron	Ionizing particle
Very high energy β and γ radiation	20–30 million volt betatron radioactive elements	8.5	0.28	Electron
X radiation	"Supervoltage" 1,000 kv	15	0.49	Electron
	"Deep therapy" 200 kv	80	2.6	Electron
	From copper (K) 8 kev	145	4.7	Electron
	From Al (K) 1.5 kev	460	15.0	Electron
Neutron radiation	Cyclotrons at 12 Mev	290	9.5	Proton
	Cyclotrons at 8 Mev	380	12.4	Proton
	400 kv deuterium ions bombarding deuterium	1,100	35.8	Proton
α radiation	Natural disintegration of radon	3,700	120	α particle
	Slow neutrons	9,000	292	Nuclear particle
Atomic rays	Uranium fission	130,000	4,240	Nuclear particle

*After Gray (56).
†Ions per micron of tissue.

micron and going up to 130,000 for very dense tracks made by uranium fission fragments. Since each ionization is estimated to require 32.5 ev, it is apparent that the denser the ionizations along the track the higher the rate of energy transfer from the ionizing particles to the atoms or molecules of the tissue. This concept known as linear energy transfer (LET) is defined as the energy released by the radiation per unit length of track in the absorbing material. It is usually expressed in units of kev/μ and the LET varies directly with the square of the charge and inversely as the square of the velocity (or energy) of the particles. The charge on an ionizing particle does not change along its path, but as velocity (energy) remaining is gradually reduced, the frequency of ionization increases up to a maximum. The rate of energy loss for electrons in water is shown in Figure 3. The denser parts of the tracks are known as tails and are the most effective

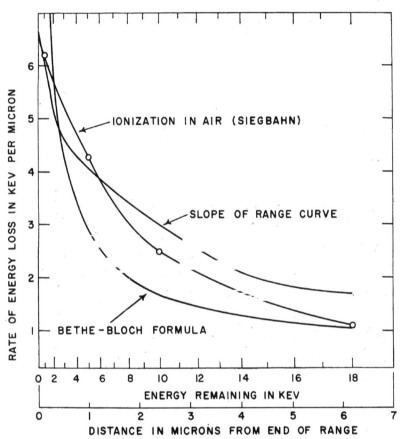

FIGURE 3.—*Rate of energy loss for electrons in water. After Robertson and Hughes (129).*

portions for any radiobiological event requiring more than two ionizations. The radiobiological significance of differences in ionization density or LET is discussed in Section VIII.

II. Units of Radioactivity and of Dosage Measurement

Genetic effects can be produced by ionizing radiation regardless of the origin of the radiation. Today many geneticists use radiations from radioisotopes or nuclear reactors as well as from X-ray machines or accelerators. A list of isotopes used for biological experimentation is given in Table 3. These isotopes emit alpha, beta, or gamma rays, or a combination of two or more.

It is imperative that careful dosage measurements be made if reproducible results are to be obtained in radiation experiments. Generally dosimetry should be done by someone well qualified to

TABLE 3.—RADIOISOTOPES USED IN BIOLOGICAL STUDIES, LISTED ACCORDING TO HALF-LIFE, WITH RADIATIONS EMITTED AND, FOR SOME ISOTOPES, DOSE RATE PER CURIE AT 1 METER.

Radioisotope	Half-life*	Radiation(s) emitted	Dose rate at 1 meter, r/hr per curie†
Manganese–56................	2.56 h	β, γ	
Potassium–42.................	12.7 h	β, γ	
Sodium–24·..................	15.00 h	β, γ	1.93
Radon......................	3.82 d	α	
Iodine–131..................	8.05 d	β, γ	0.231
Phosphorus–32..............	14.3 d	β	
Rubidium–86.......	18.6 d	β, γ	
Iron–59....................	45.1 d	β, γ	0.65
Strontium–89.....	53 d	β	
Sulfur–35..................	87.1 d	β	
Polonium–210...............	138.3 d	α, ι	
Calcium–45.................	164 d	β	
Zinc–65..........	245 d	$\beta+$, γ	0.30
Cobalt–60.....	5.27 y	β, γ	1.32
Barium–133.................	7.2 y	γ	
Hydrogen–3 (tritium)........	12.46 y	β	
Strontium–90...............	28 y	β	
Cesium–137—barium–137......	30 y	β,	0.356
Radium–226·...............	1,622 y	α,	0.84
Carbon–14··················	5,568 y	β	
Uranium–238...............	4.4×10^9 y	α	

*h = hour; d = day; y = year.
†These values taken from Kinsman (82, page 139).

handle the appropriate apparatus. There are many pitfalls for the unwary, and, especially so, with internally deposited isotopes. A good idea of the complexities involved can be obtained from the book by Hine and Brownell (72). The details of dosimetry measurements are not appropriate to this article, but suffice it to say that the amount of radioactivity is determined by counting the number of particles emitted in a given time and that it is commonly designated in curies, millicuries, or microcuries. A curie (abbreviation: c) is defined as

3.7×10^{10} disintegrations per second, without regard to the number of particles released per disintegration, their energies, or their properties. It is possible to convert a known number of disintegrations into a standard unit of dosage measurement if both the number of particles emitted and their average energy is known. Dose rates obtained from several gamma emitting isotopes are shown in Table 3. Note that the dose rate per curie varies widely with different isotopes.

There are several different units currently in use for the measurement of amounts of radiation produced in air or the amount of energy absorbed in tissue or similar material. The oldest unit still in general use is the *roentgen* (abbreviation: r). It is defined as *"that quantity of X or gamma radiation such that the associated corpuscular emission per 0.001293 gram of air produces, in air, ions carrying 1 e.s.u. of electricity of either sign"*. A roentgen is equivalent to the absorption of approximately 98 ergs/gm in water or tissue. One roentgen of X or gamma rays will produce 2.083×10^9 ion pairs per cc of air at standard temperature and pressure, 1.6×10^{12} ion pairs per gram of tissue or about 1.8 ion pairs per μ^3. For specific values of different radiations at various energies see Table 2 of Lea (93).

A second unit, less commonly used, is known as the *roentgen-equivalent-physical* (abbreviation: rep) and can be used for any ionizing radiation. A rep is defined as that quantity of corpuscular radiation which produces in tissue, per gram of tissue, an amount of ionization equivalent to that produced by 1 r of gamma radiation in air.[5] Both of these units have the disadvantage that they do not measure the energy absorbed but depend on the amount of ionization produced in air. In order to avoid this difficulty, a newer unit, the rad, has been adopted recently. It is a unit of absorbed dose and 1 rad equals 100 ergs/gm. For X-rays, one rad equals the amount of energy released by 1.08 r in water.

More detailed information about the dosage units and measurements of ionizing radiation are given in many books and articles (43, 68, 72, 83, 98, 124, 193, for example). A summary titled "Recommendations of the International Commission on Radiological Protection" has recently been published (74). This or a similar article

---

[5]Unfortunately, the definition of the rep has undergone some revision and its value has appeared in the literature both as energy absorbed per unit mass or per unit volume. The rad unit avoids this difficulty. See glossary (111) for further comments.

should be carefully read by anyone contemplating work with ionizing radiation.

III. Methods of Exposure and Facilities Used

Methods of exposing suitable material to radiation and the size and nature of the facilities used vary so widely that it is difficult to summarize them. For instance, experimental procedures used with plants vary from the treatment of a few milligrams of pollen or spores with an inexpensive ultraviolet source or small portable X-ray machine to the simultaneous treatment of hundreds or thousands of large plants or trees by gamma sources of many kilocuries in size (159, 167, 171). Space limitation does not allow a full description of all the methods used and facilities available, but representative radiation sources are listed in Table 4. A more complete description of a few types is given below. These have been selected largely on the basis of their actual usage or potential significance in cytogenetics or plant breeding.

A. X-ray Machines

X-ray machines are probably the most widely available source of ionizing radiation and have as their chief advantages availability, versatility, and, generally, ease of operation. They are also easier to shield than radioisotopes emitting highly energetic gamma rays such as Co^{60} and Cs^{137} and do not require such frequent calibration or calculations of dose as is necessary with isotopes, especially those with shorter half-lives.

X-ray machines suitable for biological experiments are usually in the 50- to 300-kv range and their output may exceed 1,000 r per minute for small samples. Newer types may go much higher. As compared to Co^{60} gamma sources, however, X-ray machines have as their chief limitations: (a) their requirement for considerable electric power; (b) that the maximum intensities and penetration of their radiation are in most cases much less than that of a Co^{60} source; (c) that the size of the object to be irradiated is definitely limited with most installations, except for very low doses; (d) the fact that X-ray machines cannot be operated economically for continuous long exposures; and (e) maintenance costs are considerable for machines operated near full capacity.

B. Gamma Sources

The availability of large amounts of long-lived isotopes emitting gamma radiation has given plant radiobiology an unequalled opportunity for exposure of plants or plant parts to large amounts of ionizing radiation and created new opportunities for comparing chronic exposure with the more usual acute methods of treatment. A list of common gamma-emitting isotopes and the intensity of gamma radiation per curie per hour at one meter is given in Table 3. The general relationship between the energy of gamma photons and their dose rate at 1 meter is given in Figure 4.

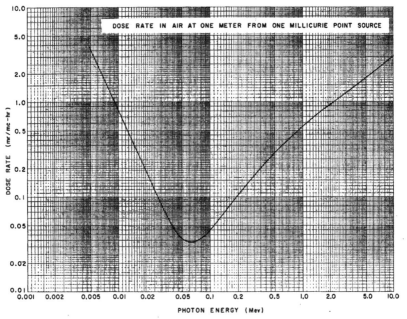

FIGURE 4.—*Dose rate in milliroentgens per millicurie per hour at 1 meter in air (without absorption) from a point source. After Slack and Way (150).*

A large number of gamma ray sources (mostly Co[60]) are now available for biological use. These vary from sources of less than a curie used for treatment of small samples in shielded containers up to those of many kilocuries. Some of these are located in large but

TABLE 4.—REPRESENTATIVE RADIATION SOURCES.*

Source	Major radiations emitted	Contaminating radiation	Half-life	Method of production	Source strength available	Remarks
Gamma Rays						
Cobalt-60	1.17- and 1.33-Mev γ	0.3-Mev β	5.3 yr	(n, γ)	Kilocuries available; 10^6 r/hr easily obtained over small volume	Contaminating β-rays are easily filtered out, resulting in pure monochromatic γ-ray sources
Cesium-137	0.669-Mev γ	0.81- and 1.2-Mev β	33 yr	Uranium fission	Multicurie sources	
Gold-198	0.408-Mev γ (99%)	0.96-Mev β (99%)	2.7 da	(n, γ)	Multicurie sources	
Thulium-170	0.084-Mev γ	0.886- and 0.197- Mev β	129 da	(n, γ)	Multicurie sources	
Beta Rays						
Phosphorus-32	Continuous β spectrum; mean energy, 400–500 kv, 1.72-Mev max	None	14.3 da	$P^{31} (n, \gamma) P^{32}$; $S^{31} (n, p) P^{32}$	Multicurie sources. 5×10^3 rep/hr easily obtained from P^{32}-Bakelite plaques	Mean β energy is given for thick source
Yttrium-91	Continuous β spectrum; 1.55-Mev max	Weak γ-rays	57 da	Uranium fission		

	Energy	γ-ray	Half-life	Yield	Remarks
Alpha Particles					
Polonium-210	5.3-Mev α	None	138 da	Multicurie sources	Ideal for micro sources
Plutonium-239	5.1-Mev α	None	2.4×10^4 yr		Ideal for very low-level α-irradiation
Fast Neutrons					
Polonium-beryllium	Mean energy, ~4.5 Mev	Small	138 da	3×10^6 neutrons/sec/curie	
D-D (accelerator)	Monochromatic neutrons, ~2.5 Mev	Small		10^9 neutrons/sec	These yields are approximate figures for 500-μa beam current at 250 kv
D-T (accelerator)	Monochromatic neutrons, ~14 Mev	Small		10^{11} neutrons/sec	
p-Be (accelerator)	Continuous distribution, peak at ~1.5 Mev	γ-ray present		Very high yields	
D-Be (accelerator)	~4 Mev	1 γ photon per neutron		Very high yields	

*After Kirby-Smith (83). All neutron energies are dependent on energy of bombarding particle.

carefully protected open fields for chronic or acute treatment of plants of almost any desired size. Some uses of gamma sources in botanical research have been summarized by Sparrow (159). While much has been written about the value of these large gamma sources in plant breeding (146, 147, 167), their usefulness in this respect seems to be generally no greater than the more conventional radiation sources such as X-ray machines or shielded radioisotopes used under laboratory conditions. They are, however, very versatile and may be especially suited for certain specialized crops which are ordinarily reproduced by vegetative methods. The main advantage of these large gamma sources is the ease with which chronic treatments can be made at any desired dose rate and the very high intensities of radiation available at least in a restricted area near the source. Daily dose rates available at various distances from three Co^{60} sources are given in Table 5. As shown in Figure 4, dose rates from a given size of source

TABLE 5.—DOSE-DISTANCE RELATIONSHIP FOR THREE COBALT-60 SOURCES IN USE IN THE BIOLOGY DEPARTMENT AT BROOKHAVEN NATIONAL LABORATORY.*

| Distance from source, meters | Greenhouse | | 10-acre field |
	0.21 curie (effective) r/20-hr day	8.2 curies (effective) r/20-hr day	2,040 curies (effective) r/20-hr day
0.5	21.0	670	154,000
1.0	5.0	203	44,000
3.0	0.8	24	5,750
5.0	0.26	8.7	2,175
7.0	0.14	4.4	1,130
10.0		2.2	550
20.0			130
40.0			31
80.0			6.4
160.0			1.3

*After Sparrow (159).

varies with the energy of the radiation. It is the opinion of the author that these sources should be regarded as versatile radiobiological tools and not primarily as facilities for the average plant breeder.

While most installations to date have used Co^{60} because of its availability, it would seem that Cs^{137} would be more suitable for future installations. The main advantage of Cs^{137} is its long half-life (30 years) compared to the 5.3 year half-life of Co^{60}. Another advantage of the Cs^{137} is its less energetic gamma which makes shielding easier. This is a potential disadvantage too, of course, wherever deep penetration into tissue is desired, and more curies are required to give the same output in r/hr.

C. Accelerators and Cyclotrons

The number of accelerators of various types available today is bewildering to the average biologist and no attempt will be made to enumerate all of them. The smaller accelerators for electrons cost only a few thousand dollars, but those more recently developed cost many millions of dollars. While the more expensive types can sometimes be scheduled for biological experiments, one can hardly visualize having such machines built at this time for our exclusive use.

In addition to their great expense, accelerators are generally difficult to use for biological experiments and are in great demand by physicists. For these reasons they are used much less in biological experiments than other types of radiation sources. They are a major source now for biological experiments with fast electrons, protons, neutrons, and stripped nuclei (44). It is possible by appropriate design to obtain microbeams of a few microns in diameter from various accelerators (205). Protons or deuterons usually are used in such microbeams to reduce the amount of scatter which would occur with X-ray or electron beams. Mesons can be produced artificially by the acceleration of alpha particles or protons to energies in the Bev range. As far as the author is aware no genetic experiments with mesons have yet been performed owing to their elusive nature and the relatively low intensity available. With the development of the extremely high energy accelerators now under test or construction (45), it should soon be possible to obtain enough of these energetic particles to test them for their relative biological effectiveness in producing breakage and mutation.[6] However, since we know that the ionization density produced along their tracks varies from extremely low to extremely high values, it is predictable that their effectiveness will cover a broad range depending on their energy at the time they traverse a given nucleus.

---

[6] The Brookhaven Alternating Gradient Synchrotron will attain energies up to 30 billion volts.

D. Nuclear Reactors

Unfiltered radiation from nuclear reactors is a complex mixture of radiations and is not suitable for well-defined biological experiments. Facilities with fairly complicated shielding are required to provide thermal or fast neutron beams with a small amount of gamma contamination (29).

Some radioactivity is induced in the biological specimens treated with thermal neutrons, but the amount of radioactivity is usually small and of fairly short half-life. The radioactivity induced constitutes no serious hazard with reasonable precautions. For most plant material exposure times in the Brookhaven facility are in hours or days. A considerable amount of work has been done in the last decade in the treatment of seeds and other plant material with thermal neutrons for comparison with X- or gamma-ray treatment (35, 141, 142, 160, 186, 190, 203). The main rationale behind this is that the relatively high ion density obtained with thermal neutron exposures may give results different from those one would get with the less densely ionizing radiation.

E. Small Portable Irradiators

In addition to the portable X-ray machines and ultraviolet sources, many other relatively inexpensive small sources have been designed and used for various purposes. Beta irradiators of various sizes and types can be used for localized exposures of suitable material either with small beams through minute pores or by surrounding the material to be irradiated with a radioisotope suitably bound to prevent corrosion or flaking (167). Maximum penetration of biological material with beta rays is usually only a few millimeters or less, depending upon the radioisotope used. Dosimetry of beta sources is difficult and complex (95). Similar irradiators can be made using alpha emitters and as these penetrate only one or a few cell layers, treatment of very localized areas is possible. Encapsulated radioactive materials in various sizes and shapes are now available commercially and can be used in a great variety of ways for experimental work.

F. Internal Emitters

Whenever radioisotopes are located inside cells or tissues, they are called internal emitters. Many internal isotopes have been studied with respect to their cytogenetic effect, but the four most commonly used are probably phosphorus–32, sulfur–35, carbon–14, and tritium. These and others less frequently used are listed in Table 3 along with their respective half-lives and radiations emitted. Of special interest to geneticists are those radioactive elements or compounds which are selectively concentrated in chromosomes and nuclei. (See Caldecott (19) for references.) Tritiated thymidine is an example (94). Important but less extensive use also has been made of various precursors labeled with C^{14}, P^{32}, S^{35}, etc. (179, 181, 204).

Tritiated thymidine was initially used by Taylor, Woods, and Hughes (187) because of its localization in the nuclei and high resolution in autoradiography. Because of its concentration in the nucleus and the short, dense, ionization tracks produced, its energy is highly localized (Figure 5) and obviously can be expected to produce changes fairly efficiently in nuclei (129). A number of such studies have been made and some results obtained by Wimber (196) are shown in Figure 6.

Other studies in which mutations have been induced by internal radioisotopes, including potentially useful ones, are too great to list here. The use of labeled chemical mutagens reported by Moutschen–Dahmen and co-workers (107) and by Smith (153) promises to be a very interesting and valuable new technique. The genetic effect will result from both chemical and physical activity.

IV. General Survey of Cellular and Nuclear Changes Induced by Ionizing Radiation

In the interests of completeness, we shall first list here briefly most of the known changes which occur in nuclei following exposure to ionizing radiations. (See also Section VI.) A few are secondary effects, e.g., polyploidy which results from a primary effect on the spindle. Many of these are not too well-understood and the significance of some with respect to genetic changes may be obscure or unknown.

1. Gene mutation

The author will not attempt to define gene mutation but uses it as a category of genetic change different from those involving some kind of chromosome aberration. Gene mutation almost without exception increases linearly with increasing dose. (See Muller (108) and other pertinent papers in this volume.)

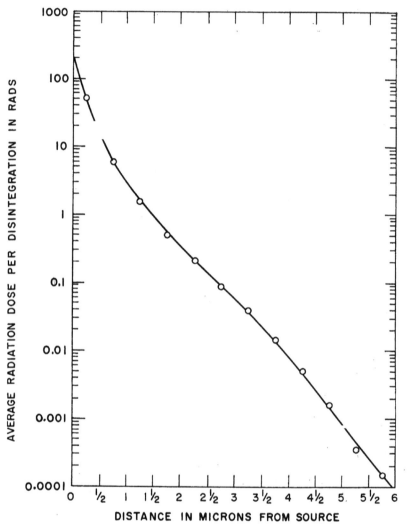

FIGURE 5.—*Radiation doses about a point source of tritium. After Robertson and Hughes (129).*

2. Chromosome, chromatid or subchromatid aberrations

The type of aberration produced depends upon the number of strands present in the chromosome, on the length or density of the

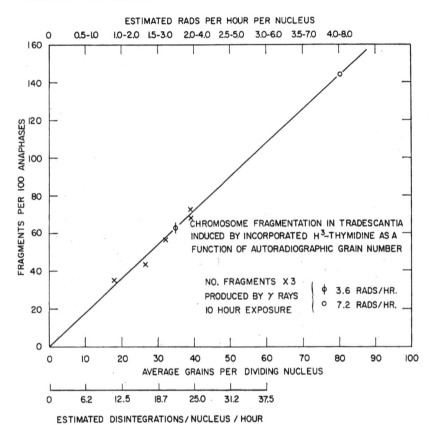

FIGURE 6.—*Increase in chromosome fragmentation in* Tradescantia palu-dosa *exposed to 1 μc/ml H³-thymidine as a function of autoradio-graphic grain number over dividing nuclei. 200 or more cells were scored for each point. Disintegrations per nucleus calculated on basis of 1 grain per 143 disintegrations. After Wimber (196).*

ionizing track which traverses the strand or strands, and on a great number of other factors. (See sections on chromosome breakage and modifying factors.)

3. Changes in chromosome number

Changes in the basic chromosome number vary from the addi-tion or subtraction of one or a few chromosomes to the alteration of whole chromosome sets. Haploidy, triploidy, and tetraploidy have all been reported to result from treatment with ionizing radiation (158). Some cases of polyploidy apparently result from inhibition or destruction of spindles as in Trillium (156), others presumably from restitution caused by multiple bridge formation. Aneuploidy could result either from nondisjunction or from abnormal or multipolar spindles.

4. Binucleate or multinucleate cells

Either of these effects may result from the inhibition of the spindle. They are commonly seen after irradiation of meiotic or post-meiotic cells in many plants and less frequently in somatic cells.

5. Effects on centromeres

One frequently sees lagging chromosomes which behave as if the centromere were nonfunctional or at least delayed in functioning. A second effect of radiation on centromeres is misdivision of the centromere during which it splits at right angles to the longitudinal axis of the chromosome (70). Ideally, one should get two isochromosomes each time misdivision occurs but both do not always survive. Nondisjunction frequently is reported in irradiated cells. It could also result from an effect on the centromere.

6. Effect on chiasma frequency and crossing-over

Radiation can increase or decrease the chiasma frequency, depending upon the time of exposure (92, 99). An effect on crossing-over has also been reported in Drosophila (108).

7. Inhibition of cell division

Retardation or inhibition of cell division occurs whenever suitable doses of ionizing radiations are delivered to a tissue which would normally undergo further cell division. Such inhibition may result from a number of causes which may include genetic damage, inhibition of DNA synthesis, or other unknown physiological disturbances. At low doses the effect may be transient but at higher doses recovery of the ability to undergo further division may be long delayed or may not occur.

8. Induction of mitotic activity

This sometimes occurs after irradiation in tissues or cells which would not normally undergo further cell division (62).

9. Death of nuclei or cells

Death of a nucleus can lead to phenotypic change in binucleate cells or heterokaryons. It could also lead to the production of haploids if either a male or female nucleus had been sufficiently damaged prior to fertilization but survived long enough for pseudo-fertilization to occur and for development to begin. Such an effect, known as androgenesis, has been reported in Habrobracon (192) and, as already mentioned, some cases of haploidy resulting from exposure to ionizing radiation have been reported in plants (158, 160). More extensive cell death has been cited by Sagawa and Mehlquist (135) as the cause of the reversion of a periclinal chimera of Dianthus back to a non-chimera. Phenotypically such a reversion resembles a mutant but obviously it should not be so classified. Very high doses will also cause sufficient damage to kill whole meristems, organs, or even whole plants.

10. Sterility or partial sterility of various types may be produced

Some of these are the result of genic or chromosomal change, but other types are not. (See Section **V**.)

11. Miscellaneous nuclear or chromosomal changes

There are a number of changes known to occur in irradiated cells whose significance with respect to genetic integrity is vague or unknown. These are listed below along with possible modes of effects in some cases.

a. *Chromosome paling.*—The production of pale regions (also called gaps) has been studied particularly in certain animal cells by means of microbeam irradiation (205). An increase in the number of understained regions in plant chromosomes following exposure to ionizing radiation has also been reported (31).

b. *Chromosome stickiness and clumping.*—These effects occur at late prophase, metaphase, and anaphase after sufficiently high doses shortly after irradiation and can lead to secondary consequences such as fragmentation and possibly polyploidy (183).

c. *Abnormal spiralization of chromonemata.*—Abnormal spiralization has been reported in meiotic chromosomes of Trillium (169) and could lead to breakage due to entanglement of unspiralled chromosome arms or possibly to restitution at anaphase due to the long unspiralled arms behaving like bridges. Either of these effects would be more likely to occur when stickiness is also present.

d. *Persistent nucleoli.*—These sometimes occur in irradiated cells and since nucleoli are known to modify chromosome mechanics (40, 101), it seems quite likely that cells with persistent nucleoli might behave differently with respect to terminalization of chiasmata and separation of chromosomes at anaphase. Irradiation of nucleoli by microbeams of ultraviolet for periods as short as 3 seconds can cause permanent stopping of cell division in Chortophaga neuroblasts (48).

e. *Nuclear enlargement (without polyploidy).*—This change sometimes occurs in irradiated cells and may have cytogenetic consequences if additional exposure occurs. (See Sparrow and Forro (164) for references.)

f. *Biochemical, cytochemical, or histochemical changes.*—Many changes falling in this category are known. Some may be the result of primary damage to the genetic system but others almost certainly are not. Many examples are known (7, 38, 53, 118, 125).

V. Effects on Growth Rate, Growth Habit, Reproductive Capacity, and/or Phenotype

1. Growth rate

Growth rate may be retarded, completely stopped, or, in some cases, stimulated. Reduced growth, one of the commonest effects seen in plants, increases with increasing doses. However, different species differ greatly in the dose required to produce a given effect. The exact cause of the growth inhibition is difficult to pin down since the amount of inhibition can be correlated with both cytogenetic damage (18, 125, 163) and with a drop in level of auxin (53). In some cases buds may be inhibited for very long periods (161) without showing visible breakdown. Accelerated growth, growth stimulation, or precocious maturation have also been reported in many plants (14, 62, 125, 160). The nature of these growth-enhancing effects are not understood, but probably result from physiological disturbance rather than direct cytogenetic alterations.

2. Abnormal growth habit including tumor induction

This category is distinguished from the previous one by an abnormal growth pattern or growth habit rather than inhibition, and may occur in a great variety of forms (59, 61, 62, 63). In the extreme, tumors may be formed as a result of the radiation treatment (166). It is not known whether the tumors originate as mutant cells or because of a general physiological disturbance. The latter seems most likely (151, 170).

3. Reproductive capacity

Reproductive capacity can be reduced to zero by gene or chromosomal changes and also in several ways which may or may not result from direct cytogenetic damage, i.e., some may be due to physiological disturbances. Sterility may arise (a) by severe stunting or growth inhibition which prevents flowering; (b) flowers form but lack the necessary reproductive structures; (c) reproductive structures are present but pollen is aborted; (d) fertilization occurs but embryos are aborted before maturity; or (e) seeds form but fail to germinate properly or die after germination.

4. Phenotypic appearance

The phenotypic appearance of X_1 plants may change for any of several reasons: (a) in haploid plants due to mutation; (b) in diploids due to dominant mutation, or if the plant is already heterozygous, due to recessive mutation or to deletion of the dominant locus; (c) reversion of a chimera as explained above (Section IV, 9); or (d) change of a characteristic presumably due to nongenetic physiological change, e.g., short shoots apparently devoid of chlorophyll may develop on chronically irradiated Tradescantia plants; however, they often regain normal color at later stages of development (155).

VI. Radiation-induced Chromosome Aberrations, Abnormalties and Related Processes

A. Terminology

It would seem appropriate to explain briefly a number of processes which occur during the production of or as a result of chromosome breakage and a description of the various types of induced aberrations. (See also Section IV.)

1. Fragmentation or breakage

The process by which a chromosome, chromatid, or subchromatid is broken into two or more pieces. Less commonly used synonyms are fracture and shattering.

2. Restitution

The process in which broken ends unite back into the original configuration. The result leaves no cytologically detectable evidence of the break.

3. Reunion or rejoining

These words are used more or less interchangeably to describe any union of broken ends other than the type referred to above as restitution.

4. Fragment

Any portion of a chromosome separated from the main chromosome by breakage. With the exception of a few organisms which have diffuse centromeres, fragments are commonly *acentric*. *Terminal* and *interstitial* fragments are frequently referred to as terminal and interstitial *deletions* or *deficiencies*. Small fragments or *dot deletions* which are the same in length as in width are also called *isodiametric fragments* and extremely small fragments are generally called *minutes*.

5. Chromosome break

The prerequisite to the production of fragments or other types of aberration. It occurs when a chromosome is irradiated before it has become effectively double (2-stranded).

6. Chromatid break

A break in one of the two chromatids of the duplicated chromosome, usually seen at late prophase or metaphase.

7. Subchromatid break

A break in a subchromatid or half-chromatid at a time when the chromosome is 4-stranded.

8. Isochromatid break

A break in both chromatids of a chromosome at or near the same position. The broken ends of the chromatids may rejoin in such a fashion as to form a dicentric and a U-shaped fragment. The dicentric then usually forms a bridge at the first anaphase following its formation. If both centromeres go to the same pole the dicentric will persist until the next anaphase or possibly longer.

9. Gaps or achromatic lesions

Unstained or poorly stained short regions often seen in chromatids after irradiation. They may represent incipient breakage.

10. Misdivision of the centromere

A break in the centromere at right angles to the longitudinal axis of the chromosome. Such a division is in contrast to the usual lengthwise splitting. Misdivision of the centromere may result in the formation of isochromosomes, i.e., two chromosomes in each of which both arms are identical genetically and cytologically.

11. Transposition

The movement of a piece of chromosome or chromatid to a new position.

12. Exchange

Exchange occurs when broken ends produced by two separate breaks rejoin in a new combination. When two separate chromosomes are involved it is called *interchange,* and when the two breaks involved are in the same chromosome an *intrachange.* A ring-shaped chromosome is usually such an intrachange. Isolocus chromatid exchanges usually result in a dicentric and a fragment.

13. Simple translocation

This results from the movement of a piece of one chromosome to another followed by union of the broken ends.

14. Reciprocal translocation

The structure arising when portions of two nonhomologous chromosomes are exchanged. These are much more common than simple translocations and are one type of *exchange.*

15. Inversion

An inversion occurs when a piece of a chromosome is inverted 180° and inserted either in its original position or in a new position in the same chromosome. Inversions are of two types, depending upon the position of the centromere relative to the two breaks. *Paracentric inversions* are those confined to a single arm of the chromosome. If the centromere is included, it is a *pericentric inversion.*

16. Duplication

An aberration in which one or more segments are duplicated. They are also called *repeats*. In a *triplication* a segment is present three times.

17. Deficiency or deletion

Alternate names for fragments but usually restricted to mean small fragments which often are lost from the daughter nuclei by virtue of the fact that they are acentric. (See also 4 above.)

18. Dicentric

Any chromosome or chromatid which contains two centromeres. Tricentrics have three centromeres, quadri-centrics four. Those with more than two can be referred to as multi-centrics or poly-centrics. A chromosome with two or more centromeres may form a bridge (or bridges) at anaphase.

19. Micronucleus

A small nucleus which usually lies in the cytoplasm at some distance from the main nucleus and in irradiated cells usually results from one or more lagging acentric fragments. There may be one to several per cell. Micronuclei can also result from other causes, such as abnormal spindle behavior, and from the lagging of centric chromosomes.

20. Spontaneous breakage

This and resultant aberrations are know to occur in cells exposed only to background levels of radiation. Suitable experiments indicate that it is very doubtful if more than a small fraction of these spontaneously appearing chromosome aberrations are the result of naturally occurring radiation (108). They presumably have their origin in the inherent instability of the nucleoprotein structural framework of the chromosome. There are wide fluctuations and variations in the rate of spontaneous breakage.

The above description of cytological effects is necessarily brief. More detailed considerations of the aberrations are given in many places (21, 30, 39, 49, 78, 125, 137, 183, 198, 199). These sources should be consulted for descriptive diagrams and pertinent references. The "Bibliography on the Effects of Ionizing Radiations on Plants" (160) is also useful in finding original references.

B. Production of Aberrations

1. General

Although chromosome breakage normally appears in dividing cells shortly after exposure to ionizing radiation, in certain cases breakage does not occur for a relatively long period after treatment. For instance, the irradiation of meiotic first metaphase (or of later stages during meiosis) produces no apparent breakage at meiosis, but breakage and aberrations do appear during the subsequent microspore division (157). It is obvious that some kind of lesion must have been produced at the time of exposure, but that it did not develop to the point that it could be recognized as a break until a later stage or even until the nucleus had passed through an interphase. Such lesions or incipient damage are usually referred to as "potential breaks." Irradiation of certain stages, such as diplotene, may produce both immediate and delayed breakage. As one can readily imagine, this delayed expression of radiation damage has led to considerable confusion since chromosomes carrying potential breaks could be considered normal if observations were not continued long enough to determine that delayed breakage also occurs in the next cell division.

Chromosome or chromatid breaks may either restitute in the original position, rejoin in a new combination, or remain open. While much effort has been devoted to all aspects of this problem, it is difficult to generalize concerning the time required for each of these events to occur or to define explicitly how and why restitution or reunion occurs. There is evidence that, in certain cases, breakage and reunion occur in a very few minutes, whereas in other cases, breakage does not occur for a very long period after exposure to radiation, and, under certain conditions, broken ends may remain open for long periods before union or reunion occurs (23). It is generally assumed that the initial pattern of chromosome breakage is completely random within a given karyotype, but there are several nonrandom cases reported (39, 183). These are generally thought to result from secondary factors such as (a) differences in the freedom of movement of different regions of the chromosome, (b) the presence or absence of heterochromatin, (c) the position of heterochromatin within the karyotype, (d) the position of the nucleolus or nucleolar-organizing regions, and/or (e) the position of the centromere relative to any given locus.

Although chromosome breakage predominates in the first division following exposure, the cytological consequences of breakage frequently also can be seen much later. In addition to chromatin bridges and micronuclei which may persist and be visible for more than one cell division, certain types of aberrations, such as inversions, translocations, and rings, may persist indefinitely and express themselves by characteristic cytological configurations and by their concomitant genetic or physiological effect. Small duplications in the form of centric fragments may also persist (15) and these, of course, can produce genetic effects. In certain cases a series of phenomena known as the breakage-fusion-bridge cycle can result whether the initial breakage is of spontaneous origin, produced by ionizing radiation or by a chemical mutagen. As reported by McClintock (100), this cycle may go on for many cell generations. Other examples of long-persisting effects are small dicentrics reported by Morrison (106) and certain conditions such as translocations found in hexaploid wheat by MacKey (96). For unknown reasons these are sufficiently unstable to lead to secondary changes such as deficiency-duplication or simple deficiencies. These, of course, can result in phenotypic changes.

2. The mechanisms of chromosome breakage

Although much is known about the chemical changes produced by ionizing radiation (7, 8, 38, 73, 91, 113), our knowledge of the exact mechanisms by which chromosomes are broken is unfortunately rather meager. This lack of understanding is due partly to our inadequate knowledge of the composition and structure of the chromosome threads themselves, and partly to a lack of adequate definition and resolution which would enable us to see changes which are produced at the submicroscopic level in the interval between the initial chemical events and the final biological event recognized as a chromosome or chromatid break.

A summary of the possible interrelationships of various events which occur preceding, during, and after chromosome breakage is given in Figure 7. Possible pathways of various modifying factors or processes are also given. The relationship of gene mutation to other types of chromosome damage is obscure.

There has been much speculation concerning the manner in which damage to DNA protein or nucleoprotein molecules could lead

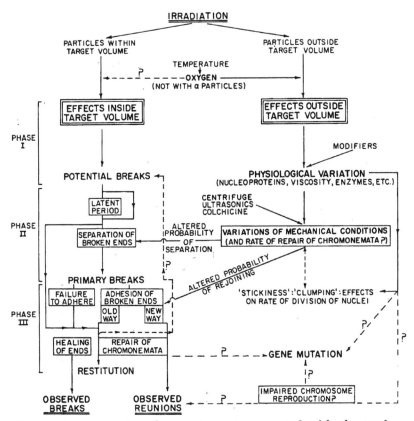

FIGURE 7.—*Suggested pathways of events concerned with the production of chromosome aberrations and mutagenesis. After Thoday (188).*

to chromosome rupture and several possible hypotheses have been proposed (3, 4, 23, 24, 79, 80, 177, 178, 185, 197, 198). Regardless of the nature of the events which occur at the submicroscopic level, it is now fairly clear that not only can chromatids be broken independently of each other but also that half chromatid or subchromatid breakage can occur (39, 195). Since the potentialities for recombination vary with the number of strands present and since the number of strands varies with the stage of meiosis and mitosis, it is apparent that the stage of cell division irradiated can have a great effect on the outcome of the initial breakage events which occur (194). The number of nucleoprotein strands in a Tradescantia chromosome, according to recent work with the electron microscope, may be as high as 32 before synthesis (177). One would expect that the number of strands or fibrils in a a chromatid, as well as their diameter, should have a bearing on the probability of breakage of that chromatid by a given ionization track or cluster.

3. Dosage response curve for one-hit events

Chromosome or chromatid breakage is considered to result from a cluster of 15 to 20 ionizations either within or very close to the chromatin strand in question (93). A sufficient density of ionization may occur at any position along the more dense tracks, such as those of alpha particles, or only near the end, or "tails", of less dense tracks. While each ionization is an independent event, in radiobiological parlance it is also possible to regard any large cluster or tail as a single event. When such a cluster or track passes through a chromosome (or chromatid), the event is called a "hit". Such clusters or tracks have a certain probability of producing a break (in contrast, it is often considered that gene mutation can be caused by as little as one ion pair). It has been shown that various one-hit chromosomal events (one-hit interstitial deletions, terminal deletions) are independent of the dose rate, of the time of exposure, and of dose fractionation. They, therefore, increase linearly with dose (Figure 8). In general, the kinetics of such events are similar to those of gene mutations, but the relative biological efficiency of different radiations in producing these end results would be expected to differ. (See also Section VIII A.)

4. Dosage response curves for two-hit events

Certain types of chromosome or chromatid aberrations (exchanges, inversions, rings, etc.) are two-hit aberrations as far as X-rays and gamma rays are concerned and behave in a different fashion with respect to several of the variables mentioned above from one-hit chromosome aberrations. Whereas with one-hit aberrations the frequency is strictly proportional to dose, with two-hit aberrations the frequency increases disproportionately with increasing dose and with sufficiently high intensities becomes almost proportional to the square of the dose (Figure 8). The dose-squared relationship is best observed at the higher intensities when the exposure time is kept constant but the intensity varied (Figure 9). From these and similar

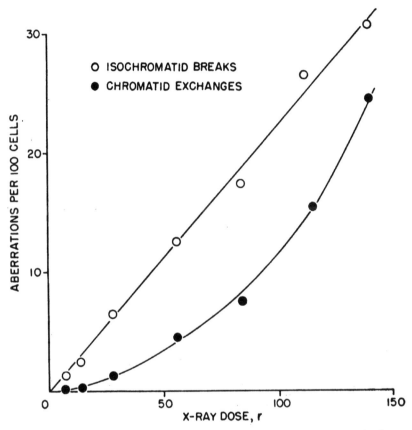

FIGURE 8.—*Relation between X-ray dosage and frequency of isochro-matid breaks (one-hit) and chromatid exchanges (two-hit). Time of exposure constant. After data of Sax modified by Giles (49).*

observations, it can be concluded that the yield of two-hit exchange aberrations increases roughly as the square of the dose and that each of the two breaks is separately induced. Since the two breaks must be present simultaneously to permit the exchange to occur, it is obvious that such factors as intensity (Figure 9), exposure time, and dose frac-tionation should all have detectable effects on the frequency of two-hit events. Oxygen concentration also is a factor of considerable importance (39, 49, 80, 125).

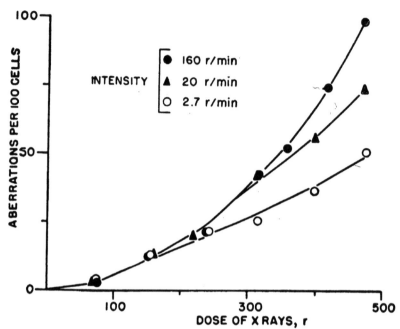

FIGURE 9.—*Effect of dosages of X-rays at different intensities upon the yield of chromosome exchanges. Data of Sax, after Giles (49).*

Certain types of chromosomal aberrations behave in a fashion intermediate between one-hit and two-hit events. For instance, the yield of interstitial deletions increases as the 1.5 power of the dose as shown by Rick (127), whereas isochromatid deletions have a value between 1.0 and 1.5. The exact value depends on the particular radiation used (84, 188).

It has been known for some time that some types of two-hit aberrations fail to show the nonlinear response when cells are exposed to certain densely ionizing particles, such as neutrons, alpha rays, protons, etc. (Figure 10). The reason for this situation is that these particles have sufficient density of ionization and sufficient length of track to produce more than one break for each passage through a nucleus. Thus, what would for X- or gamma rays be a two-particle event (two separate ionization "tails") is, in actual fact, a one-particle event (one ionization track) for the densely ionizing particles mentioned above. (See Swanson (183) for references.)

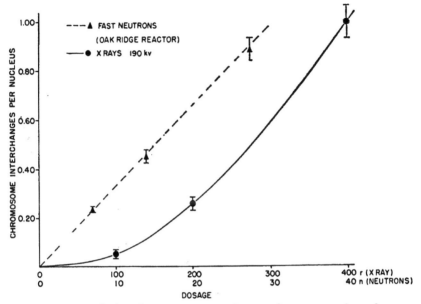

FIGURE 10.—*Relation between frequencies of chromosome interchange (dicentric and centric rings) and dosages of fast neutrons and X-rays. 1 n unit = ca. 2.5 r. After Giles (49).*

VII. Nature of Mutational Events Induced by Ionizing Radiation

The relationship between the frequency of induced mutations and the amount of ionizing radiation depends in part upon the nature of the mutational event. Phenotypic change which results from chromosomal breakage, or the resultant aberrations, will have the same dosage response curves as the breakage or aberration event or events which caused it. For instance, simple one-hit deletions (of markers) will show a linear increase with increasing dose, whereas losses resulting from two-hit deletions will have a dosage response curve which is nonlinear and will approach the dose-squared relationship at high dosages or high intensities (163).

It has been shown in a wide variety of organisms that point mutations increase linearly with increasing dose (Figure 11) and are generally independent of intensity (108). Occasional departures from linearity are found (see 110), but in this case it has been suggested

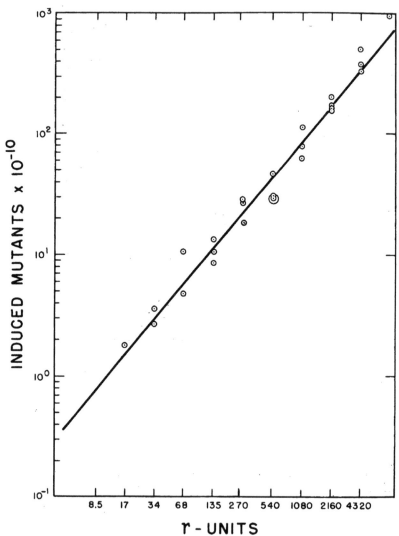

FIGURE 11.—*Reverse mutations induced in* met-2 *auxotrophs by various doses of X-rays. After Demerec and Sams (34).*

that a selective mechanism was responsible for the nonlinearity. A recent case of nonlinearity has been found in mice in which mutation rate is dose-rate dependent (133). It was also reported in this paper, and confirmed later (134), that acute X- and gamma radiations are equally mutagenic to spermatagonia and that both are more effective than chronic gamma irradiation. No adequate explanation for this difference has been given, but it seems possible that such an effect would be expected if some of the "mutations" from the acute exposure were two-hit deletions and not true point mutations. Evidence of a dose-rate effect as well as a nonlinear response has been reported for marker losses in several different plant species. Such an effect for chronically irradiated *Lilium testaceum* is shown in Figure 12. The explanation given is that some of the marker losses result from two-hit deletions and thus the dosage-squared component contributes significantly to nonlinear response (32, 162, 163).

Classically, it has been considered that so-called point mutations or intragenic mutations are a valid category of genetic change. This may be true for Drosophila and for many extensively studied microorganisms. However, it was the view of Stadler (173, 174), who did early outstanding work on the induction of mutations in maize with ionizing radiations, that all the so-called point mutations in maize were actually minute chromosomal aberrations or deficiencies and that point mutations as defined above did not actually exist in this material. As far as this author is aware, it is questionable whether anyone has since analyzed mutations in higher plants with sufficient detail to disprove Stadler's conclusion. It is further of considerable interest that recent work of Demerec (33) has shown that most mutations induced by ionizing radiation in *E. coli* appear to be deficiencies, although point mutations apparently can be produced by ultraviolet radiation or arise spontaneously in this organism. In actual practise, the distinction between deletions and true mutations often becomes one of resolution, since only the most critical kind of test could reveal any distinction. Such tests are rarely applicable in higher plants.

If radiation-induced point mutations actually exist, one would expect that they could revert back to the original form by back mutation. For example, Giles (50) has shown in Neurospora that reverse mutation does occur at certain loci. These mutations would appear to meet the criterion for point mutation, but an equally clear case has not yet been found in higher plants. While the exact nature or even the actual existence of radiation-induced point muta-

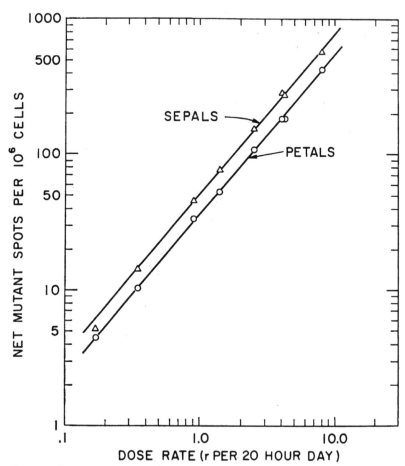

FIGURE 12.—*Somatic mutation rate in sepals and petals of chronically irradiated* Lilium testaceum. *After Sparrow, Cuany, Miksche, and Schairer (163).*

tions seems in doubt in higher plants, there is no corresponding doubt about mutational events derived from chromosomal breakage and rearrangement. The various kinds of such aberrations have been outlined above (Section VI) and almost all of these have been identified with phenotypic changes of some type. For instance, duplications, deletions, translocations, inversions, isochromosomes, and breakage-fusion-bridge cycles have all been well documented elsewhere (16, 39,

64, 126, 158, 198, 199). In addition to these, position effect has been described by Catcheside (20) in *Oenothera blandina*. In spite of some of the practical difficulties sometimes associated with certain types of aberrations, e.g., reduced fertility, the record clearly shows that appropriately designed (or even accidental) cases of chromosome engineering can yield results of great potential value. An outstanding example is the work of Sears (140) in which leaf rust resistance was transferred from the wild grass, *Aegilops umbellatum* L., to common wheat, *Triticum aestivum* L., by means of a short interstitial translocation. A similar transfer of stem rust resistance has since been reported by Elliott (36) from *Agropyron elongatum* to *T. aestivum*. A more extensive discussion of the usefulness of chromosomal aberrations in plant genetics and plant breeding is given in earlier publications and elsewhere in this volume (46, 47, 64, 86, 97, 123, 158).

VIII. Factors which Modify Cytogenetic or Other Radiobiological Responses

One of the difficulties with the status of modern radiobiology is a plethora of facts and a deficiency of general principles. This regretable state of affairs is well illustrated in the case of the literature on factors which modify radiation response. It is possible to catalogue these facts (Tables 6, 7, 8), but unfortunately the exact reason or mechanism by which modification comes about is often unknown.

Since it is important that radiation experiments be as reproducible as possible, the maximum number of variables should be controlled carefully. Some things which may seem to be trivial details often turn out to be important modifying factors, i.e., moisture content of seeds, exact age of seeds or seedlings, growth rate, and temperature. A diagram of some possible post-radiation events and of various stages at which modifying factors may act is given in Figure 13.

In the interest of providing a brief summary of the known modifying factors and to provide a guide to investigators desiring uniformity of results, most of the major modifying factors are described or listed below.

A. Physical Factors

1. Miscellaneous factors

Many physical factors other than total dose can modify the degree

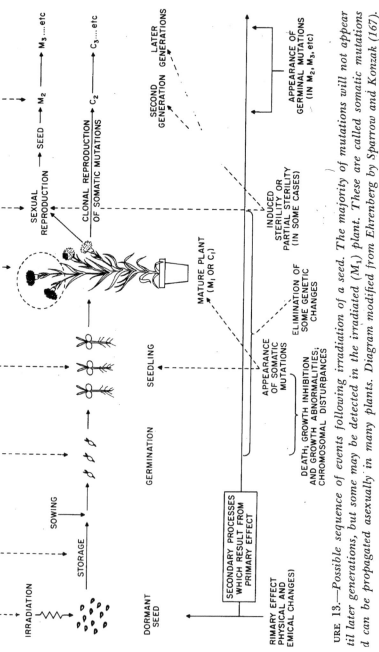

URE 13.—*Possible sequence of events following irradiation of a seed. The majority of mutations will not appear til later generations, but some may be detected in the irradiated (M_1) plant. These are called somatic mutations d can be propagated asexually in many plants. Diagram modified from Ehrenberg by Sparrow and Konzak (167).*

of the radiation response. Dose rate, dose fractionation (see Section VI, 4), linear ion density (see below), and temperature are probably the most important of these variables, but several others also can cause a significant modification under appropriate circumstances (Table 6). For instance, small doses of ultraviolet in combination with 1 rad of X-rays can give a yield of aberrations roughly equal to that normally obtained from 100 rads of X-rays alone (85). Similar but much less dramatic synergistic effects have been known for some time (see references in Table 6). It is difficult to make generalized

TABLE 6.—PHYSICAL FACTORS AFFECTING THE RADIOSENSITIVITY OF PLANTS.*

Factor	References
A. Dose fractionation...........................	21,39,49,93,125,137,199,202
B. Dose rate....................................	21,39,86,93,125,137,183,199
C. Linear ion density...........................	21,26,57,88,93,125,202
D. Previous exposure to ionizing radiation.........	202
E. Exposure to other radiations	
1. Ultraviolet............................	183(page 382)
2. Infrared..............................	49(page 739),202(page 374)
3. Visible light.........................	160,202(page 358)
F. Exposure to ultrasonic energy................	125,137(page 19)
G. Bioelectrical potential...	160
H. Temperature................................	49(page 741),202(page 360)
I. Centrifugation..............................	137(page 19)
J. Pressure, hydrostatic...	160
K. Phase state.................................	202(page 353)

*After Gunckel and Sparrow (62). The references in this table are held to a minimum and are in many cases to review papers rather than to the original research. References to many original papers can be found in the reviews cited or in the bibliography by Sparrow, Binnington, and Pond (160).

statements concerning these factors. Other conditions of the experiment and the effect studied will determine, in part, the degree and direction of modification. The articles referred to in Table 6 contain extensive discussions of the various physical factors.

2. Ionization density, LET and RBE

It is well known that different radiations or the same radiation at different energies behave differently with respect to the average number of ionizations produced per micron of tissue traversed. (See Figure 14 and Table 2.) Since each ionization is estimated to require 32.5 ev, it is apparent that the denser the ionizations along the track, the higher the rate of energy transfer from the ionizing particles to

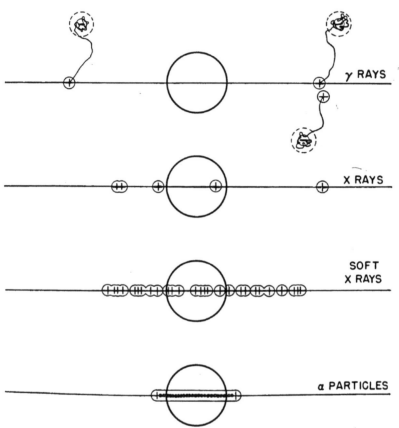

FIGURE 14.—*Separation of ion clusters in relation to the size of virus particles 27 mμ in diameter. After Gray (54).*

molecules of the tissue, i.e., the higher the linear energy transfer (LET). (See Section I.)

The biological consequences of variations in LET have been studied for many years and are of great importance in radiogenetics. Certain qualitative differences and many quantitative differences in cytogenetic responses can be attributed to differences in the ionization density of the radiation used. The relative biological effect (RBE) for many responses increases with increasing LET but may go through a maximum (Figure 15) and then decrease. This saturation effect can be explained, in some cases at least, by the production of surplus

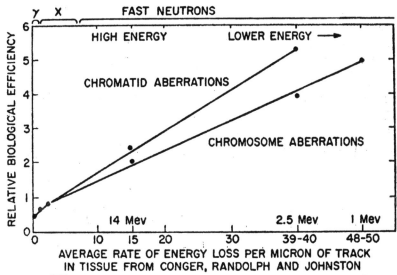

Figure 15—*The effect of ionizing radiation on different energies on their relative efficiency in inducing chromosomal aberrations in Tradescantia microspores. After Sax (137). [See Conger, Randolph, and Johnston (25).]*

ionizations, e.g., if a sensitive target volume requires only x ionizations to produce an effect, then $2x$ ionizations in the same volume will not increase the detectable effect (Figure 14). Certain effects, such as those produced by single ionizations, are less efficiently produced with increasing LET (Figure 16). It is generally assumed that point mutation can be induced by one ion pair or one ion cluster (108, 199). For chromosome breakage, however, a different relationship apparently exists.

From early work, Lea concluded that chromosome breakage (in Tradescantia) required an estimated minimum of 15 to 20 ionizations to occur within a cross section of a chromatid thread not thicker than 0.1 μ (93, page 280). This means that breakage of such chromosomes can only be produced by relatively dense ion clusters. The response of Tradescantia chromosomes to radiations of different ion density has recently been studied in detail by Conger, Randolph, Sheppard, and Luipold (26). By comparing the efficiency of various radiations to that of Co⁶⁰ gamma rays, they showed that the RBE climbs to a

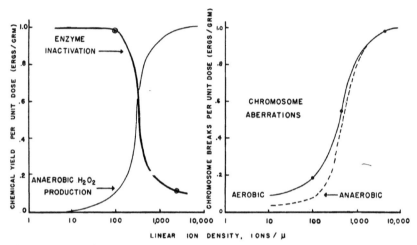

FIGURE 16.—*The influence of ion density on chemical change and chromosome breakage. After Gray (55).*

peak of 8 to 11 between approximately 50 to 70 kev/μ. It then falls considerably and flattens out in the LET range between roughly 120 and 220 kev/μ. Less extensive but somewhat similar observations were made by Larter and Elliott (89) who found thermal neutrons more effective in producing translocations than X-rays, or beta rays from S^{35} and P^{32}. It also has been found that the dense ionization associated with thermal neutron treatment caused more breakage per unit inhibition of seedling growth than did X-rays (18).

In addition to the LET effect on RBE's for chromosome breakage, there is evidence that breaks caused by high-ion density tracks, such as alphas, are less likely to rejoin than those caused by less dense tracks of X- or gamma rays (87). The RBE's of densely ionizing radiation with respect to various types of mutations also have been studied by several investigators. Theoretically, for point mutations low-ion density should be more efficient; however, the effect depends upon the kind of mutation produced (1, 13, 108, 109, 199). For example, Ehrenberg and Nybom (35) found a higher percentage of *erectoides* mutations in barley with thermal neutrons than with X rays. Other studies show that radiations of different ionization densities sometimes produce characteristically different spectrums of mutations, at least in plants (46, 67, 86). An extensive bibliography (88) on RBE is available for those interested.

B. Chemical Factors

A very extensive literature exists relating to the modification of control of radiosensitivity by chemical substances. Variations in the amount of naturally occurring chemical constituents as well as in those which can be added experimentally to the internal or external cellular environment are all potentially suspect and many already are known to be effective (Table 7). While many of the results obtained could have been predicted, the reasons for some of the modifications are still obscure or unknown. However, the study of these factors is important since such studies may suggest some of the chemical steps involved in chromosome breakage and mutagenesis. Moreover, a better understanding should lead to more adequate control of the most important environmental variables and this, in turn, should yield more reproducible data. This also could have practical applications, e.g., one could plan experiments in which the maximum yield of a certain kind of cytogenetic effect could be obtained with the minimum amount of sterility or lethality. Such mutagenic specificity is summarized by Smith (153).

Due to the complexity of the relationships involved and lack of space, it is impossible to explain here even the best known mechanisms by which protection or enhancement is brought about. However, many authors have discussed at length chemical modification and the mechanisms of action involved (2, 5, 12, 22, 38, 71, 91, 125, 153, 201).

C. Biological Factors

1. Cytological and genetic factors

a. *Chromosome number and polyploidy.*—The chromosome number in diploids and the degree of polyploidy are both known to affect radiosensitivity (160, 163, 202). The degree of polyploidy and the kind of polyploidy are also known to affect the type of mutation and the rate (96). Likewise, the higher the degree of polyploidy, the more generations which must be grown to ensure the appearance of recessive mutations.

b. *Chromosome size or nuclear volume.*—The size of the nucle- •

TABLE 7.—CHEMICAL FACTORS AFFECTING THE RADIOSENSITIVITY OF PLANTS. *

Chemical	References	Chemical	References
A. Antioxidants (lipid)	69(pages 81,91), 139		
B. Auxins (indoleacetic acid)	53(page 10),105		
C. Carcinogens			
1. Benzopyrene	157(page 1522)	2. Urethane	
D. Chelating and complexing agents			
1. Cupferron	202(page 366)	3. Ethylenediamine	
2. Dowex–50	160	tetraacetic acid	
		(EDTA or Versene)	137,202
E. Enzymes and cofactors			
1. Adenosinetriphosphate (ATP)	197,200	4. Diphosphopyridine nucleotide (DPN)	184(page 242)
2. Catalase	145,176	5. Peroxidase	180
3. Cytochrome C	180		
F. Metabolic inhibitors			
1. Carbon monoxide	125, 137(page 23)	4. Triiodobenzoic acid (TIBA)	170
2. Cyanide	125,200	5. Uranyl nitrate	157(page 1520)
3. Maleic hydrazide	155		
G. Metabolites			
1. Carbon dioxide	137(page 23)	5. Fat	157(page 1523)
2. Carboxylic acids:		6. Saponin	160
malonic and maleic acids	160	7. Sodium	
3. Desoxypentose nucleic		ribose-nucleate	144
acid (DNA)		8. Vitamins	
(Bacteria, U.V.)	65	a. Ascorbic acid	27
4. Digitalis glycoside	160	b. Synkavite	81
H. Miscellaneous			
1. Casein hydrolysate	191	3. Megaphen	119
2. "Kollidon"	57(page 389)	4. Yeast extract	202(page 361)
I. Mitotic poisons			
1. Acenaphthene	160	3. 3,5–dinitro–o–cresol	132
2. Colchicine	157,160		
J. Mutagens (chemical)			
1. Mustard oil	130(page 207)	2. Nitrogen mustard	160

*After Gunckel and Sparrow (62). Also, see footnote to Table 6.

TABLE 7.—CONTINUED.

Chemical	References	Chemical	References
K. Oxidizing substances			
1. Ferric sulfate	157(page 1520)	5. Sodium	
2. Hydrogen peroxide	91,125,180	peroxydisulfate	128
3. Nitric oxide	2,122	6. Sodium pyrosulfate	128
4. Oxygen	57(page 378), 91,125,143,202		
L. Neutron absorbers			
1. Boron	86(page 38)	3. Lithium	86(page 38)
2. Cadmium	86(page 38)		
M. Nutrients	73		
1. Boron	149	6. Phosphorus	157(page 1521)
2. Calcium	175,176,177	7. Potassium	9
3. Cobalt	60	8. Sulphur	157(page 1521)
4. Magnesium	60	9. Zinc	60
5. Nitrogen	157(page 1521)		
N. pH	160		
O. Reducing substances (without –SH groups)			
1. Ascorbic acid	27,86(page 38)	4. Sodium hydrosulfite	57(page 388)
2. Hydrogen	57(page 379)	5. Sodium hyposulfite	57(page 388)
3. Hydrogen sulfide	117		
P. Sulfhydryl compounds			
1. Cysteamine	73, 202(page 362)	4. Glutathione	57(page 388), 202(page 357)
2. Cysteine	137(page 24), 202(page 362)	5. Thiourea	57(page 388)
3. 2,3–dimercaptopropanol (BAL)	57(page 388), 125		

us or chromosome is very important in determining the frequency of chromosome breakage per roentgen of exposure (116, 161, 163, 165). Data collected from several species of plants with different nuclear volumes indicate that, under comparable conditions of chronic irradiation, somatic nuclei yield a more or less standard number of chromosome breaks per cubic micron of nuclear volume per roentgen of exposure (155). It is thus apparent that any nuclear volume difference between species, between stages in development, or between tissues in either plants or animals should have a great effect on the yield (per r of exposure) of breaks or on any of the known genetic changes derived from breakage.

c. *DNA content per nucleus.*—The DNA content is related to average nuclear volume, and it appears that the larger the amount of DNA per complement, the higher the radiosensitivity (164, 168).

d. *Amount of heterochromatin.*—The amount and distribution of heterochromatin has been shown to affect the radiosensitivity of nuclei in various ways (39, 157).

e. *Number of nucleolar organizing regions.*—The number of such regions is often related to the amount of heterochromatin and may be related to the average amount of nucleolar material per nucleus. Both of these factors are known to affect the radiosensitivity of cells (41, 57).

f. *Number of nuclei.*—It has been shown, for example in fungi, that the number of nuclei can affect the radiosensitivity (81). It would also be expected that multinucleate cells would be more resistant than the normal type of mononucleate cell.

g. *Stage of the nuclear cycle.*—The stage of the nuclear cycle has been known for many years to affect the amount of chromosome breakage (14, 125, 138, 157) and the amount of mutation produced per roentgen of exposure. In some cases the extent of the differences is quite large. For instance, in *Trillium erectum* there is about a 60-fold difference in amount of fragmentation between the least sensitive and most sensitive stages of meiosis (39, 157). In *Zea mays* mutation rates also vary widely during microsporogenesis (148). Changes in length of chromonemata at different stages of meiosis in Trillium also has been correlated with sensitivity (169).

h. *Average length of nuclear cycle or of intermitotic time.*—It has been shown, especially for chronically irradiated material, that the longer the average nuclear cycle, the more radiation damage is accumulated under a given dose rate (121, 163).

i. *Position of a gene in the chromosome.*—It has been clearly shown that the position of a gene with reference to the centromere has a significant effect on the frequency of deletion (42, 163).

j. *Position of the centromere and length of chromosome arm.*—Both of these factors are known to affect radiation response, although more work is required to establish clearly the degree of the effect (41, 42).

k. *Portion of the cell irradiated.*—Microbeam irradiation of cells has clearly illustrated that there are wide differences in sensitivity of different portions of cells (205). For instance, the nucleus is much more sensitive than the cytoplasm and for ultraviolet irradiation, at least, the nucleolus is an extremely sensitive spot in the nucleus (48).

l. *Genotype.*—Genetic differences as small as single gene differences can cause significant changes in sensitivity (2). Different varieties of a single species also may show different sensitivities (114, 152, 202).

2. Morphological organization and stage of development

a. *Type of cell or tissue.*—In plants, as in animals, the type of cell or tissue in question may have a considerable influence on the radiosensitivity (8). For instance, meiotic cells are usually much more sensitive than somatic cells (138, 169). Shoot and scutellum are also less sensitive than roots in barley during early developmental stages (104).

b. *Stage of differentiation and age of plant.*—There is a considerable amount of evidence which indicates that the tolerance of a plant, as well as the number of mutations or chromosome aberrations produced, may vary with the stage of differentiation of the seeds, seedling, or plant (11, 14, 103, 104, 136).

c. *Portion or amount of plant irradiated.*—The radiosensitivity of a whole plant is greater than the sensitivity of a small area or volume of a plant (58, 102, 172).

3. Physiological conditions

The radiosensitivity is influenced by various physiological states of the cell, tissue, or plant. Space does not allow a detailed discussion of these, but they are summarized in Table 8. Of the various factors listed, water content of spores and seeds has been most extensively studied and is known to have a large effect on radiosensitivity and on yield of chromosome aberrations and mutations. Recent studies have shown that the water content during post-irradiation storage of seeds is also of great importance (28). It is therefore a factor which should be held constant whenever possible during radiation experiments.

The other factors listed in Table 6 will not be mentioned, except to point out that it is becoming increasingly clear that the yield of mutations is dependent upon the time of exposure with regard to

TABLE 8.—PHYSIOLOGICAL FACTORS WHICH AFFECT SENSITIVITY.*

Factor	References	Factor	References
1. Age of cell, tissue or seed	86(page 37), 114(page 155)	5. Diseased tissue	160,171
2. Synthetic activity		6. Concentration in	
General (growth rate)	73	cell suspensions	121(page 76)
	90	7. Dormancy	160
Specific		8. Vernalization	37
of DNA	66,73	9. Leaching	75,77
of protein		10. Water content	17,115,
(Bacteria, U.V.)	65(page 878), 66		160,202
		11. Post irradiation	—
of auxin	53(page 10)	storage	28
3. Respiration	90		
4. Nutritional state	90,121		

*After Gunckel and Sparrow (62). Also see footnote to Table 6.

certain physiological or biochemical steps involved in protein and nucleic acid synthesis (2, 66). Various factors listed above under chemical factors could perhaps have also been included in this section. Also, please refer to Table 7.

Summary and Conclusions

This paper attempts to review certain aspects of the physical nature of ionizing radiations and their effects upon nuclei, chromosomes, and genes. The major physical characteristics of a number of the ionizing radiations most important to radiobiologists are presented and some methods of treatment and facilities commonly used are described. The major biological effects, including nuclear and chromosomal changes known to be induced by ionizing radiation in plants or plant cells, are listed and explained briefly. Special attention is given to the problem of chromosome breakage and aberration. The kinetics of the dosage response curves obtained under different conditions for various types of aberrations and mutations are explained and discussed. The possible relationship between chromosome aberration and mutation, and the factors which influence radiosensitivity are considered in some detail.

The review is mainly concerned with fundamental aspects of radiation cytogenetics and does not attempt to relate this knowledge to practical problems of plant breeding. It is concluded that further comparative studies of the effects of different physical and chemical mutagens and of various techniques of exposure are urgently needed. In spite of rapid progress being made, our knowledge of radiobiology and radiation genetics leaves much to be desired with respect to our ability to define or predict the most efficient methods for the production of useful mutations in plants. Expansion of the research effort at both basic and applied levels is desirable.

References

1. Alexander, M. L. 1958. Radiation damage in the developing germ cells of *Drosophila virilis* from fast neutron treatment. *Genetics*, **43:** *458–469*.

2. Alper, T. 1960. Cellular radiobiology. *Ann. Rev. Nuclear Sci.*, **10:** *489–530*.

3. Ambrose, E. J. 1956. The structure of chromosomes. *Prog. Biophysics and Biophys. Chem.*, **6:** *25–55*.

4. Anderson, N. G. 1956. Cell division: Part One. A theoretical approach to the primeval mechanism, the initiation of cell division, and chromosomal condensation. *Quart. Rev. Biol.*, **31:** *169–199*.

5. Atwood, K. C. 1959. Cellular radiobiology. *Ann. Rev. Nuclear Sci.*, **9:** *553–592*.

6. Auerbach, C. 1958. Radiomimetic substances. *Radiation Res.*, **9:** *33–47*.

7. Augenstine, L. G., ed. 1960. Bioenergetics. *Radiation Res. Suppl.* **2.**

8. Bacq, Z. M., and Alexander, P., ed. 1955. Fundamentals of Radiobiology. *New York: Academic Press; London: Butterworths Sci. Pub.*

9. Bair, W. J., and Hungate, F. P. 1958. Synergistic action of ethylene-diamine-tetra-acetate (EDTA) and radiation on yeast. *Science*, **127:** *813*.

10. Behrens, C. F. 1953. Atomic Medicine. *Baltimore: Williams and Wilkins Co.*

11. Biebl, R. 1958. The radiosensitive phase in plant germination. *Proc. Intern. Conf. Peaceful Uses of Atomic Energy, 2nd Conf., Geneva. New York: United Nations*, **27:** *299–304*.

12. Blois, M. S. Jr., Lindblom, R. O., Brown, H. W., Weissbluth, M., and Lemmon, R. M. 1960. Free Radicals in Biological Systems. *New York: Academic Press.*

13. Bora, K. C. 1958. Relative biological efficiencies of 220 kvp X-rays and 14.1 Mev fast neutrons in the induction of aberrations in

chromosomes of *Tradescantia* at low temperature. *Proc. Intern. Conf. Peaceful Uses of Atomic Energy, 2nd Conf., Geneva. New York: United Nations,* **22:** *342–350.*

14. Breslavets, L. B. 1946. Plants and X rays. *Moscow: Academy of Sciences.* English translation published by *Amer. Inst. Biol. Sci., Washington 6, D. C. (1960).*

15. Brewbaker, J. L., and Natarajan, A. T. 1960. Centric fragments and pollen-part mutation of incompatibility alleles in Petunia. *Genetics,* **45:** *699–704.*

16. Burnham, C. R. 1956. Chromosomal interchanges in plants. *Bot. Rev.,* **22:** *419–552.*

17. Caldecott, R. S., and North, D. T. 1961. Factors modifying the radio-sensitivity of seeds and the theoretical significance of the acute irradiation of successive generations. *This Symposium 365–404.*

18. ————, Frolik, E. F., and Morris, R. 1952. A comparison of the effects of X-rays and thermal neutrons on dormant seeds of barley. *Proc. Nat. Acad. Sci.,* **38:** *804–809.*

19. ———— and Snyder, L. A., ed. 1960. A Symposium on Radio-isotopes in the Biosphere. *Minneapolis: Univ. of Minnesota.*

20. Catcheside, D. G. 1939. A position effect in Oenothera. *Jour. Genetics,* **38:** *345–352.*

21. ————. 1948. Genetic effects of radiations. In *Advances in Genetics. New York: Academic Press.,* **2:** *271–358.* (M. Demerec, Editor.)

22. Claus, W. D., ed. 1958. Radiation Biology and Medicine. *Reading, Mass.: Addison-Wesley Pub. Co.*

23. Cohn, N. S. 1956. An analysis of the rejoining of X-ray-induced broken ends of chromosomes. *Genetics,* **41:** *639.*

24. ————. 1957. The effect of carbon monoxide on the restitution of X-ray-induced chromosome breaks in *Allium cepa. Genetics,* **42:** *366.*

25. Conger, A. D., Randolph, M. L., and Johnson, A. H. 1956. Chromosomal aberration production by X-rays and by monochromatic 2.5-Mev and 14-Mev neutrons. *Genetics,* **41:** *639.*

26. ————, ————, Sheppard, C. W., and Luippold, H. J. 1958. Quantitative relation of RBE in Tradescantia and average LET of gamma-rays, X-rays, and 1.3-, 2.5-, and 14.1–Mev fast neutrons. *Radiation Res.,* **9:** *525–547.*

27. Cooke, A. 1953. Effect of gamma irradiation on the ascorbic acid content of green plants. *Science,* **117:** *588–589.*

28. Curtis, H. J., Delihas, N., Caldecott, R. S., and Konzak, C. F. 1958. Modification of radiation damage in dormant seeds by storage. *Radiation Res.*, 8: *526–534.*

29. ————, Person, S. R., Oleson, F. B., Henkel, J. E., and Delihas, N. 1956. Calibrating a neutron facility for biological research. *Nucleonics,* 14: *26–29.*

30. Darlington, C. D., and LaCour, L. F. 1945. Chromosome breakage and the nucleic acid cycle. *Jour. Genetics,* 46: *180–267.*

31. ————, ————. 1953. The classification of radiation effects at meiosis. *Heredity,* 6 *(Suppl.): 41–55.*

32. Davies, D. R., and Wall, E. T. 1960. The effect of acute, chronic, and fractionated doses of gamma radiation on the induction of somatic mutations in *Trifolium repens.* In *Large Radiation Sources in Industry. Vienna: Intern. Atomic Energy Agency,* 2: *223–237.*

33. Demerec, M. 1960. Frequency of deletions among spontaneous and induced mutations in Salmonella. *Proc. Nat. Acad. Sci.,* 46: *1075–1079.*

34. ———— and Sams, J. 1959. Induction of mutations in individual genes of *Escherichia coli* by low X radiation. In *Immediate and Low Level Effects of Ionizing Radiations. Intern. Jour. Radiation Biol. Spec. Suppl., 283–291.*

35. Ehrenberg, L., and Nybom, N. 1954. Ion density and biological effectiveness of radiations. *Acta Agr. Scand.* 4: *396–418.*

36. Elliott, F. C. 1957. X-ray-induced translocation of Agropyron stem rust resistance to common wheat. *Jour. Heredity,* 48: *77–81.*

37. Engel, O. S. 1950. Effect of winter wheat seed vernalization on sensitivity of sprouts to X-ray action (Russian). *Dokl. Akad. Nauk. SSSR,* 71: *1151–1153.*

38. Errera, M. 1957. Effets biologiques des radiations: Aspects biochimiques. *Protoplasmatologia,* 10: *1–241.*

39. Evans, H. J. 1961. Aberrations of chromosome structure induced by physical and chemical agents. *Intern. Rev. Cytol.,* (in press).

40. ————. 1961. The frequency and distribution of interchange and isochromatid aberrations induced by the irradiation of diploid and tetraploid cells. In *Effects of Ionizing Radiation on Seeds and Its Significance for Crop Improvement. Proc. Sci. Symposium, Intern. Atomic Energy Agency and The Food and Agr. Organization United Nations. Karlsruhe, 259–270.*

41. ———— and Sparrow, A. H. 1961. Nuclear factors affecting sensitivity: II. *Brookhaven Symp. in Biol., Brookhaven Nat. Lab., Upton N. Y.,* 14. In press.

42. Fabergé, A. C. 1957. The possibility of forecasting the relative rate of induced loss for endosperm markers in maize. *Genetics*, 42: *454–472.*

43. Fano, U. 1954. Principles of radiological physics. In *A. Hollaender, ed., Radiation Biology. New York: McGraw-Hill,* 1: *1–144.*

44. Fluke, D. J., Brustad, T., and Birge, A. C. 1960. Inactivation of dry T–1 bacteriophage by helium ions, carbon ions, oxygen ions: Comparison of effect for tracks of various ion density. *Radiation Res.,* 13: *788–808.*

45. Fowler, P. H., and Perkins, D. H. 1961. The possibility of therapeutic application of beams of negative π mesons. *Nature,* 189: *524–528.*

46. Gaul, H. 1958. Present aspects of induced mutations in plant breeding. *Euphytica,* 7: *275–289.*

47. ————. 1961. Use of induced mutants in seed-propagated species. *This Symposium, 206–251.*

48. Gaulden, M. E., and Perry, R. P. 1958. Influence of the nucleolus on mitosis as revealed by ultraviolet microbeam irradiation. *Proc. Nat. Acad. Sci.,* 44: *553–559.*

49. Giles, N. H. 1954. Radiation-induced chromosome aberrations in Tradescantia. In. *A. Hollaender, ed., Radiation Biology. New York: McGraw-Hill,* 1: *713–761.*

50. ————. 1955. Forward and back mutation at specific loci in Neurospora. *Brookhaven Symp. in Biol., Brookhaven Nat. Lab., Upton, N. Y.,* 8: *103–125.*

51. Glasser, O. 1960. Medical Physics. *Chicago: The Year Book Publishers, Inc.*

52. ————, Quimby, E. H., Taylor, L. S., and Weatherwax, J. L. 1950. Physical Foundations of Radiology. *New York and London: Paul B. Hoeber, Inc.*

53. Gordon, S. A. 1957. The effects of ionizing radiation on plants: Biochemical and physiological aspects. *Quart. Rev. Biol.,* 32: *3–14.*

54. Gray, L. H. 1952. The energy transfer from ionizing particles to an aqueous medium and its bearing on the interpretation of radio-chemical and radiobiological change. *Jour. Cellular Comp. Physiol.,* 39, Suppl. 1: *57–74.*

55. ————. 1954. Some characteristics of biological damage induced by ionizing radiations. *Radiation Res.,* 1: *189–213.*

56. ————. 1955. Aspects physiques de la radiobiologie. In *M. Haïssinsky, ed., Actions Chimiques et Biologiques des Radiations. Paris: Masson et Cie., Ser. 1: 4–91.*

57. ———. 1956. Cellular radiobiology. *Ann. Rev. Nuclear Sci.*, 6: *353–422.*

58. ——— and Scholes, M. E. 1951. The effect of ionizing radiations on the broad bean root: VIII. Growth rate studies and histological analyses. *Brit. Jour. Radiol.*, 24: *82–92, 176–180, 228–236, 285–291, 348–352.*

59. Gunckel, J. E. 1957. The effects of ionizing radiation on plants: Morphological effects. *Quart. Rev. Biol.*, 32: *46–56.*

60. ———. 1961. Modifications of growth and development induced by ionizing radiations. *Encyclopedia of Plant Physiol.*, 15: *1–23.*

61. ——— and Sparrow, A. H. 1954. Aberrant growth in plants induced by ionizing radiation. *Brookhaven Symp. in Biol., Brookhaven Nat. Lab., Upton, N. Y.*, 6: *252–279.*

62. ———, ———. 1961. Ionizing radiations: Biochemical, physiological and morphological aspects of their effects on plants. *Encyclopedia of Plant Physiol.*, 16: *555–611.*

63. ———, ———, Morrow, I. B., and Christensen, E. 1953. Vegetative and floral morphology of irradiated and non-irradiated plants of *Tradescantia paludosa*. *Amer. Jour. Bot.*, 40: *317–332.*

64. Gustafsson, Å. 1961. The induction of mutations as a method in plant breeding. In *M. Errera and A. Forssberg, ed., Mechanisms in Radiobiology. New York: Academic Press,* 2. In press.

65. Haas, F. L., and Doudney, C. O. 1957. A relation of nucleic acid synthesis to radiation-induced mutation frequency in bacteria *Proc. Nat. Acad. Sci.*, 43: *871–883.*

66. ———, ———, and Kada, T. 1961. Effects of preirradiation and postirradiation cellular synthetic events on mutation induction in bacteria. *This Symposium, 145–172.*

67. Hagberg, A., Gustafsson, Å., and Ehrenberg, L. 1958. Sparsely contra densely, ionizing radiations and the origin of erectoid mutations in barley. *Hereditas, 44: 523–530.*

68. Handloser, J. S. 1959. Health Physics Instrumentation. *New York: Pergamon Press.*

69. Hannan, R. S. 1956. Research on the Science and Technology of Food Preservation by Ionizing Radiations. *New York: Chemical Pub. Co.*

70. Haque, A. 1953. The irradiation of meiosis in Tradescantia *Heredity,* 6: *(Suppl.): 57–75.*

71. Hennessy, T. G., Howton, D. R., Levedahl, B. H., Mead, J. F., Myers, L.S., and Schjeide, O. A., ed. 1959. Radiobiology at the Intra-cellular Level. *New York and London: Pergamon Press.*

72. Hine, G. J., and Brownell, G. L. 1956. Radiation Dosimetry. *New York: Academic Press.*

73. Holmes, B. E. 1957. Biochemical effects of ionizing radiation. *Ann. Rev. Nuclear Sci.,* 7: *89–134.*

74. International Commission on Radiological Protection. 1959. Radiation Protection. *New York and London: Pergamon Press.*

75. Jackson, W. D. 1959. The life-span of mutagens produced in cells by irradiation. In. *J. H. Martin, ed., Radiation Biology, New York: Academic Press; London: Butterworths Sci. Pub., 190–208.*

76. Johns, H. E. 1953. The Physics of Radiation Therapy. *Springfield, Ill.: Charles C. Thomas.*

77. Kamra, O. P., Kamra, S. K., Nilan, R. A., and Konzak, C. F. 1960. Radiation response of soaked barley seeds: II. Relation of radiobiological damage to substances lost by leaching. *Hereditas,* 46: *261–273.*

78. Kaufmann, B. P. 1954. Chromosome aberrations induced in animal cells by ionizing radiations. In *A. Hollaender, ed., Radiation Biology, New York: McGraw-Hill,* 1: *627–711.*

79. ————, Gay, H., and McDonald, M. R. 1960. Organizational patterns within chromosomes. In *G. H. Bourne and J. F. Danielli, ed., International Review of Cytology. New York: Academic Press,* 9: *77–127.*

80. Kihlman, B. A. 1961. Biochemical aspects of chromosome breakage. *Advances in Genetics,* 10. In press.

81. Kimball, R. F. 1957. Nongenetic effects of radiation on microorganisms. *Ann. Rev. Microbiol.,* 11: *199–220.*

82. Kinsman, S., ed. 1957. Radiological Health Handbook. *Cincinnati: Robert A. Taft Sanitary Engineering Center, U. S. Dept. of Health, Education, and Welfare.*

83. Kirby-Smith, J. S. 1956. The measurement and properties of ionizing radiation. In. *G. Oster and A. W. Pollister, ed., Physical Techniques in Biological Research. New York: Academic Press,* 2: *57–110.*

84. ———— and Daniels, D. S. 1953. The relative effects of X rays, gamma rays, and beta rays on chromosomal breakage in Tradescantia. *Genetics,* 38: *375–388.*

85. ————, Nicoletti, B., and Gwyn, M. L. 1960. Synergistic action of X-rays and ultraviolet radiation on chromosomal breakage in Tradescantia pollen. *This Symposium, abstracts of contributed papers.*

86. Konzak, C. F. 1957. Genetic effects of radiation on higher plants. *Quart. Rev. Biol.*, 32: 27–45.

87. Kotval, J. P., and Gray, L. H. 1947. Structural changes produced in microspores of Tradescantia by alpha-radiation. *Jour. Genetics*, 48: 135–154.

88. Langham, W. H., ed. 1960. Literature Search on the Relative Biological Effectiveness (RBE) of Ionizing Radiations. *Los Alamos Sci. Lab., Los Alamos, New Mexico. LAMS-2343.*

89. Larter, E. N., and Elliott, F. C. 1956. An evaluation of different ionizing radiations for possible use in the genetic transfer of bunt resistance from Agropyron to wheat. *Can. Jour. Bot.*, 34: 817–823.

90. Laser, H. 1956. The influence of oxygen on radiation effects. In *G. E. W. Wolstenholme and C. M. O'Connor, ed., Ciba Foundation Symposium on Ionizing Radiations and Cell Metabolism. Boston: Little, Brown and Co., 106–119.*

91. Latarjet, R., ed. 1958. Organic Peroxides in Radiobiology. *New York: Pergamon Press.*

92. Lawrence, C. W. 1961. The effect of the irradiation of different stages in microsporogenesis of chiasma frequency. *Heredity*, 16: 83–89.

93. Lea, D. E. 1947. Actions of Radiations on Living Cells. *Cambridge: Univ. Press; New York: MacMillan.*

94. Lima-De-Faria, A. 1959. Bibliography on autoradiography with special reference to tritium labeled DNA precursors. *Hereditas*, 45: 632–648.

95. Loevinger, R., Japha, E. M., and Brownell, G. L. 1956. Discrete radioisotope sources. In *G. J. Hine and G. L. Brownell, ed., Radiation Dosimetry. New York: Academic Press, 693–799.*

96. MacKey, J. 1954. Neutron and X-ray experiments in wheat and a revision of the speltoid problem. *Hereditas*, 40: 65–180.

97. ———. 1961. Methods of utilizing induced mutations in crop improvement. *This Symposium, 336–364.*

98. Marinelli, L. D., and Taylor, L. S. 1954. The measurement of ionzing radiations for biological purposes. In *A. Hollander, ed., Radiation Biology. New York: McGraw-Hill, 1: 145–190.*

99. Marquardt, H. 1951. Die Wirkung der Röntgenstrahlen auf die Chiasmafrequenz in der Meiosis von *Vicia faba. Chromosoma*, 4: 232–238.

100. McClintock, B. 1939. The behavior in successive nuclear divisions of a chromosome broken in meiosis. *Proc. Nat. Acad. Sci.*, 25: 405–416.

101. ————. 1941. Spontaneous alterations in chromosome size and form in *Zea mays*. *Cold Spr. Harb. Symp. Quant. Biol.*, **9**: 72–81.

102. Mericle, L. W., and Mericle, R. P. Unpublished data.

103. ————, ————. 1957. Irradiation of developing plant embryos: I. Effects of external irradiation (X rays) on barley embryogeny, germination, and subsequent seedling development. *Amer. Jour. Bot.*, **44**: 747–756.

104. ————, ————. 1961. Radiosensitivity of the developing plant embryo. *Brookhaven Symp. in Biol., Brookhaven Nat. Lab., Upton, N. Y.*, **14**. In press.

105. Mika, E. S. 1952. Effect of indoleacetic acid on root growth of X-irradiated peas. *Bot. Gaz.*, **113**: 285–293.

106. Morrison, J. W. 1954. A dicentric wheat chromosome in division. *Can. Jour. Bot.*, **32**: 491–502.

107. Moutschen-Dahmen, J., Moutschen-Dahmen, M., Verly, W. G., and Koch, G. 1960. Autoradiograms with tritiated myleran. *Exp. Cell Res.*, **20**: 585–588.

108. Muller, H. J. 1954. The nature of the genetic effects produced by radiation; The manner of production of mutations by radiation. In *A. Hollaender, ed., Radiation Biology. New York: McGraw-Hill*, **1**: 351–473; 475–626.

109. ————. 1956. On the relation between chromosome changes and gene mutations. *Brookhaven Symp. in Biol., Brookhaven Nat. Lab., Upton, N. Y.*, **8**: 126–147.

110. ————, Herskowitz, I. H., Abrahamson, S., and Oster, I. I. 1954. A non-linear relation between X-ray dose and recovered lethal mutations in Drosophila. *Genetics*, **39**: 741–749.

111. National Research Council. 1958. Glossary of Terms in Nuclear Science and Technology. *2nd Printing. New York: American Society of Mechanical Engineers. (ASA N1.1–1957)*.

112. Newell, H. E., and Naugle, J. E. 1960. Satellites and space probes are revealing the kinds and amounts of radiation men will encounter in space. *Science*, **132**: 1465–1472.

113. Nickson, J. J., ed. 1952. Symposium on Radiobiology: The Basic Aspects of Radiation Effects on Living Systems. *Oberlin College, 1950. New York: John Wiley and Sons*.

114. Nilan, R. A. 1956. Factors governing plant radiosensitivity. In *A Conference on Radioactive Isotopes in Agriculture, Michigan State Univ. Washington: U. S. Govt. Printing Office, 151–164. TID 7512*.

115. ———— and Konzak, C. F. 1961. Increasing the efficiency of mutation induction. *This Symposium, 437–460.*

116. Nybom, N. 1956. Some further experiments on chronic gamma irradiation of plants. *Bot. Notiser,* **109:** *1–11.*

117. ————, Gustafsson, Å., and Ehrenberg, L. 1952. On the injurious action of ionizing radiation in plants. *Bot. Notiser,* **105:** *343–365.*

118. Ord, M. G., and Stocken, L. A. 1959. Biochemical effects of ionizing radiation. *Ann. Rev. Nuclear Sci.,* **9:** *523–552.*

119. Peters, K. 1954. Die Beeinflussung des Radium Effektes in den Wurzelspitzenzellen von *Vicia faba equina* durch Megaphen. *Naturwissenschaften,* **41:** *89–90.*

120. Piore, E. R., Beams, J. W., Bethe, H. A., Haworth, L. J., and McMillan, E. M. 1959. An explanatory statement on elementary particle physics and a proposed federal program in support of high energy accelerator physics. In *A Report of a Special Panel Appointed by The President's Science Advisory Committee and The General Advisory Committee to the Atomic Energy Commission.*

121. Powers, E. L. 1957. Cellular radiobiology. *Ann. Rev. Nuclear Sci.,* **7:** *63–88.*

122. ————, Webb, R. B., and Ehret, C. F. 1960. Storage, transfer, and utilization of energy from X-rays in dry bacterial spores. *Radiation Res., Suppl.* **2:** *94–121.*

123. Prakken, R. 1959. Induced mutation. *Euphytica,* **8:** *270–322.*

124. Price, W. J. 1958. Nuclear Radiation Detection. *New York: McGraw-Hill.*

125. Read, J. 1959. Radiation Biology of *Vicia faba* in Relation to the General Problem. *Oxford: Blackwell Scientific Publications.*

126. Rhoades, M. M. 1955. The cytogenetics of maize. In *G. F. Sprague, ed., Corn and Corn Improvement. New York: Academic Press, 123–219.*

127. Rick, C. M. 1940. On the nature of X-ray-induced deletions in Tradescantia chromosomes. *Genetics,* **25:** *467–482.*

128. Riley, H. P. 1957. Chemical protection against X-ray damage to chromosomes. *Genetics,* **42:** *593–600.*

129. Robertson, J. S., and Hughes, W. L. 1959. Intranuclear irradiation with tritium-labeled thymidine. *Proc. First Nat. Biophysics Conf., Columbus, Ohio, 1957. New Haven: Yale Univ. Press, 278–283.*

130. Rosen, G. v. 1954. Radiomimetic reactivity arising after treatment employing elements of the periodic system, organic com-

pounds, acids, electric current in an electrolyte, temperature shocks, and ionizing radiation: Comparison with the mutagen effect. *Socker Handl. II,* 8: *157–273.*

131. Rossi, H. H. 1960. Spatial distribution of energy deposition by ionizing radiation. *Radiation Res., Suppl.* 2: *290–299.*

132. Russell, M.A., and Michelini, F. J. 1951. Relation of mitotic activity to the effects of X-rays and nitrogen mustard as indicated by the growth of corn seedlings. *Cancer Res.,* 11: *687–693.*

133. Russell, W. L., Russell, L. B., and Kelly, E. M. 1958. Radiation dose rate and mutation frequency. *Science,* 128: *1546–1550.*

134. ————, ————, and Oakberg, E. F. 1958. Radiation genetics of mammals. In *W. D. Claus, ed., Radiation Biology and Medicine. Reading, Mass.: Addison-Wesley Pub. Co., 189–205.*

135. Sagawa, Y., and Mehlquist, G. A. L. 1957. The mechanism responsible for some X-ray-induced changes in flower color of the carnation, *Dianthus caryophyllus. Amer. Jour. Bot.,* 44: *397–403.*

136. Sarić, M. R. 1957. The radiosensitivity of seeds of different ontogenetic development: I. The effects of X-irradiation on oat seeds of different phases of ontogenetic development. *Radiation Res.,* 6: *167–172.*

137. Sax, K. 1957. The effect of ionizing radiation on chromosomes. *Quart. Rev. Biol.,* 32: *15–26.*

138. ———— and Swanson, C. P. 1941. Differential sensitivity of cells to X-rays. *Amer. Jour. Bot.,* 28: *52–59.*

139. Schjeide, O. A., Mead, J. F., and Myers, L. S., Jr. 1956. Notions on sensitivity of cells to radiation. *Science,* 123: *1020–1022.*

140. Sears, E. R. 1956. The transfer of leaf-rust resistance from *Aegilops umbellulata* to wheat. *Brookhaven Symp. in Biol., Brookhaven Nat. Lab., Upton, N. Y.,* 9: *1–22.*

141. Shapiro, S. 1956. The Brookhaven radiations mutation program. *Conf. on Radioactive Isotopes in Agriculture, Michigan State Univ. Washington: U. S. Govt. Printing Office, 141–150. TID–7512.*

142. ————. 1961. Applications of Radioisotopes and Radiation in the Life Sciences. *Hearings before the Subcommittee on Research and Development, Joint Committee on Atomic Energy, Congress of the United States, U. S. Govt. Printing Office, 195–227.*

143. Shchepot'yeva, E. S., Ardashnikov, S. N., Lur'ye, G. E., and Rakhmanova, T. B. 1959. Effect of Oxygen in Ionizing Radiation. *Moscow: State Publishing House for Medical Literature.* (English translation available from the Office of Technical Ser-

vices, Department of Commerce, Washington 25, D. C. AEC–tr–4265.)

144. Sieburth, L. R. 1951. The interaction of nucleic acid and X radiations on the chromatin of rye and pea. *Northw. Sci.*, **25:** *132–136.*

145. Singh, B. N. 1941. The relationship of catalase ratio to germination of X-rayed seed as an example of pretreatments. *Jour. Amer. Soc. Agron.*, **33:** *1014–1016.*

146. Singleton, W. R. 1955. The contribution of radiation genetics to agriculture. *Agronomy Jour.*, **47:** *113–117.*

147. ————, ed. 1958. Nuclear Radiation in Food and Agriculture. *Princeton, N. J.: D. van Nostrand Co., Inc.*

148. ————, Konzak, C. F., Shapiro, S., and Sparrow, A. H. 1956. The contribution of radiation genetics to crop improvement. *Proc. Intern. Conf. on the Peaceful Uses of Atomic Energy, Geneva, 1955. New York: United Nations,* **12:** *25–30.*

149. Skok, J. 1957. Relationship of boron nutrition to radiosensitivity of sunflower plants. *Plant Physiol.*, **32:** *648–658.*

150. Slack, L., and Way, K. 1959. Radiations from Radioactive Atoms in Frequent Use. *Washington: U. S. Govt. Printing Office.*

151. Smith, H. H. 1958. Genetic plant tumors in Nicotiana. *Ann. N. Y. Acad., Sci.,* **71:** *1163–1177.*

152. ————. 1958. Radiation in the production of useful mutations. *Bot. Rev.,* **24:** *1–24.*

153. ————. 1961. Mutagenic specificity and directed mutation. *This Symposium, 413–436.*

154. Soderholm, L. H., and Walker, E. R. 1955. Effect of cathode rays on germination and early growth of wheat. *Bot. Gaz.,* **116:** *281–290.*

155. Sparrow, A. H. Unpublished data.

156. ————. 1944. X-ray sensitivity changes in meiotic chromosomes and the nucleic acid cycle. *Proc. Nat. Acad. Sci.,* **30:** *147–155.*

157. ————. 1951. Radiation sensitivity of cells during mitotic and meiotic cycles with emphasis on possible cytochemical changes. *Ann. N. Y. Acad. Sci.,* **51:** *1508–1540.*

158. ————. 1956. Cytological changes induced by ionizing radiations and their possible relation to the production of useful mutations in plants. In *Work Conference on Radiation Induced Mutations, Brookhaven Nat. Lab., Upton, N. Y., 76–113.*

159. ————. 1960. Uses of large sources of ionizing radiation in botanical research and some possible practical applications. In

Large Radiation Sources in Industry. Vienna: Intern. Atomic Energy Agency, **2:** *195–219.*

160. ———, Binnington, J. P., and Pond, V. 1958. Bibliography on the Effects of Ionizing Radiations on Plants, 1896–1955. *Upton, N. Y.: Brookhaven Nat. Lab.*

161. ——— and Christensen, E. 1953. Tolerance of certain higher plants to chronic exposure to gamma radiation from cobalt-60. *Science,* **118:** *697–698.*

162. ——— and Cuany, R. L. 1959. Radiation-induced somatic mutations in plants. *Conf. on Radioactive Isotopes in Agriculture, Oklahoma State Univ. USAEC Document No. TID–7578, 153–156.*

163. ———, ———, Miksche, J. P., and Schairer, L. A. 1961. Some factors affecting the responses of plants to acute and chronic radiation exposures. In *Effects of Ionizing Radiation on Seeds and Its Significance for Crop Improvement. Proc. Sci. Symposium sponsored by the Intern. Atomic Energy Agency and the Food and Agriculture Organization of the United Nations. Karlsruhe, 1960, 289–320.*

164. ——— and Forro, F. 1953. Cellular radiobiology. *Ann. Rev. Nuclear Sci.,* **3:** *339–368.*

165. ——— and Gunckel, J. E. 1956. The effects on plants of chronic exposure to gamma radiation from radiocobalt. *Proc. Intern. Conf. on the Peaceful Uses of Atomic Energy, Geneva, 1955. New York: United Nations,* **12:** *52–59.*

166. ———, ———, Schairer, L. A., and Hagen, G. L. 1956. Tumor formation and other morphogenetic responses in an amphidiploid tobacco hybrid exposed to chronic gamma irradiation. *Amer. Jour. Bot.,* **43:** *377–388.*

167. ——— and Konzak, C. F. 1958. The use of ionizing radiation in plant breeding: Accomplishments and prospects. In *E. C. Tourje, ed., Camellia Culture. New York: MacMillan Co., 425–452.*

168. ——— and Miksche, J. P. 1961. Correlation of nuclear volume and DNA content with the tolerance of plants to chronic radiation exposures. *Science. In press.*

169. ———, Moses, M. J., and DuBow, R. J. 1952. Relationships between ionizing radiation, chromosome breakage and certain other nuclear disturbances. *Exp. Cell. Res., Suppl.* **2:** *245–267.*

170. ——— and Schairer, L. A. 1958. Some factors influencing radioresistance and tumor induction in plants. *Proc. Intern. Conf. on the Peaceful Uses of Atomic Energy, 2nd .Conf., Geneva. New York: United Nations,* **27:** *335–340.*

171. ——— and Singleton, W. R. 1953. The use of radiocobalt as a source of gamma rays and some effects of chronic irradiation on growing plants. *Amer. Naturalist,* **87:** *29–48.*

172. ———, Sparrow, R. C., and Schairer, L. A. 1960. The use of X-rays to induce somatic mutations in Saintpaulia. *African Violet Mag.,* 13(4): *32–37.*

173. Stadler, L. J. 1932. On the genetic nature of induced mutations. *Proc. 6th Intern. Congr. Genet.,* 1: *274–294.*

174. ———. 1954. The gene. *Science,* **120:** *811–819.*

175. Steffensen, D. 1955. Breakage of chromosomes in Tradescantia with a calcium deficiency. *Proc. Nat. Acad. Sci.,* **41:** *155–160.*

176. ———. 1957. Effects of various cation imbalances on the frequency of X-ray induced chromosomal aberrations in Tradescantia. *Genetics,* **42:** *239–252.*

177. ———. 1959. A comparative view of the chromosome. *Brookhaven Symp. in Biol., Brookhaven Nat. Lab., Upton, N. Y.,* **12:** *103–124.*

178. ———. 1961. Structure and stability of chromosomes in the irradiated cell. *Proc. IX Intern. Bot. Congr.* In *Recent Advances in Botany; Lectures and Symposia Presented to the Congress. Toronto: Univ. of Toronto Press.* In press.

179. Stent, G. S., and Fuerst, C. R. 1960. Genetics and physiological effects of the decay of incorporated radioactive phosphorus in bacterial viruses and bacteria. *Advances in Biol. and Med. Phys., ics,* **7:** *1–75.*

180. Stone, W. S. 1956. Indirect effects of radiation on genetic material. *Brookhaven Symposia in Biology, Brookhaven Nat. Lab., Upton, N. Y.,* **8:** *171–190.*

181. Strauss, B. S. 1958. The genetic effect of incorporated radioisotopes: The transmutation problem. *Radiation Res.,* **8:** *234–247.*

182. Sun, K.- H. 1954. Effects of atomic radiation on high polymers. *Modern Plastics,* 32(1): *141–150; 229–239.*

183. Swanson, C. P. 1957. Cytology and Cytogenetics. Englewood Cliffs, N. J.: *Prentice-Hall.*

184. ——— and Kihlman, B. 1956. The induction of chromosomal aberrations by ionizing radiations and chemical mutagens. In G. E. *Wolstenholme and C. M. O'Connor, ,ed., Ciba Foundation Symposium on Ionizing Radiations and Cell Metabolism. Boston: Little, Brown and Co.,* 239–274.

185. ———, Merz, T., and Cohn, N. 1959. Metabolism and the stabil-

ity of chromosomes. *Proc. Xth Intern. Congr. Genetics*, 1: *300–305.*

186. Symposium on Effects of Ionizing Radiation on Seeds and Its Significance for Crop Improvement. *Proc. Sci. Symposium sponsored by the Intern. Atomic Energy Agency and the Food and Agriculture Organization of the United Nations. Karlsruhe, 1960.* (in press).

187. Taylor, J. H., Woods, P. S., Hughes, W. L. 1957. The organization and duplication of chromosomes as revealed by autoradiographic studies using tritium-labeled thymidine. *Proc. Nat. Acad. Sci.,* 43: *122–128.*

188. Thoday, J. M. 1953. Sister-union isolocus breaks in irradiated *Vicia faba:* The target theory and physiological variation. *Heredity,* 6 *(Suppl.): 299–309.*

189. Tobias, C. A., Anger, H. O., and Lawrence, J. H. 1952. Radiological use of high energy deuterons and alpha particles. *Amer. Jour. Roentgenol. and Radium Therapy,* 67: *1–27.*

190. U. S. Atomic Energy Commission. 1960. Genetics Research Program of the Division of Biology and Medicine. *Washington: Office of Technical Services, Dept. of Commerce, TID–4041.*

191. Wainwright, S. D., and Nevill, A. 1955. Some effects of post-irradiation treatment with metabolic inhibitors and nutrients upon X-irradiated spores of *Streptomyces. Jour. Bact.,* 70: *547–551.*

192. Whiting, A. R. 1948. Incidence and origin of androgenetic males in X-rayed Habrobracon eggs. *Biol. Bul.,* 95: *354–360.*

193. Whyte, C. N. 1959. Principles of Radiation Dosimetry. *London and New York: John Wiley.*

194. Wilson, G. B., and Sparrow, A. H. 1960. Configurations resulting from iso-chromatid and iso-subchromatid unions after meiotic and mitotic prophase irradiation. *Chromosoma,* 11: *229–244.*

195. Wilson, G. B., Sparrow, A. H., and Pond, V. 1959. Sub-chromatid rearrangements in *Trillium erectum* L.: I. Origin and nature of configurations induced by ionizing radiation. *Amer. Jour. Bot.,* 46: *309–316.*

196. Wimber, D. E. 1959. Chromosome breakage produced by tritium-labeled thymidine in *Tradescantia paludosa. Proc. Nat. Acad. Sci.,* 45: *839–846.*

197. Wolff, S. 1960. Problems of energy transfer in radiation-induced chromosome damage. *Radiation Res., Suppl.* 2, *122–132·*

198. ———. 1961. Chromosome aberrations. In *A. Hollaender, ed., Radiation Protection and Recovery. London: Pergamon Press, 157–174.*

199. ————. 1961. Selected aspects of radiation genetics. In *M. Errera and A. Forssberg, ed., Mechanisms in Radiobiology*, 2: *(in press)*.

200. ———— and Luippold, H. E. 1956. The biochemical aspects of chromosome rejoining. In *J. S. Mitchell, B. E. Holmes, and C. L. Smith, ed., Progress in Radiobiology. Edinburgh: Oliver and Boyd, 217–221*.

201. Wolstenholme, G. E. W., and O'Connor, C. M., eds. 1956. Ciba Foundation Symposium on Ionizing Radiations and Cell Metabolism. *Boston: Little, Brown and Co.*

202. Wood, T. H. 1958. Cellular radiobiology. *Ann. Rev. Nuclear Sci.,* 8: *343–386.*

203. Work Conference on Radiation-induced Mutations. 1956. *Brookhaven Nat. Lab., Upton, N. Y.*

204. Zelle, M. R. 1960. Radioisotopes and the genetic mechanism: Mutagenic aspects. In *R. S. Caldecott and L. A. Snyder, ed., Radioisotopes in the Biosphere. Minneapolis: Univ. of Minnesota, 160–180.*

205. Zirkle, R. E. 1957. Partial-cell irradiation. *Advances in Biol. and Med. Physics,* 5: *103–146.*

Comments

MEHLQUIST: I am commenting with reference to your slide showing increased tolerance with increased size of the chromosomes and/or cell size. As you know, we tried some of our Delphinium hybrids along with carnations planted directly in the gamma field at Brookhaven and found that the Delphiniums were so sensitive that they were killed to the ground with the dosage at which the carnations showed little effect. Yet the Delphinium chromosomes are many times the size of carnation chromosomes.

Chemicals and Their Effects

CHARLOTTE AUERBACH

Institute of Animal Genetics,
Edinburgh, Scotland

IT IS NOW almost exactly 20 years since the first highly effective mutagens were detected. This symposium provides an opportunity for taking stock of developments in this new field of research. There can be no question of its expansion; an ever-increasing number of chemicals has been found to be mutagenic, and there is no reason to expect that this expansion will come to a standstill unless we halt it purposely because we feel that there is no good reason for testing more and more chemicals for mutagenic ability. This is one of the points which, I hope, we shall discuss at this meeting. But let us ask a different question: To what extent has work on chemical mutagens fulfilled the expectations with which it was started?

Expectations are largely a personal matter, but I think that the main expectations of mutation workers at that time may be classed under three headings. Chemical mutagens were expected to help elucidate (a) the chemical nature of the genetic material and (b) the relationship between intragenic and intergenic changes, and (c) to open the way for the production of specific types of genetic change. How far have these expectations been fulfilled?

The Nature of the Genetic Material

The original, somewhat naive, idea was that by studying the chemical group or groups that produce mutations we would learn something about the nature of the other partner in the reaction, the genetic material. This, I am afraid, has proved an illusion. We now know that there is not one magic group which confers mutagenic ability on a compound. Instead, vastly different compounds may have mutagenic abilities, and closely related ones may differ in this respect. On the other hand, during these 20 years our knowledge of the chemical nature of the genetic material has advanced spectacularly, but the advance did not come from mutation research. The situation now is almost exactly the reverse of what we expected 20 years ago. Instead of inferring the chemical nature of the gene from the nature of the substances which make it mutate, we tend to interpret the action of mutagenic compounds on the basis of what we know about the nature of the genetic material. In some cases, notably in mutation experiments on virus and bacteriophage, this approach has already given highly interesting results, but as a general attitude to mutation research it narrows the area of investigation and is likely to result in misleading interpretations. Even if we accept it as highly probable that, in higher organisms as well as in bacteria and bacteriophages, the essential genetic specificity resides in the DNA of the chromosomes, it still remains naive to imagine that a chemical—and often a highly reactive one—introduced into a cell will react with the DNA without at the same time producing a variety of chemical changes in the remaining constituents of the cell. Mutation, as we now know, is a complicated process of which we usually only see the beginning and the end: the introduction into the cell of a mutagenic agent, and the emergence of an organism or a clone of cells with altered properties. In between are many steps. The chemical has to pass barriers of permeability; it may produce the actual mutagen by interaction with the cytoplasm; the initial lesion in the genetic material may be restored or made permanent; chromosome breakage may be followed by restitution or rearrangement; the mutated gene has to create new biochemical pathways; the mutated cell has to multiply in the face of competition from non-mutated cells. The same chemical that produces a change in the DNA will often affect the course of these intermediary events by reactions outside the DNA.

What we know at present about the action of most mutagens is little more than speculation. I think this is partly due to the concentration of effort towards the one question, how does the mutagen react with the DNA? I shall briefly discuss what is known about the action of some groups of mutagens.

Alkylating Agents

These include some of the most potent chemical mutagens—the "mustards", characterized by the presence of one or several chloroethyl groups, the epoxides, ethylene imines, methane sulphonates, and β–propiolactone. A great number of alkylating agents have been synthesized for use in cancer therapy. The successful ones were subsequently tested on Drosophila, mainly by Fahmy and Fahmy (35)[1] in

---

[1]See References, page 136.

London. Most of them were found to be mutagenic. The close correlation between carcinostatic and mutagenic abilities among the alkylating agents is therefore in large part due to a bias in the selection of mutagens for testing. It is probable that chromosome breakage is the main mechanism by which alkylating agents kill dividing tumour cells. In view of the connection between mutation and chromosome breakage it would therefore be expected that alkylating agents with carcinostatic ability will be mutagens. The fact that many of them are also carcinogenic is less readily interpreted. It may be taken as support for the theory that the primary event in carcinogenesis may be a somatic mutation.

Oncologists have found that all effective carcinostatic alkylating agents carry two or more active groups, e.g., ethylene imine or chloroethyl groups, and they put forward the theory that cross-linkage between biologically important molecules is a prerequisite for carcinostatic as well as mutagenic ability. Yet monofunctional mustards have been found to be very effective in the production of mutations in Neurospora (88), and the monofunctional compounds ethylene oxide and ethylene imine are among the most effective mutagens for barley (31, 32), where they also produce many chromosome breaks. There still remained the possibility that polyfunctional compounds are *relatively* more effective than monofunctional ones in breaking chromosomes. If this were true, they should produce higher ratios of translocations to lethals in Drosophila. This was tested recently in our laboratory for the two compounds ethylene oxide and diepoxybutane (65). The results suggested that cross-linkage plays no significant role in the production of chromosome breakage by alkylating agents.

Alkylating mutagens have been effective in all tested organisms from bacteria to mammals. Data on mammals are not easy to obtain. Two methods have been used successfully on mice. The first consists of injecting the chemical intraperitoneally and examining the F_1 for heritable semisterility caused by induced translocations. By this method, translocations have been detected after treatment with nitrogen mustard (37), but toxicity was so strong that even with the highest tolerated dose the number of translocations was small. Possibly, treatment of semen for artificial insemination will be more successful and we intend to try this in our laboratory. Triethylene melamine (TEM) proved a much more suitable mutagen for mice (16). Very high frequencies of dominant lethals and translocations were obtained after intraperitoneal injection. One translocation is of particular interest because it produces a position effect on two recessive coat color genes in the neighborhood of the break. Both of these genes, when present in heterozygous combination with the translocation, give a flecked pattern (17). Position effects had pre-viously not been known in mice. Curiously enough, at the same time, several similar ones were found among X-ray-induced translocations (80).

A different method for testing mutagens on mammalian cells has been developed by Klein and Klein (52) and, independently, by Mitchison (61). It has been used in our Institute for testing TEM (27). Tumors are induced in hybrids between two inbred strains of mice which are isogenic apart from one allelic difference at the histo-compatibility locus. The hybrid tumors, possessing both antigens, will not take in either parental strain unless they have lost one of the antigens. The frequency with which this happens could be increased by TEM. Whether the underlying event is mutation proper, chromo-some loss, deficiency, or somatic crossing-over could not be decided.

It is customary to interpret the effects of alkylating agents in terms of interaction with DNA. This may be correct, but it is not proved. The fact that the DNA, the transforming principle, is exceedingly sensitive to the destructive action of mustard gas (45) does not necessarily imply that the primary attack of mustard gas as a mutagen must be on DNA. Some evidence in favor of this assumption has recently been provided by the finding (58) that several alkylating agents, in particular ethyl methanesulphonate, can produce mutations in bacteriophage treated *in vitro*.

Whatever its point of attack, mustard gas, like X-rays, act in a "hit"-wise fashion, that is, it produces mutations and chromosome breaks by independent reactions that affect separate points in a more or less random fashion. For X-rays this was established by stud-ies of dose-effect relations. Evidence of this type is not easy to obtain for the action of chemicals on a complex organism. For certain muta-gens a linear relationship between dose and effect has been shown to hold good within rather narrow ranges of cencentration (33, 42), but this finding is open to different interpretations. In particular, it

suffers from the fact that the dose of a chemical as applied exter-
nally bears only a remote relation to the amount that reaches the
genes. We used a more indirect approach in experiments on Drosophila (68). The effective dose of mustard gas was measured in terms
of sex-linked lethals, and against this was plotted the frequency of
translocations in the same tests. We found that translocation frequency increased almost exactly as the square of lethal frequency,
and we took this to mean that mustard gas, like X-rays, acts by independent "hits" in or near the genetic material, one hit being sufficient to produce a lethal while two hits are required for a
translocation.

Corroborative evidence came from experiments in which treatment of Drosophila ♂ ♂ by mustard gas was immediately preceded or
followed by X-rays (73). With both arrangements, the frequency
of lethals after the combined treatment was the sum of the frequencies produced by either treatment separately. That of translocations was higher and calculation showed that it agreed well
with what would be expected if rearrangements were formed indis-
criminately from broken ends whether these had been produced by
X-rays or mustard gas. It is interesting to note that Wolff and Luippold (93) obtained a different result when they exposed plant chromosomes to a combined treatment with neutrons and X-rays. In
these experiments the frequency of interchanges was the sum of the
frequencies that had been produced separately by either radiation.
This was attributed to the fact that breaks produced by the closely
ionized neutron track are not scattered at random throughout the
nucleus, but lie close together and therefore tend to rejoin with
each other rather than with independently produced X-ray breaks.
It may be concluded that mustard gas, unlike neutrons, produces
breaks which *are* scattered randomly.

Urethane

The ability of urethane to produce translocations in flowering plants was discovered by Oehlkers (71) at the beginning of the last
war. Subsequently, the mutagenic action of urethane was confirmed for Drosophila (92). Neurospora, on the other hand, has
shown itself wholly refractory to its action (46). This may be connected with the fact that urethane shows an unusual degree of
organism specificity also in its carcinogenic action. Rogers (77) found

that the active principle in the production of lung tumors by urethane is a metabolite which is produced in mice but not in guinea-pigs. It is, therefore, possible that Neurospora is resistant to the mutagenic action of urethane because it cannot form the mutagenically effective metabolite. Oehlkers attributes the chromosome-breaking action of urethane in plant cells to general disturbances of metabolism. Rogers (78) has shown a close connection between urethane and nucleic acid metabolism. In Drosophila breaks produced by urethane and X-rays interact freely with each other to give the number of rearrangements expected from the overall number of breaks by both treatments (73).

Alkaloids

Oehlkers and his collaborators (72) have found that a number of pharmacologically important alkaloids, such as morphine and scopolamine, produce chromosome rearrangements in plants. Quite recently, Clark in Australia (20) has established that certain alkaloids, e.g., heliotrin and related substances, are highly effective mutagens for Drosophila. These substances occur in Senecio and some other plants and cause liver disease in sheep.

Peroxides

This group forms a link between chemical and radiation mutagenesis, for peroxides have been shown to play the role of intermediates in mutagenesis by ultraviolet light and, probably, X-rays. Hydrogen peroxide is a weak mutagen for micro-organisms (46), but is wholly ineffective in Drosophila where it is quickly destroyed by catalase. Certain organic peroxides, on the other hand, are effective mutagens for both Drosophila and Neurospora (3, 28, 82).

Formaldehyde

The mutagenic action of formaldehyde was discovered by Rapoport (74) soon after the last war. He obtained high mutation frequencies by mixing formaldehyde with the food of Drosophila larvae. Mutations can also be produced by injecting aqueous solutions of formaldehyde into the abdomen of adult flies (6). This is a less effective method but will be discussed first because there are reasons to believe that it acts *via* the formation of organic peroxides from formaldehyde and metabolically produced hydrogen perox-

ide. Mutation frequency is enhanced in Drosophila ♂ ♂ whose catalase has been poisoned by KCN (83, 84). Under these conditions,
formaldehyde injection even produces some mutations in ♀ ♀, which
otherwise are quite refractory to it. Solutions of formaldehyde are
but weakly mutagenic for conidia of Neurospora; so are solutions
of hydrogen peroxide (46). A mixture of both compounds in solution
is strongly mutagenic (28), and addition of dihydroxydimethyl
peroxide compound is a good mutagen for Drosophila (82).

When formaldehyde is applied as an admixture to the food
of Drosophila, it probably acts in quite a different way. While
injection produces mutations mainly in mature spermatozoa, formaldehyde food acts preferentially or exclusively on the long-
growth stage of the primary spermatocyte that, in the larvae, pre-
cedes meiosis (9). Female larvae are wholly resistant to formaldehyde
food; so are adults. Experiments with isotopically labelled formaldehyde (47) have shown that lack of penetration into certain cells
cannot be responsible for these results; for large amounts of labelled
formaldehyde were found inside the germ cells of adults, of female
larvae, and of spermatogonia. The fact that bunches of identical
or complementary cross-overs were obtained from treated ♂ larvae
(85) shows that the effective substance enters the nucleus of spermatogonia; yet it hardly, if ever, produces mutations at this stage,
for several hundred cross-tests between autosomal lethals that
appeared as clusters in the progeny of individual ♂ ♂ yielded no
clusters of identical lethals (9).

It is tempting to assume that the preferential mutagenic action
of formaldehyde food on the growth stage of the primary spermatocyte is somehow connected with the synthetic processes going on
during this stage. This assumption is supported by various observations. Thus, any condition that slows down larval development
reduces the mutagenic effectiveness of formaldehyde food (8). More
recently, Alderson found that, in a synthetic and sterile medium,
formaldehyde requires adenosine riboside for its mutagenic action
(1, 2).

Substances Related to Nucleic Acid

Even before DNA acquired its predominant role in our model
of the genetic material, substances related to nucleic acid metabolism were tested for mutagenic activity, many of them with success.

Acridines, which react with nucleic acid, have produced mutations in at least one series of experiments with barley (22). One of them, proflavine, is a very effective mutagen for bacteriophage (15). Pyronin, which reacts with ribonucleic acid, is mutagenic for Drosophila (18). Its overall effect is weak, but certain treated individuals yield very high mutation frequencies. It seems that very special conditions have to be fulfilled for its effective action (7, 19).

Many purines produce mutations in micro-organisms (41, 70), and mutations and chromosome breaks in plants (49). For many years, no mutagenic pyrimidines were found in spite of attempts to detect them. This has changed since it was discovered that it is possible to force bacterial cells to take in analogues of the normal pyrimidine bases by closing the normal pathway leading to pyrimidine synthesis. Bacteria that are unable to synthesize thymine may replace it quantitatively with offered bromouracil. Mutations occur in such bacteria and in bacteriophages grown on them (13, 57, 79).

The fact that bromouracil is so readily incorporated into bacterial DNA makes it tempting to assume that it is mutagenic through incorporation, and ingenious schemes of mutagenesis through base changes in the genetic DNA have been based on this assumption (40). Some mutagenic purines, such as caffeine, are either not incorporated at all or only in indetectably small amounts (69) and proflavine probably does not act by incorporation. For purines, interference with enzymes concerned with nucleic acid metabolism has been suggested as an alternative cause of mutagenic ability (69). Particularly interesting, though still unexplained, is the fact that in bacteria mutagenesis by purines like caffeine and theophylline can be prevented by "antimutagens" in particular adenosine riboside (69). Radiation-induced mutation frequency is not depressed by these antimutagens, and a certain proportion of spontaneous mutations is likewise refractory to them. A search for similar systems of mutagen-antimutagen balance seems a promising approach to a study of the complex biochemical interactions in mutagenesis.

Nitrous Acid

In 1939, Thom and Steinberg (90) produced variants in *Aspergillus* grown on food with an admixture of mannitol-nitrite. Ten year later, admixture of sodium nitrite to the food of Drosophila produced mutations in one experiment, but repetitions gave negative results (76). These tests had been undertaken with the idea that nitrite, like formaldehyde, might produce mutations by acting on the proteins, which then were thought to be carriers of genetical specificity. Quite recently, nitrite has been used again and with spectacular success. Starting with very precise chemical predictions about desamination of the bases in nucleic acid, a group of German workers (63) found a method for the production of mutations in tobacco mosaic virus treated *in vitro*. In this case, there can be little, if any, doubt that mutations were produced by direct chemical action on the genetical material, for treatment was equally effective when applied to the naked RNA of the virus. Subsequently, similar results have been obtained with the RNA of polio virus (14) and the DNA of the pneumococcus-transforming principle (56). Nitrite has also been used successfully for the production of mutations in bacteriophage (91) and bacteria (48). Treatment of bacteriophage produced many sectored mutants, as would be expected from chemical reaction with one of the two strands of DNA. When an exceptional phage strain with single-stranded DNA was treated, there were no sectored mutants (89). Although it seems plausible to assume that nitrite also acts directly on DNA in bacteria, proof is still lacking.

In conclusion of this section I would like to say this. Work with chemical mutagens has not thrown new light on the nature of the gene, except insofar as the effects of certain mutagens can be understood best from our present model of gene structure. In my opinion, it would be regrettable if further research were directed wholly towards the study of this particular class of mutagens. The very variety of mutagenic chemicals shows that the stability of the hereditary mechanism is under complex control. Analysis of many different types of mutagen may help us unravel some of these complexities. It may also teach us how to loosen the control in ways that produce desirable mutations for applied purposes.

The Relation Between Intergenic and Intragenic Changes

The question whether there is an essential difference between intergenic and intragenic changes has been often discussed (62), sometimes heatedly. There are good reasons, theoretical as well as observational, for answering it in the affirmative, but decisive experimental evidence would be desirable. One possibility of obtaining such evidence was to look for a chemical that would produce gene mutations but no chromosome breakage. No clear case of this kind has been encountered. Although mutagens differ widely in their chromosome-breaking ability, most of them can be shown to possess this ability to some extent and none can be shown not to possess it. The opposite is not true. Several compounds that break the chromosomes of plants have failed to produce mutations in Drosophila. This, however, cannot be taken as evidence that these substances act on intergenic bonds rather than on genes, unless it can be shown that they can also break the chromosomes of Drosophila. *A priori,* it is probable that substances that can do this will also produce small deficiencies that appear as recessive lethals, and position effect rearrangements that appear as visible mutations. There are good reasons for supposing that a substance that breaks plant chromosomes may not have the same effect on Drosophila chromosomes. Not only are the metabolic differences between these organisms likely to affect the action of introduced chemicals. It should also be kept in mind that plant chromosomes, in contrast to the very stable chromosomes in the germ cells of Drosophila, are easily broken, e.g., by low doses of X-rays, excessive oxygen pressure (21) or, conversely, anoxia (60). Thus, only a substance that fails to produce mutations in the same organism in which it produces chromosome breaks can be used as evidence in favor of distinct intergenic bonds. No clear case of such a substance has so far been found.

Again chemical mutation research has failed to provide a decisive answer to a definite question. An answer, albeit still a partial one, has come from a different field of genetics. Analysis of the fine structure of the genetic material in bacteriophage has shown that a mutation may affect only one or at most a few nucleotides (12), surely a change that must be called intragenic. Since, however, the chromosomes of higher organisms are much more highly structured than those of bacteriophage and, in particular, have various types of protein intimately associated with their DNA, the relation between intergenic and intragenic changes remains an important subject of investigation. Crystallographic, electronmicroscopic, biochemical, and biophysical investigations are likely to play a major role in this research. Analysis of the effects of chemical mutagens will, I think, be mainly important by showing the complexity of the processes that result in chromosome breakage and by separating out some of the relevant factors.

One very important factor has been discovered in radiation studies. Wolff and Luipold (93), found that the reunion of broken .chromosome ends is an energy-requiring process which proceeds only under conditions that permit protein synthesis. This makes it necessary to distinguish between the effects of mutagens on the cohesion of the chromosomes and on the mechanism of reunion. A chemical may affect these processes separately, and the effects may differ in dependence on the cellular environment. Thus, chromosomes that have been broken by 8-ethoxycaffeine readily form rearrangements in Vicia, but tend to remain as free fragments in Allium (50). Conversely, urethane produces mainly fragments in Vicia and mainly rearrangements in Oenothera (26). Discrepancies between the reports of cytologists on the effects of a given mutagen will often be due to differences in the choice of organism or cell stage.

Only barley and Drosophila have been used for systematic attempts at comparing the intergenic and intragenic effects of chemicals. In Drosophila, most chemicals produce fewer rearrangements for a given frequency of mutations (sex-linked lethals) than do X-rays. That this is not due to a shortage of breaks but to failures of reunion has been shown for two substances, mustard gas (68) and TEM (34). Compared with a dose of X-rays giving the same frequency of sex-linked lethals, both these substances produce too few translocations but at least as many chromosome fragmentations resulting in zygotic lethals. The main cause for the relative shortage of reunions is apparently the delayed effect of these compounds; for breaks that open in different cell cycles cannot form rearrangements. Indeed, when mustard gas-treated spermatozoa were stored in the seminal receptacles of the ♀, there was an increase in the frequency of large rearrangements (5).

The tendency of chemical mutagens to produce intrachromosomal rather than interchromosomal changes may be likewise a consequence of their delayed action, if it is assumed that latent breaks in the same chromosome have a greater chance than latent breaks in different chromosomes to open in the same cell cycle (81).

Storing of mustard gas-treated spermatozoa in the seminal receptacles of ♀ ♀ did not affect the frequency of sex-linked lethals, although calculations showed that the increase should have been perceptible if delayed mutation had occurred at the same rate as delayed breakage (10). Yet it is known that mutation is often delayed after mustard gas treatment (4). It appears, therefore, that delayed gene mutation requires replication of the chromosome, while delayed breakage does not. If this can be firmly established, it will form a distinguishing feature between chemically induced intergenic and intragenic changes in Drosophila.

Even if the mechanisms, by which mustard gas produces gene mutations and chromosome breaks should differ in their final stages, they must have a common initial step. This follows already from the previously mentioned experiment in which it was shown that the frequency of translocations increases as the square of the frequency of lethals (68). It is further substantiated by the finding that during the late spermatogonial stage, which is the most highly sensitive one to the mutagenic effects of mustard gas, lethals and breaks leading to translocations increase proportionally in frequency (87).

Minute deficiencies and other minute rearrangements occupy a somewhat intermediate position between gene mutations and large chromosome rearrangements. This is so in radiation mutagenesis, where the frequency of minute rearrangements increases linearly with dose, that of large rearrangements more nearly as the square of dose. It is also so in chemical mutagenesis, where compounds that are inferior to X-rays in the production of large rearrangements may yet be superior in the production of minute ones (36, 81). It is possible that a proportion of minute deficiencies is produced not by breakage and reunion but by some other mechanism such as unequal crossing-over, or errors in chromosome replication; but this is certainly not true for all of them, and probably only for a small minority. Reverse repeats, e.g., cannot arise without previous chromosome breakage, and they have been found repeatedly after exposure of Drosophila to chemical mutagens (59, 67, 81). The tendency of chemical mutagens to produce minute rather than large rearrangements may be due to some spreading effects of the chemical damage.

Of particular interest is the ability of a few—and possibly of more—chemical mutagens to produce high frequencies of small duplications. Many duplications were found in Vicia chromosomes that had been treated with nitrogen mustard (38), and formaldehyde food produced many duplications in Drosophila (81).

While there is good evidence that a common mechanism is responsible for the production of gene mutations, chromosome breaks, and minute rearrangements by chemicals, crossing-over seems to be induced in some different way. Formaldehyde food produces crossing-over but not mutations in spermatogonia (83, 86). For a given frequency of mutations, mustard gas produces considerably fewer cross-overs than does formaldehyde (85). The relative frequencies of lethals and induced cross-overs were markedly changed when mustard gas was given to male pupae rather than imagines (10). These findings agree with observations on plants which show that there is no correlation between the ability of a chemical to produce translocations and its effect on chiasma frequency (55). Further, chemical treatment of plants can produce translocations in pachytene when crossing-over is completed (54).

In barley, gene mutations are scored as segregants for visible abnormalities, mainly chlorophyll defects, in F_2, while chromosomal changes are recognized by their effect on the fertility of the F_1. Judged by these criteria, chemical mutagens differ widely in their relative abilities to produce intergenic and intragenic changes (30). Curiously enough, the extremes are formed by two purines—8-ethoxycaffeine, which produces almost exclusively chromosome changes, and nebularine (purine–9–riboside) which produces almost exclusively mutations. Unfortuntately, it is not possible to determine how many of the segregating chlorophyll mutations are caused by small deficiencies or rearrangements. Even so, the striking difference between the effects of these two substances, and similar differences between other chemicals, lend support to the model of a chromosome in which the genes are connected by links of non-genic material.

The Production of Specific Genetical Changes

From its early stages, work with chemical mutagens brought evidence for the possibility that these substances might be more specific than X-rays in their effects on the genetic material. During many years, however, the detectable specificity was regional rather than genic. The distribution of chemically induced chromosome breaks in plants was found not to be random (38, 51, 72, 75). After treatment with certain chemicals, it was highly localized (23). Most chemicals appear to produce more breaks in heterochromatic than euchromatic regions. In Drosophila, too, the distribution of sex-linked lethals over the X-chromosomes was found to differ between X-rays and certain chemicals (11, 35).

The discovery of specific effects of mutagens on individual loci meant a great step forward in mutation research. Many cases of "mutagen specificity" are now known in a variety of organisms, but it is very doubtful whether all of them have a similar basis. The most striking example is that of one particular region in the chromosome of bacteriophage T4 (39), where mutations induced by 5–bromouracil or 2–aminopurine are crowded together into preferred sites. While it is at least probable that these "hot spots" arise from specific interactions between mutagen and genetic site, this is not necessarily—and not even plausibly—so in many other cases of mutagen specificity. In bacteria (24) and fungi (53), different loci and different alleles of the same locus show different mutational reponses to a variety of physical and chemical mutagens. In extreme cases, a gene may be "mutagen stable", i.e., it may fail to respond to some or all tested mutagens, while yet being able to mutate spontaneously (25). These "mutation spectra" can be profoundly changed by introduction of an additional mutant gene into the same nucleus (43). We do not yet know how this is brought about, but various explanations can be envisaged and tested. An influence of the residual genotype on the reaction between gene and mutagen is the least likely one. We have come to realize that mutation is a complicated process of which the reaction between mutagen and gene is only one step, although it is the essential one. Interactions between mutagen and cytoplasm, conditions of the cell preceding and following treatment, degree and type of competition between the newly mutated cell and the remaining non-mutated ones, as

well as other circumstances which it would lead too far to discuss here, all these decide whether the essential step will take place
and whether it will result in a detectable mutation. Specificity may
occur at any one of these levels; at which one, cannot be decided
without special analysis of each individual case.

In higher organisms, apparent mutagen specificity at individual loci may be an expression of regional specificity along the chromosome. In Drosophila the distribution of visible mutations over
the X-chromosome depends to some extent on the mutagenic treatment (36). In the silkworm, two linked genes determining the color
of the embryonic membranes appear to mutate in different ratios
after treatment with X-rays and several chemical mutagens (66).
Most or all of these apparent mutations are due, however, to deficiencies, and the specific responses observed are therefore properties
of the chromosome regions in which the genes are located. Also, specificity may occur at some later step in the mutagenic process, such
as the rejoining of broken chromosome ends. This is suggested by
the observation that the differential response of the two loci to mutagens varies with the sex of the treated animal (64).

The mutagen specificities that have been observed in barley
give rise to similar ambiguities of interpretation. The mutation
spectra of both chlorophyll and erectoides mutations have been
shown to vary according to the mutagen used (44). Some of these specificities have been correlated with detectable chromosomal aberrations. How many may be due to differences in the frequencies and
types of undetected chromosome rearrangements cannot at present
be decided. Moreover, in barley as in the silkworm, the mutation
spectrum is under the influence of physiological conditions (29),
such as hydration or degree of pregermination, and thus is unlikely
to reflect specificities of chemical interaction at the level of the
gene itself.

These doubts concerning the origin of specific responses to
mutagens do in no way detract from the great theoretical and practical importance of the observed phenomenon of mutagen specificity.
Theoretically, the recognition that mutagen specificity need not
always result from specific chemical reactions between mutagens and
nucleotides should make the geneticist more rather than less interested in chemical mutagens. It should make him realize that chemical

mutation research, in addition to and often instead of supplying the chemist with food for speculation, is one of the best tools for analyzing the fascinating series of events that culminate in induced mutation. The practical value of mutagen specificity depends on economical considerations and not on theories about the origin of specificity. If a treatment can be found that profitably produces a desired type of mutation, this will be a tremendous practical success, whatever its theoretical basis. Indeed, it seems to me that specificity of the reaction between mutagen and DNA is least likely to lead to such a success. Research with micro-organisms strongly suggests that this type of specificity differs between alleles and sites of the same locus; in general, it will therefore not appear as locus specificity. The aim of mutation breeding, on the other hand, is phenotypical specificity, that is the specific production of mutations that yield a desired phenotypical effect. If such specificity exists at all, it is more likely to occur at a late step in mutagenesis, when mutations leading to similar final effects may have a common biochemical pathway. A judicious combination of mutagenic and anti-mutagenic treatment may be one approach to its detection.

If we now look back at the questions with which we started this stocktaking of progress in chemical mutagenesis, it turns out, surprisingly, that the greatest progress has been made along the path that at first seemed by far the steepest, the path directed towards the detection of mutagen specificity. Progress along the two other paths, directed towards analysis of the chemical nature of the gene and of the relationship between intergenic and intragenic changes, has been somewhat disappointing. I believe this has been due mainly to two opposing tendencies on the part of investigators. One is the tendency to disperse efforts by hunting for more and more mutagens. The other, which comes into play when a new mutagen has been detected, is the tendency to channel all efforts at analysis into the field of nucleic acid chemistry. Outside these two methods of approach lie many neglected possibilities of biological analysis, whose exploration almost certainly will contribute to our understanding of the structure of higher chromosomes and of the complexities of the mutagenic processes.

References

1. Alderson, T. 1960. Significance of ribonucleic acid in the mechanism of formaldehyde-induced mutagenesis. *Nature,* **185:** *904–907.*

2. ————. 1960. Mechanism of formaldehyde-induced mutagenesis: The uniqueness of adenylic acid in the mediation of the mutagenic activity of formaldehyde. *Nature,* **187:** *485–489.*

3. Altenburg, L. S. 1954. The production of mutations in Drosophila by tertiary-butyl hydroperoxide. *Proc. Nat. Acad. Sci.,* **40:** *1037–1040.*

4. Auerbach, C. 1950. Difference between effects of chemical and physical mutagens. *Pub. Staz. Zool. Napoli,* **22:** *(Suppl): 1–21.*

5. ————. 1951. Problems in chemical mutagenesis. *Cold Spr. Harb. Symp. Quant. Biol.,* **16:** *199–213.*

6. ————. 1952. Mutation tests on *Drosophila melanogaster* with aqueous solutions of formaldehyde. *Amer. Nat.,* **86:** *330–332.*

7. ————. 1955. The mutagenic action of pyronin-B. *Amer. Nat.,* **89:** *241–245.*

8. ————. 1956. Analysis of the mutagenic action of formaldehyde food: III. Conditions influencing the effectiveness of the treatment. *Zeit. indukt. Abstamm. -u-Vererb. Lehre,* **87:** *627–647.*

9. ————, and Moser, H. 1953. Analysis of the mutagenic action of formaldehyde food: I. Sensitivity of Drosophila germ cells. *Zeit. indukt. Abstamm. -u-Vererb. Lehre,* **85:** *479–504.*

10. ———— and Sonbati, E. M. 1960. Sensitivity of the Drosophila testis to the mutagenic action of mustard gas. *Zeit. indukt. Abstamm. -u-Vererb. Lehre,* **91:** *237–252.*

11. Belitz, H. J. 1957. Vergleichende Untersuchungen der Verteilung spontaner und ducrh Chinon I (Bayer 4073) induzierter Mutationen über die genetische Karte des X-Chromosoms von *Drosophila melanogaster. Zeit. indukt. Abstamm. -u-Vererb. Lehre,* **88:** *434–442.*

12. Benzer, S. 1957. The elementary units of heredity. In *The Chemical Basis of Heredity (edited by McElroy and Glass) Baltimore: Johns Hopkins Press, 70–93.*

13. ———— and Freese, E. 1958. Induction of specific mutations with 5-bromouracil. *Proc. Nat. Acad. Sci.,* **44:** *112–119.*

14. Boeye, A. 1959. Induction of a mutation in poliovirus by nitrous acid. *Virology,* **9:** *691–700.*

15. Brenner, S., Benzer, S., and Barnett, L. 1958. Distribution of proflavine-induced mutations in the genetic fine structure. *Nature,* **182:** *983–985.*

16. Cattanach, B. M. 1957. Induction of translocations in mice by triethylenemelamine. *Nature,* 180: *1364–1365.*

17. ————. 1961. A chemically induced variegated-type position effect in the mouse. *Zeit. indukt. Abstamm. -u-Vererb. Lehre.* In press.

18. Clark, A. M. 1953. The mutagenic activity of dyes in *Drosophila melanogaster. Amer. Nat.,* 87: *295–306.*

19. ————. 1958. The mutagenic action of pyronin in Drosophila. *Zeit. indukt. Abstamm. -u-Vererb. Lehre,* 89: *123–130.*

20. ————. 1959. Mutagenic activity of the alkaloid heliotrine in Drosophila. *Nature,* 183: *731–732.*

21. Conger, A. D. 1952. Breakage of chromosomes by oxygen. *Proc. Nat. Acad. Sci.,* 38: *289–299.*

22. d'Amato, F. 1950. Mutazioni clorofilliane nell'orzo indotte da derivativi acridinici. *Caryologia,* III: *211–220.*

23. Darlington, C. D., and McLeish, J. 1951. Action of maleic hydrazide on the cell. *Nature,* 167: *407–409.*

24. Demerec, M. 1953. Reaction of genes of *Escherichia coli* to certain mutagens. *Symp. Soc. Expt. Biol.,* 7: *43–54.*

25. ————, Witkin, E. M., Labrum, E. L., Galinsky, I., Hanson, J., Monsees, H., and Fetherston, T. H. 1952. Bacterial genetics. *Carnegie Inst. Wash. Year Book No. 51: 193–205.*

26. Deufel, J. 1951. Untersuchungen über den Einfluss von Chemikalien und Röntgenstrahlen auf die Mitose von *Vicia faba. Chromosoma,* 4: *239–272.*

27. Dhaliwal, S. S. 1961. Studies on histocompatibility mutations in mouse tumour cells using isogenic strains of mice. *Genet. Res.* In press.

28. Dickey, F. H., Cleland, G. H., and Lotz, C. 1949. The role of organic peroxides in the induction of mutations. *Proc. Nat. Acad. Sci.,* 35: *581–586.*

29. Ehrenberg, L. 1955. Factors influencing radiation induced lethality, sterility, and mutations in barley. *Hereditas,* 41: *123–146.*

30. ————, Gustafsson, A., and Lundqvist, U. 1956. Chemically induced mutation and sterility in barley. *Acta Chem. Scandinav.,* 10: *492–494.*

31. ———— and Gustafsson, A. 1957. On the mutagenic action of ethylene oxide and di-epoxybutane in barley. *Hereditas,* 43: *594–602.*

32. ————, Lundqvist, U., and Ström, G. 1958. The mutagenic action of ethylene imine in barley. *Hereditas,* 44: *330–336.*

33. Fahmy, O. G., and Fahmy, M. J. 1953. Chromosome breaks among recessive lethals induced by chemical mutagens in *Drosophila melanogaster*. *Heredity*, 6 *(Suppl.): 149–159.*

34. ————, ————. 1954. Cytogenetic analysis of the action of carcinogens and tumour inhibitors in *Drosophila melanogaster:* II. The mechanism of induction of dominant lethals by 2:4:6– tri (ethyleneimino) – 1:3:5–triazine *Jour. Genet., 52: 603–619.*

35. ————, ————. 1956. Cytogenetic analysis of the action of carcinogens and tumour inhibitors in *Drosophila melanogaster:* V. Differential genetic response to the alkylating mutagens and X-radiation. *Jour Genet., 54: 146–164.*

36. ————, ————. 1957. Comparison of chemically and X-ray-induced mutations in *Drosophila melanogaster*. In *Advances in Radiobiology, Edinburgh and London: Oliver and Boyd, 437–447.*

37. Falconer, D. S., Slizynski, B. M., and Auerbach, C. 1952. Genetical effects of nitrogen mustard in the house mouse. *Jour. Genet., 51: 81–88.*

38. Ford, C. E. 1948. Chromosome breakage in nitrogen mustard treated *Vicia faba* root tip cells. *Proc. 8th Intern. Congr. Genetics, 270–271.*

39. Freese, E. 1959. The difference between spontaneous and base-analogue induced mutations of phage T4. *Proc. Nat. Acad. Sci., 45: 622–633.*

40. ————. 1959. On the molecular explanation of spontaneous and induced mutations. *Brookhaven Symp. in Biol., Brookhaven Nat. Lab., Upton, N. Y., 12: 63–75.*

41. Fries, N., and Kihlman, B. 1948. Fungal mutations obtained with methyl-xanthines. *Nature, 162: 573–574.*

42. Gibson, F. G., Brink, R. A., and Stahmann, M. A. 1950. The mutagenic action of mustard gas on *Zea mays*. *Jour. Hered., 41: 232–238.*

43. Glover, S. W. 1956. A comparative study of induced reversions in *Escherichia coli*. In *Genetic Studies with Bacteria, Carnegie Inst. Wash. Pub. 6, 12: 121–136.*

44. Gustafsson, A. 1960. Chemical mutagenesis in higher plants. *Abhandl. Deutsche Akad. Wissenschaften Berlin, Klasse Medizin, 1960,* I: *14–29.*

45. Herriot, R. M. 1948. Inactivation of viruses and cells by mustard gas. *Jour. Gen. Physiol., 32: 221–230.*

46. Jensen, K. A., Kirk, I., Kølmark, G., and Westergaard, M. 1951.

Chemically induced mutations in Neurospora. *Cold Spr. Harb. Symp. Quant. Biol.*, 16: 245–261.

47. Kaplan, W. D., and Pelc, S. R. 1956. Autoradiographic studies of Drosophila gonads following the feeding of ^{14}C labelled formaldehyde. *Zeit. indukt. Abstamm. -u-Vererb. Lehre*, 87: 356–364.

48. Kaudewitz, F. 1959. Production of bacterial mutants with nitrous acid. *Nature*, 183: 1829–1830.

49. Kihlman, B. 1952. A survey of purine derivatives as inducers of chromosome changes. *Hereditas*, 38: 115–127.

50. ———. 1955. Oxygen and the production of chromosome aberrations by chemicals and X-rays. *Hereditas*, 41: 384–404.

51. ——— and Levan, A. 1951. Localized chromosome breakage in *Vicia faba*. *Hereditas*, 37: 382–388.

52. Klein, G., and Klein, E. 1958. Histocompatability changes in tumors. *Jour. Cell. Comp. Phys.*, 52 *(Suppl. 1)*: 125–168.

53. Kølmark, G. 1953. Differential response to mutagens as studied by the Neurospora reverse mutation test. *Hereditas*, 39: 270–276.

54. Linnert, G. 1951. Die Einwirkung von Chemikalien auf die Meiosis. *Zeit. indukt. Abstamm. -u-Vererb. Lehre*, 83: 422–428.

55. ———. 1953. Der Einfluss von Chemikalien auf Chiasmenbildung und Mutationsauslösung bei Oenothera. *Chromosoma*, 5: 428–453.

56. Litman, R., and Ephrussi-Taylor, H. 1959. Inactivation et mutation des facteurs, génétiques de l'acide desoxyribonucléique du Pneumocoque par l'ultraviolet et par l'acide nitreux. *C. R. Acad. Sci., USSR*, 249: 838–840.

57. ——— and Pardee, A. B. 1960. The induction of mutants of bacteriophage T2 by 5–bromouracil: IV. Kinetics of bromouracil-induced mutagenesis. *Biochem. and Biophys. Acta*, 42: 131–140.

58. Loveless, A. 1958. Increased rate of plaque-type and host-range mutation following treatment of bacteriophage *in vitro* with ethyl methane sulphonate. *Nature*, 181: 1212–1213.

59. Mehtab, M. 1953. Chromosomal rearrangements in the progeny of Drosophila males treated with mustard gas. *Nature*, 171: 262–263.

60. Merz, T. 1959. The effect of extended anaerobic treatments on the chromosomes of *Vicia faba*. *Jour. Biophys. Biochem. Cytol.*, 5: 135–142.

61. Mitchison, N. A. 1956. Antigens of heterozygous tumours as material for the study of cell heredity. *Proc. Roy. Phys. Soc.*, 250: 45–48.

62. Muller, H. J. 1955. On the relation between chromosome changes and gene mutations. *Brookhaven Symp. in Biol., Brookhaven Nat. Lab., Upton, N. Y.*, **8**: *126–147.*

63. Mundry, K. W., and Gierer, A. 1958. Die Erzeugung von Mutationen des Tabakmosaik-Virus durch chemische Behandlung seiner Nukleinsäure *in vitro. Zeit. indukt. Abstamm. -u-Vererb. Lehre,* **89**: *614–630.*

64. Nakao, Y. 1957. Difference between male and female treatments in visible X-ray induced mutation rates in the silkworm. *Intern. Symp. Genetics, Tokyo, 260–264.*

65. ————. Unpublished communication.

66. ————, Tazima, Y., and Sakurai, Y. 1958. Specificity of interaction between the individual gene locus and the structure of chemical mutagens. *Zeit. indukt. Abstamm. -u-Vererb. Lehre,* **89**: *216–220.*

67. Nasrat, G. F. 1954. Some cytological observations on the delayed effect of mustard gas. *Nature,* **174**: *968–969.*

68. ————, Kaplan, W. D., and Auerbach, C. 1954. A quantitative study of mustard gas induced chromosome breaks and rearrangements in *Drosophila melanogaster. Zeit. indukt. Abstamm. -u-Vererb. Lehre,* **86**: *249–262.*

69. Novick, A. 1955. Mutagens and antimutagens. *Brookhaven Symp. in Biol., Brookhaven Nat. Lab., Upton, N. Y.*, **8**: *201–215.*

70. ———— and Szilard, L. 1951. Experiments on spontaneous and chemically induced mutations of bacteria growing in the chemostat. *Cold Spr. Harb. Symp. Quant. Biol.,* **16**: *337–343.*

71. Oehlkers, F. 1943. Die Auslösung von Chromosomenmutationen in der Meiosis durch Einwirkung von Chemikalien. *Zeit. indukt. Abstamm. -u-Vererb. Lehre,* **81**: *313–341.*

72. ————. 1953. Chromosome breaks influenced by chemicals. *Heredity,* **6**: *(Suppl.): 95–106.*

73. Oster, I. I. 1958. Interactions between ionizing radiation and chemical mutagens. *Zeit. indukt. Abstamm. -u-Vererb. Lehre,* **89**: *1–6.*

74. Rapoport, I. A. 1946. Carbonyl compounds and the chemical mechanism of mutations. *C. R. Acad. Sci., USSR,* **54**: *65–67.*

75. Revell, S. H. 1953. Chromosome breakage by X-rays and radiomimetic substances in *Vicia. Hereditas,* **6**: *(Suppl.): 107–124.*

76. Robertson, A., and Rendel J. M. Unpublished communication.

77. Rogers, S. 1955. Studies on the mechanism of action of urethane in initiating pulmonary adenomas in mice: I. The indirect nature of its oncogenic influence. *Jour. Nat. Cancer Inst.,* **15**: *1675–1683.*

78. ————. 1957. Studies on the mechanism of action of urethane in initiating pulmonary adenomas in mice: II. Its relation to nucleic acid synthesis. *Jour. Exp. Med.*, **105**: *279–306.*

79. Rudner, R., and Balbinder, E. 1960. Reversions induced by base analogues in *Salmonella typhimurium. Nature,* **186**: *180.*

80. Russell, L. B., and Bangham, J. W. 1959. Variegated-type position effects in the mouse. *Genetics,* **44**: *532.*

81. Slizynska, H. 1957. Cytological analysis of formaldehyde induced chromosomal changes in *Drosophila melanogaster. Proc. Roy. Soc. Edin.,* **B. 66**: *288–304.*

82. Sobels, F. H. 1956. Mutagenicity of dihydroxydimethyl peroxide and the mutagenic effects of formaldehyde. *Nature,* **177**: *979–980.*

83. ————. 1956. Studies on the mutagenic action of formaldehyde in Drosophila: II. The production of mutations in females and the induction of crossing-over. *Zeit. indukt. Abstamm. -u-Vererb. Lehre,* **87**: *743–752.*

84. ———— and Simons, J. W. I. M. 1956. Studies on the mutagenic action of formaldehyde in Drosophila: I. The effect of pretreatment with cyanide on the mutagenicity of formaldehyde and of formaldehyde-hydrogen peroxide mixtures in males. *Zeit. indukt. Abstamm. -u-Vererb. Lehre,* **87**: *735–742.*

85. ———— and van Steenis, H. 1957. Chemical induction of crossing-over in Drosophila males. *Nature,* **179**: *29–31.*

86. ————, Bootsma, D., and Tates, A. D. 1959. The induction of crossing-over and lethal mutations by formaldehyde food in relation to stage specificity. *Drosophila Information Service (DIS),* **33**: *161–162.*

87. Sonbati, E. M., and Auerbach, C. 1960. The brood pattern for intragenic and intergenic changes after mustard gas treatment of Drosophila males. *Zeit. indukt. Abstamm. -u-Vererb. Lehre,* **91**: *253–258.*

88. Stevens, C. M., Mylroie, A. Auerbach, C., Jensen, K. A., Kirk, I., and Westergaard, M. 1950. Biological action of "mustard gas" compounds. *Nature,* **166**: *1016–1022.*

89. Tessman, J. 1959. Mutagenesis in phages ΦX 174 and T4 and properties of the genetic material. *Virology,* **9**: *375–385.*

90. Thom, C., and Steinberg, R. A. 1939. The chemical induction of genetic changes in fungi. *Proc. Nat. Acad. Sci.,* **25**: *329–333.*

91. Vielmetter, W., und Wieder, C. M. 1959. Mutagene und inaktivierende Wirkung salpetriger Säure auf freie Partikel des Phagen T2. *Zeit. Naturforsch;* **14 b**: *312–317.*

92. Vogt, M. 1948. Mutationsauslösung bei Drosophila durch Äthyl-
    urethan. *Experienta, 4: 68–69.*
93. Wolff, S., and Luippold, H. E. 1955. Metabolism and chromosome-
    break rejoining. *Science,* **122:** *231–232.*
94. ————, Atwood, K. C., Randolph, M. L., and Luippold, H. E.
    1958. Factors limiting the number of radiation-induced chro-
    mosome exchanges. *Jour. Biophys. Biochem. Cytol., 4: 365–372.*

Comments

Ross: Is there any evidence for chemical specificity either for chromosome breaks or gene mutations at the same locus of both homologues
of a chromosome in one cell?

AUERBACH: This is a possibility that interests me very much. I hope that
it will be tried this year in our Institute by Doctor Oster who has
worked out a scheme for testing it in Drosophila.

NILAN: Results obtained at Pullman with diethyl sulfate have some
bearing on Doctor Auerbach's discussion of the delayed effect of certain chemical mutagens on chromosomes. In barley, very high frequencies of chlorophyll-deficient seedling mutations were induced by
this chemical. However, only a very few chromosome aberrations were
found. Such aberrations, resulting from chromosome interchanges, are
common following irradiation. On further study, some evidence for
an appreciable frequency of inversions has been obtained. Thus, it
appears that more intrachromosomal than interchromosomal rearrangements have occurred, indicating a delayed effect of this chemical on
chromosomes.

CALDECOTT: In regard to the relation between chemical mutagens and
ionizing radiations, I would like to re-emphasize the importance of the
physiological state of the cell and the kinds and frequencies of genetic
damage that can be detected. Clearly, it would be quite impossible to
design an experiment where there could be a precise comparison between
mutagens.

AUERBACH: I quite agree; it is too often overlooked that any mutagenic
treatment—in particular with a chemical—affects the cell, and sometimes the organism as a whole, in addition to acting on the chromosomes.

MacKey: I greatly appreciate the cautiousness shown by Dr. Auerbach in discussing results from different organisms in relation to mutagenic treatment. I believe there is great danger in trying to explain different radiobiological phenomena observed in different research objects according to one pattern. For example, the radiosensitivity in relation to polyploidy goes a completely different way in a ploidy series of wheat or any higher plant and yeast. From diploid level onwards, radioresistance gradually increases with genome number in the first case but decreases in the latter. The results cannot be explained only by substitution of cells in the multicellular organism. It seems likely that a factor not yet discussed here intervenes, viz., the mutual interaction of cells in a multicellular tissue. Work by Dr. A. M. Clark with Habrobacon favors such an interpretation. He found haploid embryos more resistant to X-rays during cleavage. No significant difference in radiosensitivity existed at blastema and early larval stages, and at late larval, prepupal, and pupal stages, the diploids showed the higher resistance.

Kramer: There is perhaps some precedent for this approach in the use of such chelating agents as ethylene diamine tetra acetic acid on Chlamydomonas and Drosophila. Some preliminary work by Nuffer, I believe, has indicated less success with corn, but this may be due to inadequate methods of treatment. Surely this approach has value both in plant breeding and in learning more about recombination.

Auerbach: The importance of induced recombination for plant breeding deserves, I think, more consideration than it has received so far. It seems that the efficiencies of mutagens in affecting crossing over, on the one hand, and producing chromosome rearrangements, on the other, are only loosely correlated. It might, therefore, be promising to undertake a screening program—first, cytologically, then genetically—in a suitable organism like maize for chemicals that enhance recombination with little concomitant sterility or other undesirable effects resulting from rearrangements.

Vallentyne: I would like to comment briefly on the first of your three initial hopes, viz., the possibility of learning something about the chemical nature of the hereditary unit from a knowledge of the chemistry of mutagenic substances. As you were talking about the action of formaldehyde and nitrous acid, it occurred to me that there is a reaction that should be considered in relation to the natural process of mutation. This is the reaction between keto and amino compounds first studied by L. C. Maillard in 1912, and subsequently known as the Maillard reaction (or the browning reaction in food chemistry). It is of interest that the reaction proceeds at measurable rates in the temperature range to which living matter is normally subjected; secondly, that the reactive groups are not only present in DNA, but present in juxtaposition; and thirdly, that the groups involved in the reaction are those that give specificity to the bases of DNA which in turn give specificity to the DNA itself. The juxtaposition of the keto and amino groups in the Watson-Crick model of DNA seems rather unusual in terms of the possibility of a Maillard-type reaction. Perhaps the groups are protected by the very existence of hydrogen-bonding in the double helical arrangement. This I do not know. It is conceivable, however, that the groups would be reactive on rupture of the double helix, either with each other or with keto and amino compounds in the surrounding nuclear plasm of cytoplasm. I wonder if the Maillard reaction could be operating *in nature* to produce mutation in a manner analogous to that of nitrous acid. Would you care to comment on this?

AUERBACH: I am afraid my knowledge of chemistry is not good enough for an answer to this interesting question.

Effects of Preirradiation and Postirradiation Cellular Synthetic Events on Mutation Induction in Bacteria[1]

FELIX L. HAAS, CHARLES O. DOUDNEY, AND TSUNEO KADA[2]

The University of Texas M. D. Anderson Hospital and Tumor Institute,
Houston, Texas

D URING THE past few years several investigations have made it evident that the processes leading to mutation induction by ultraviolet light (UV) are intimately related to gene replication, which is then followed by the influence of the modified gene in enzyme synthesis. Since these events are primarily biochemical, involving interrelations in the syntheses of deoxyribonucleic acid (DNA) ribonucleic acid (RNA), and protein, it seems reasonable that one can no longer examine such biological phenomena as mutation induction and expression without simultaneously studying their biochemical basis. Conversely, mutation induction and mutation expression are endpoints of the biochemical events of DNA replication and genetic control of enzyme synthesis respectively, and they may serve as useful tools in working out the biochemistry of these events.

This paper is concerned with several theoretical aspects of mutation induction which have been investigated by such combined biochemical and biological studies. The experimental results also have some implications for modern theories of genetic DNA replication. At the very least, it is hoped that the value and relative ease of performing such integrated studies will be demonstrated.

The importance of the physiological state of the biological material used in such investigations, and probably in all biological investigations, cannot be overemphasized. It is obvious that the same biochemical and physiological processes are not operating at all times during the life cycle of any cell. Important events may be entirely obscured when working with random populations composed of members of all ages and stages of physiological development.

[1]This investigaton is supported in part by research grant C–3323 from the National Institutes of Health and by Atomic Energy Commission contract AT–(40–1)–2139.

[2]Postdoctoral Research Fellow; Research Training Grant CRT–5047 from the National Institutes of Health.

145

The experiments presented here were carried out with an appreciation both for the simultaneous study of biochemistry and biology, and for uniformity of the biological material. Various strains of the bacterium *Escherichia coli* were employed, and in all experiments the bacteria used were synchronized as to cell division and development by a cold treatment method (17).[3] In most of the experiments the syntheses of RNA, DNA, and protein were followed in the cultures simultaneously with mutation induction and expression.

Materials and Methods

Description of bacterial strains and the mutation followed in each

E. coli strain B.—Originally obtained from Oak Ridge National Laboratory stock of Dr. A. Hollaender. *Mutation*—aberrant colonial color response on *Difco* eosin-methylene blue (EMB) agar after 2 days incubation at 37° C.

E. coli strain WP2.—Originally isolated by Witkin (26) as a tryptophan-requiring mutant of *E. coli* strain B/r. *Mutation*—reversion of the tryptophan requirement to the non-requiring state.

E. coli strain $15_{T-Me-Tyr}$.—A triple auxotroph of *E. coli* strain B/r requiring thymine, methionine, and tyrosine isolated in our laboratory from thymine-requiring *E. coli* strain 15_T- following ultraviolet light exposure. Strain 15_T- was obtained from Dr. S. Zamenhoff of Columbia University. *Mutation*—reversion of tyrosine requirement.

Description of growth media

M medium.—A salts-glucose basal growth medium to which various supplements under test, or various metabolic inhibitors, were added. The composition of the medium has previously been described (7). In experiments using auxotrophs appropriate amounts of the required growth factors were added. All preirradiation and postirradiation incubation was carried out in liquid M medium supplemented as indicated in the various experiments.

Agar plating medium

1. *For color mutants.*—*Difco* EMB agar was used as the final plating medium. Survivors and mutations were scored on the same plates.

---

[3]See References, page 169.

2. *For prototrophic reversion.*—M medium supplemented with 2.5 per cent *Difco* nutrient broth and solidified with 3 per cent *Bacto* agar was used for final platings. The low amino acid concentrations in this medium allows unreverted cells to make sufficient divisions to develop small visible colonies (26). It also enables reverted cells to become phenotypically expressed. Unreverted survivors are counted as small colonies on high-dilution platings, while revertants appear as large colonies against a background of auxotrophic growth at low dilutions.

3. *For mutation expression.*—M medium solidified with 3 per cent *Bacto* agar was used. No nutrient broth was added to this medium.

Radiation sources

Ultraviolet irradiation (UV).—A model 30600 Hanovia mercury-vapor lamp. The UV output at the position of the cells was 92.5 ergs/mm^2/sec at wave lengths below 2800 Å (determined by a Hanovia model AV–971 ultraviolet meter).

X-ray.—A General Electric "Maxitron 250" unit set at 200 KVP and 30 MA, with 1-mm aluminum filtration added. X-ray output at the locus of cell suspensions was approximately 2860 r/min.

Preirradiation and postirradiation treatments

The techniques employed have been previously described (3, 4, 7, 9). Briefly they are as follows: A 24-hour-old slant culture was used to inoculate 50 ml sterile M medium (plus growth factors in the case of auxotrophic strains). This culture was grown for 15.5 hours at 37° C with aeration, and then held at 6 C for 1 hour to synchronize cell division. The cells were centrifuged down in the cold and resuspended in 50-ml fresh M medium (plus supplements if necessary). In the preirradiation experiments supplements under test were added to this medium. The culture was grown for an additional period, aliquots taken for test at various intervals, and chilled to halt further growth. For postirradiation experiments the cells were incubated for 50 minutes before chilling. Chilled cells were washed with cold 0.9 per cent saline and M medium, then resuspended in cold M medium so as to titer approximately 6 × 10^8 colony-forming organisms per ml in the UV studies and 2 × 10^{10} organisms per ml in the X-ray studies. Aliquots of this suspension at proper dilution were plated on appropriate medium for determination of the number of organisms subjected to the radiation. Other aliquots were irradiated with UV or X-ray depending on the experiment.

In preirradiation supplementation experiments, immediately following irradiation, appropriate dilutions were plated on agar plating media by the glass rod spreader technique.

For postirradiation experiments aliquots of the irradiated culture were diluted 1:4 into appropriately supplemented growth medium and incubated at 37° C on a reciprocal shaker for time intervals indicated in the various experiments prior to plating on agar media. All plates were incubated at 37° C for 2 days in the case of color mutants and for 3 days in the cases of prototrophic reversions before scoring for survivors and mutants.

Biochemical determinations

Culture samples were taken at indicated intervals during preirradiation and postirradiation incubation for analysis of RNA, DNA, and protein. The samples were precipitated and washed with 0.5N perchloric acid, and the nucleic acids then hydrolyzed by incubation in perchloric acid for 50 minutes at 70° C (20). They were then analyzed for DNA content by the Burton (1) method. For RNA determination, the UV absorption at 260 mu and 290 mu were determined (25), and the amount of DNA, as determined by the Burton analysis, subtracted with correction for extinction coefficients. Protein was determined by the Folin (15) method.

Experimental Results and Discussion

I. Effect of Preirradiation Growth Factor Supplementation on Radiation-Induced Mutations

Early experiments were carried out using *E. coli* strain B, and consisted of attempts to identify possible extragenic factors which might be affected by UV so as to produce mutations. These experiments have been previously reported (7). The induced mutation-radiation dose curves previously obtained by many investigators suggested that radiation-sensitive material present in the cell was activated or altered by radiation as a prerequisite to mutation induction. The material also appeared to be limited in amount since the induced mutation frequency leveled off at high doses of radiation (18, 19, 29).

The mutations followed were those giving aberrant colonial color response on EMB agar (color mutants). Synchronized cell cultures were incubated for 1 hour in M medium supplemented with either casein hydrolysate (2 mg/ml), purines and pyrimidines (0.01 mg/ml each of adenine, guanine, cytosine, and uracil), or with a mixture of B vitamins (1 µg/ml of each). Following incubation, aliquots were irradiated with increasing doses of UV and then dilutions were plated on EMB agar. The results indicated that cells incubated with purines and pyrimidines yielded somewhat higher induced mutation frequencies following UV exposure. Other supplements produced no changes in the subsequent UV-induced mutation frequency from that obtained with the unsupplemented control except for the growth factors, riboflavin and p–amino benzoic acid. Further investigation, using various combinations of purines and pyrimidines and a UV dose giving maximum mutation frequency, demonstrated that maximum increase in mutation frequency was obtained when uracil, cytosine, and either adenine or guanine were present. Substitution of thymine for uracil resulted in considerable reduction in the mutation frequency obtained (7).

Experiments were next conducted to determine the duration of incubation in purines and pyrimidines necessary for attaining the maximum increase in UV–induced mutation. It was found that an initial incubation period of 20 minutes in purine-pyrimidine medium is necessary for initiation of the increase in subsequent radiation-induced mutation frequency. Following this lag a rapid increase is observed and the maximum frequency is attained at 30 to 35 minutes incubation. On the other hand, when cells are incubated in yeast extract there is no lag, and increase in susceptibility to subsequent mutation induction starts immediately with incubation. This suggested the possibility that the lag period observed with purine-pyrimidine incubation represents the time necessary for synthesizing radiation-reactive substances from · the purines and pyrimidines. Further experiments, using ribosides (adenosine, guanosine, uridine, and cytidine) supported this concept since these supplements led to reduction of the lag period to less than 5 minutes (7).

Similar experiments using X-ray suggest that a large fraction of X-ray-induced mutations are mediated through similar mechanisms (8). Incubation of cells in yeast extract or purine-pyrimidine medium prior to irradiating increases the induced mutation frequency some-what, although the effect is not as marked as with UV.

II. Effect of Postirradiation Treatments on the Fate of Potential UV-induced Mutations

Witkin (26) has suggested that the immediate postirradiation synthesis of protein is necessary for expression of induced prototrophs with certain auxotrophic strains of E. coli and Salmonella typhimuri-um. In her experiments, expression of induced prototrophic mutants was directly related to availability of a complex supply of amino acids during the first hour of postirradiation growth. Chloramphenicol (CMP), which specifically interferes with protein synthesis, prevented mutation expression if the cells were treated during the first post-irradiation hour. We have confirmed the existence of similar relation-ships for the color mutations of E. coli strain B, and have shown that the mutation frequency increase due to preirradiation incubation in purines and pyrimidines is also dependent on postirradiation amino acid supply. Involvement of postirradiation macromolecular syn-theses (protein, RNA, DNA) in the mutagenic process has been studied for both color mutants of strain B and for prototrophic muta-tions of various auxotrophic strains. While the results were essentially comparable, for technical reasons most of these postirradiation studies were accomplished with the latter system.

Mutation frequency decline

Involvement of metabolic processes in mutation induction has been studied by observing the effects of various conditions limiting to amino acid or protein synthesis on UV-induced mutation (3, 4, 8). Immediately following irradiation the cells were incubated in nitrogen-free medium for periods of time varying from 0 to 90 minutes, prior to plating on M plus 2.5 per cent nutrient broth agar. Using this technique, it was found that when nitrogen-dependent synthetic activities are limited there is a rapid decline in induced mutation frequency obtained at plating on nutrient broth-containing medium. The decline in mutation frequency is observed almost immediately and proceeds to a level unaffected by further incubation within 20 to 30 minutes at 37° C. The decline rate is similar in cul-tures incubated in M medium and in purine–pyrimidine-supple-mented medium prior to irradiation. Evidence that the processes involved in decline are enzymatic was obtained by studying the decline rate at different incubation temperatures ranging from 6° to 37° C. This established that the mutation frequency decline (MFD) process has a Q_{10} of approximately 2. The hypothesis was suggested that when amino acids or a nitrogen source were absent, some enzymatic process removes the potential mutation before it is "fixed" in the genetic structure. Amino acids might then serve in "stabilization" of the pre-mutation or mutagen, removing it from susceptibility to MFD until those processes leading to incorporation in the genetic structure could be completed.

The role of CMP in interfering with mutation induction was investigated and found to promote MFD in a manner completely analogous with that caused by nitrogen or amino acid deprivation (4). It was thus considered that CMP acts to prevent some process of "mutation stabilization" (MS) involving amino acids, thus allowing MFD to proceed. Figure 1 compares the time course and relationships of MFD and MS in regard to tryptophan-requirement reversions in *E. coli* strain WP2. When the cells are plated immediately after irradiation onto broth-supplemented M agar the mutation frequency attained at the UV dose used was about 40 prototrophs per 10^6 survivors. However, this level of nutrient broth supplementation affords amino acid concentrations much lower than optimum for maximum mutation response as is apparent from the *Min + AA* curve in Figure 1. Here the cells were incubated following irradiation in M medium supplemented with higher levels of amino acids for increasing time intervals before plating onto the limiting broth-supplemented M agar. It is apparent that when the cells are incubated for 30 to 40 minutes with excess amino acids, the induced mutation frequency obtained on subsequent plating almost doubles. Therefore, the amino acids have been taken into the cell and are involved in processes leading to mutation induction. On the other hand, when the cells are incubated in medium containing no amino acids or other nitrogen source (*Min − N* curve in Figure 1), or if CMP is added to amino acid-supplemented medium (*Min + AA + CHL* curve in Figure 1), MFD leads to a low mutation frequency within 20 minutes. Oxidative phosphorylation is probably required for MS since, if the irradiated cells are incubated in complete medium to which dinitrophenol (DNP) has been added, MS does not take place (*Min + AA + DNP*

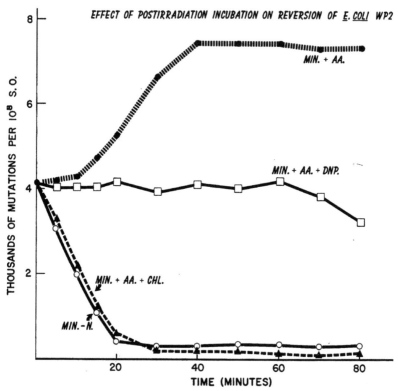

FIGURE 1.—*Comparison of "mutation stabilization" (MS), "mutation frequency decline" (MFD), and the effect of dinitrophenol on UV-induced reversion of tryptophan requirement in Escherichia coli strain WP2 The UV-irradiated cells were incubated at 37°C in the following media for the indicated time periods and then plated on M + 2.5 per cent nutrient broth agar: (1) Min + AA = M medium + casein hydrolysate (NBC, vitamin free, enzymatic, 2 mg/ml) and dl-tryptophan, 0.2 mg/ml; (2) Min + AA + DNP = medium (1) plus dinitrophenol (5 × 10⁻³ molar); (3) Min + AA + Chl = medium (1) plus chloramphenicol (20 micrograms/ml); (4) Min − N = M medium with ammonium sulfate deleted. The plates were incubated 3 days then scored for mutation and survival. No significant changes in survival level were produced by the above treatments.*

curve in Figure 1). DNP specifically uncouples oxidative phosphorylation and also prevents MFD when it is added to nitrogen-free medium. When incubation in complete medium or M minus nitrogen medium is allowed for short intervals before adding DNP intermediate levels of MS or MFD are obtained. Experiments in which high levels of amino acids were added to the incubation medium subsequent to or during MFD indicate that the process is not reversible. Once MFD has proceeded to any level, addition of amino acids will not increase the mutation frequency obtained with subsequent incubation beyond that level.

Witkin (26, 27) has suggested that delay in DNA synthesis could increase mutation since more time would be available for protein synthetic processes involved in mutation induction prior to DNA synthesis. On the other hand, Kimball, et al (13) propose that delay in DNA synthesis would decrease mutation since more time would be available for reversion of an unstable "pre-mutational" state to a stable state not inducing a mutational change.

It has previously been demonstrated that UV will cause a delay in DNA synthesis (12), and several investigators (2, 6, 10) have since shown that resumption of DNA synthesis following UV requires prior synthesis of RNA and protein. In our experiments several inhibitors which block RNA or protein synthesis prior to initiation of DNA synthesis in a UV-irradiated culture were found not only to inhibit DNA synthesis, but also to promote a decided decrease in the induced mutation frequency (5). If a quantitative correlation could be established between the effects of these agents on mutation frequency and their effect in delaying DNA synthesis, then the hypothesis advanced by Kimball, et al (13), would be supported. But if MFD occurs independently of the effect on DNA synthesis, then the hypothesis that MFD is an active enzymatically controlled process (3) would be more likely. Several experiments designed to give evidence as to the most likely of these alternatives were performed. These experiments measured the effect of increasing periods of postirradiation treatments blocking RNA or protein synthesis on the subsequent DNA synthesis and mutation frequency. Absence of nitrogen source, absence of tryptophan for the tryptophan-requiring auxotroph, presence of 6 aza uracil, and presence of CMP were used as inhibitors. With all experiments the results demonstrated that conditions sufficient in duration to lead to lower mutation frequency through blockage of RNA or protein synthesis produce no significant

effect on subsequent DNA synthesis. Only with treatment of at least
20 minutes is maximum MFD obtained and this decline is not associated with any change in the pattern of DNA synthesis. The same
holds true for RNA and protein synthesis in the culture. Net RNA
and protein syntheses begin only after 25 to 30 minutes postirradiation
incubation, and treatments during the first 20 minutes causing maximum MFD do not appear to appreciably change the subsequent
pattern of RNA and protein synthesis. It is apparent from these
experiments that delay in DNA synthesis cannot be held responsible
for MFD following UV irradiation.

Mutation fixation

The potential mutation eventually becomes established in the
cell to the extent that it is no longer susceptible to conditions promoting MFD. This process has been termed "Mutation Fixation"
(MF) (4). It should not be confused with the final process of mutation
induction involving DNA synthesis (see below). CMP challenge (see
Figure 2 for outline of technique) is most frequently used for demon-
starting MF. The CMP challenge technique is based on the principal
that, when irradiated cells are incubated in complete medium, all
processes involved in mutation induction will proceed. However,
when CMP is added after a given incubation interval, all potential
mutations remaining "unfixed" at the time of CMP addition will be
eliminated by the MFD processes. In Figure 2 comparison of the
MS curve with that for MF (determined by CMP challenge) shows
that the potential mutations are stabilized in the cell for a considerable period prior to initiation of the MF process. As late as 30 minutes
after irradiation all potential mutations remain subject to processes
promoting MFD, but after this time decreasing numbers are affected
by CMP challenge and after 75 minutes none are susceptible. Neither
MS nor MF is correlated with gross protein synthesis since this process
is not initiated until 10 minutes after MF has begun.

**Mutation fixation and mutation expression in relation to
macromolecular syntheses**

When the progress of RNA, DNA, and protein synthesis are
simultaneously followed with MF in a UV-irradiated culture, an
interesting correlation is apparent (Figure 3). Progression of MF is
closely correlated with RNA synthesis. Both processes are initiated at

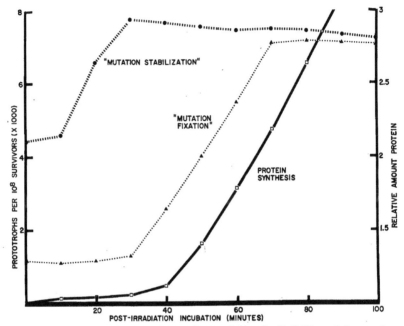

FIGURE 2.—*Comparison of "mutation stabilization" (MS) and "mutation fixation" (MF) in Escherichia coli strain WP2. Mutation followed was reversion of the tryptophan requirement. Following UV irradiation cells were_incubated at 37°C in M medium plus casein hydrolysate (2 mg/ml) and dl–tryptophan (0.2 mg/ml). Culture aliquots were plated at the indicated times (MS) on broth-supplemented M agar. Also, after these intervals of incubation chloramphenicol (CMP) (20 micrograms/ml) was added to samples which were incubated for an additional 45 minutes before plating. This latter procedure is termed "CMP challenge" and measures MF since all mutations remaining subject to the CMP-promoted MFD process at the time of CMP addition are eliminated during the subsequent 45 minutes incubation. Relative protein content of the culture at the times of CMP addition is also given.*

about the same time and MF attains its maximum when the RNA has *doubled* in amount. It thus appears that DNA synthesis *per se* may be segregated from the MF process.

The uridine analogue, 5–hydroxyuridine (5–HU), when present during postirradiation incubation, promotes marked decline in mutation frequency (Figure 4). However, this is not the MFD process

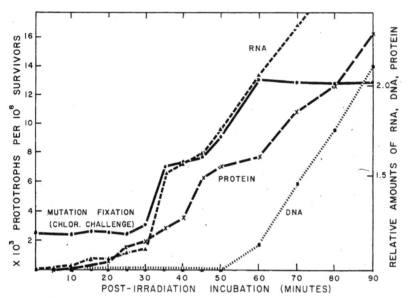

Figure 3.—*Relation of "mutation fixation" (MF), as indicated by CMP-challenge, to synthesis of RNA, DNA, and protein in a UV-irradiated culture of Escherichia coli strain WP2. Following irradiation the bacterial suspension was diluted 1:4 into M medium plus casein hydrolysate (2 mg/ml) and dl–tryptophan (0.2 mg/ml) and incubated at 37°C. At the indicated intervals aliquots were taken and CMP (20 micrograms/ml) added to them. They were then incubated for an additional 45 minutes at 37°C to allow CMP-promoted MFD to take place and plated on M + 2.5 per cent broth agar. At the time that samples were taken for CMP treatment other samples were taken for RNA, DNA, and protein determinations.*

revealed by incubation in CMP or amino acid-deficient media. With 5–HU treatment decline in mutation frequency does not start immediately but only after some 20 minutes incubation in its presence. Moreover, 5–HU promoted decline appears inversely correlated with MF and with RNA synthesis. This evidence, which suggests that 5–HU is exerting its effect through RNA synthesis, would account for the lag in initiation of the decline in mutation frequency. One possible explanation for the 5–HU promoted decline in mutation frequency is that the analogue is incorporated into new RNA, thus

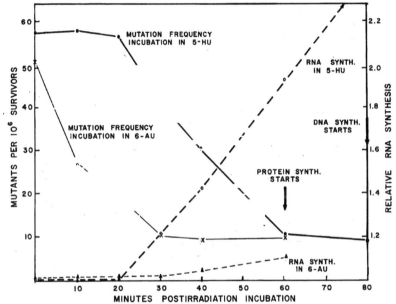

FIGURE 4.—*Comparison of the effects of 6–aza uracil (6AU) and 5–hydroxyuridine (5HU) on macromolecular syntheses and the freqeuncy of induced prototrophs in UV-irradiated Escherichia coli strain WP2. Following irradiation the bacterial suspension was diluted 1:4 into M medium plus casein hydrolysate (2 mg/ml) and dl–tryptophan (0.2 mg/ml). This was immediately divided into two portions and 6AU (0.05 micrograms/ml) was added to one portion and 5HU (0.05 micrograms/ml) to the other. Both cultures were then incubated at 37°C, and after the indicated incubation intervals samples were taken from each tube and plated on M plus 2.5 per cent nutrient broth agar. At the same times samples were taken for RNA, DNA, and protein analyses.*

producing nonfunctional RNA and using up the radiation-produced mutagenic precursors.

MFD is also obtained in the presence of the analogue 6–aza uracil (6–AU) (Figure 4), and is similar with this analogue to that obtained with CMP. Like CMP, 6–AU blocks protein synthesis, but unlike CMP, this is brought about through direct blockage of RNA synthesis. Addition of uridine will reverse the blockage to RNA and protein syntheses and also prevent 6–AU promoted MFD.

Schwartz and Strauss (21) have shown that incubation of UV-irradiated *E. coli* strain WP2 in the presence of tryptazan (a tryptophan analogue which is incorporated into protein) will result in a decline in mutation frequency. They suggest that this decline is caused by protein synthesis utilizing tryptazan. Their results constitute significant biochemical evidence implicating protein synthesis in mutation induction. If protein formed in the presence of tryptazan and subsequent to mutation fixation were intimately involved with RNA in the replication of genetic DNA, then "nonfunctional" protein would probably prevent utilization of the corresponding RNA in DNA replication and in so doing use up the mutagenic precursors.

All these experiments indicate that MF is closely correlated with the RNA synthesis, and that the mutations are established in some structure or form *before* any measurable DNA synthesis takes place. They support our previously presented hypothesis (4) that RNA and protein syntheses are intimately involved in replication of genetic DNA; and that the RNA, modified by incorporation of a radiation-altered precursor, in some manner leads to a corresponding modification in newly formed DNA. This implies that induced mutations first appear in the daughter DNA initially synthesized after irradiation, and that a period of time is thus necessary between UV treatment and establishment of the mutation in the genome. During this interval manipulations interfering with RNA and protein syntheses can prevent the potential mutation from being induced.

It is of considerable importance to determine at what point in the sequence of postirradiation events the induced mutation is first capable of being phenotypically expressed. Establishment of this point would differentiate processes involved in gene synthesis from those of enzyme synthesis involved in phenotypic expression. During the course of experiments with reversion of the tryptophan requirement of *E. coli* strain WP2, it became apparent that, during the early stages of postirradiation incubation, if the bacteria were plated on M agar rather than on broth-supplemented M agar, much lower induced mutation frequencies were obtained. After 90 minutes postirradiation incubation in complete medium, however, the mutation frequency obtained was the same on both M agar and broth-supplemented M agar. It seemed probable that the lower level of mutation obtained on M agar might be due to inability of the induced reversions in these tryptophan-deficient strains to be phenotypically expressed without some initial supplementation. If true, this behavior would indicate that mutation expression requires macromolecular synthetic events following those involved in gene replication and separable from the former when studying prototrophic mutations in an auxotrophic strain. The results of experiments to test this hypothesis have been previously reported (9).

A typical CMP challenge experiment was performed; also at the same time, samples were plated on M agar as well as on broth-supplemented M agar (Figure 5). As usual MF closely follows RNA synthesis. However, when identical aliquots are plated at the same incubation times on M agar, few mutations appear until after 40 minutes of incubation. At this point MF has practically been completed. After 40 minutes incubation, *and closely following DNA synthesis,* the mutation frequency obtained on the M agar plates rises sharply, and after 90 minutes the frequency of induced mutation measured is the same on both plating media. The results indicate that a period of protein synthesis subsequent to DNA replication is necessary for expression in the case of auxotrophic reversions. This hypothesis is susceptible to further testing. If the irradiated cells are incubated in complete medium for a time sufficient to allow initiation of DNA synthesis and then CMP is added to the incubating culture, mutation expression should be prevented in that portion of mutant cells which have not yet synthesized the requisite protein. This late CMP treatment should not promote MFD, however, since mutation fixation would already have been completed (see Figure 5). When such an experiment was performed, it was found that if CMP is added after 60 minutes incubation in complete medium, DNA synthesis continues in its presence; but further mutation expression is prevented when the cells are plated on M agar. When the culture is treated with CMP after 100 minutes of incubation and then plated on M medium, the mutation frequency is the same as that obtained in an untreated twin culture plated on broth-supplemented M agar. Other experiments were performed with an auxotroph requiring thymine, methionine, and tyrosine *(E. coli* strain $15_{\text{T-Me-Tyr}}$-). The mutation followed was reversion of the tyrosine requirement. When this strain is deprived of thymine but not amino acids during post-

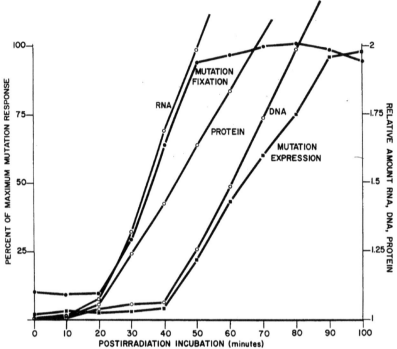

FIGURE 5.—*Relation of "mutation fixation" (MF) and "mutation expres-*
sion" to macromolecular synthesis in a UV-irradiated culture of Escheri-
chia coli strain WP2. Following irradiation the bacterial suspension was
diluted 1:4 into M medium plus casein hydrolysate (2 mg/ml) and dl–
tryptophan (0.2 mg/ml) and incubated at 37°C. MF was determined
by the "CMP-challenge" method (see Figure 2) followed by plating on
M plus 2.5 per cent nutrient broth agar. "Mutation expression" was
determined by plating directly onto M agar after the indicated periods
of incubation in M medium plus casein hyrdolysate and tryptophan.
Prototrophic mutations are expressed as the percentage of maximum
mutation frequency response (80 phototrophs/10⁶ UV survivors when
plating was on broth-supplemented M agar, and 60 prototrophs/10⁶ UV
survivors when plating was on M agar). At the same times that samples
were taken for CMP challenge and for plating on M agar, samples were
also analyzed for RNA, DNA, and protein.

irradiation incubation, no DNA synthesis takes place and mutation
expression does not occur with subsequent plating on M plus

thymine plus methionine medium. However, when thymine is restored to the incubating culture, DNA synthesis resumes and mutation expression is observed on the latter plating medium.

While processes leading to mutation induction occur prior to DNA synthesis and involve macromolecules other than the DNA, the mutation apparently comes finally to reside in new DNA but is not expressed until this DNA has functioned to establish the requisite enzyme.

III. Involvement of Postirradiation Macromolecular Syntheses in X-ray-Induced Mutation

Considerable time has been devoted recently to examining the relation of postirradiation macromolecular synthesis to X-ray-induced mutation. Earlier it was stated that preirradiation supplementation experiments indicated that a part of X-ray-induced mutations are due to incorporation of radiation-altered precursors during post-irradiation synthetic processes (7, 8). Because of this we have been examined the postirradiation synthetic events of X-irradiated bacteria for the same processes involved in UV-induced mutation. The techniques and methods elaborated for the UV work were used in the X-ray studies, except that the bacteria were at a higher concentration during exposure. The total dose in most experiments was 10 Kr given at a rate of 2860 r per minute. The cells were held in an ice bath during exposure.

Figure 6 presents results obtained when X-irradiated cells are plated on M agar and on broth-supplemented M agar after various intervals of postirradiation incubation in complete medium. It is evident from the M agar plates that an appreciable portion of the total mutational yield is expressed after only 10 minutes incubation in complete medium. With UV irradiation a much lower proportion of prototrophs were expressed prior to initiation of DNA synthesis (see Figure 5). In the case of X-ray exposed cells, during the first 10 minutes of postirradiation incubation approximately 40 per cent of the prototrophs are expressed in the absence of measurable DNA synthesis. DNA synthesis is initiated after 10 minutes and the M agar platings indicate that additional mutations are expressed following initiation of DNA synthesis.

When protein synthesis is inhibited by adding CMP immediately after irradiation, there is no appreciable effect on mutation frequency

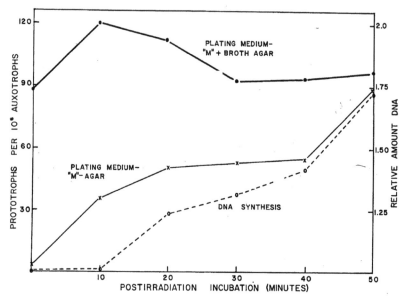

FIGURE 6.—*Comparison of mutation frequency expressed on M agar with that expressed on broth-supplemented M agar, and the relation of DNA synthesis to mutation expression in an X-irradiated culture of Escherichia coli strain WP2. Following X-ray irradiation (10 kr) the bacterial suspension was diluted 1:4 into M medium plus casein hydrolysate (2 mg/ml) and dl–tryptophan (0.2 mg/ml) and incubated at 37°C. At the indicated times aliquots were withdrawn and plated on M agar and on M plus 2.5 per cent nutrient broth agar. Identical samples were also analyzed for DNA content at each time.*

observed with plating the cells on broth-supplemented M agar after various intervals of incubation. However, when plated on unsupplemented M agar, the mutation frequency is quite low (Figure 7). While a large fraction of the potential mutations are not susceptible to CMP (are not unstable and subject to MFD as in UV irradiation), it is obvious that their expression is prevented by CMP. Therefore, *all* X-ray-induced mutations, as wll as UV-induced mutations, evidently require a period of protein synthesis following establishment in the genome if they are to attain expression. When CMP is added after 10 minutes incubation in complete medium (Figure 7) it is found that a large part of the mutations appearing on broth-supplemented

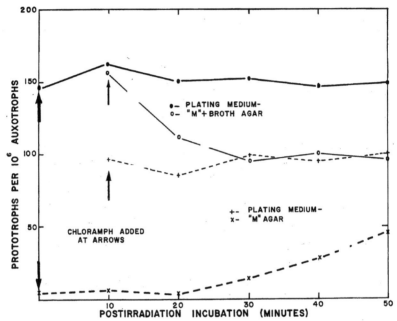

FIGURE 7.—*Effect of CMP in an X-irradiated culture of Escherichia coli strain WP2. Following X-ray irradiation (10 kr) the bacterial suspension was diluted 1:4 into M medium plus casein hydrolysate (2 mg/ml) and dl–tryptophan (0.2 mg/ml). This culture was then divided into two portions. To one portion chloramphenicol (20 micrograms/ml) was added immediately and incubation at 37°C started. The other portion was incubated for 10 minutes at 37°C before adding chloramphenicol (20 micrograms/ml) and then incubation was continued. At the indicated incubation times samples were withdrawn from each culture and plated on M agar and on M plus 2.5 per cent nutrient broth agar.*

plates have become susceptible to CMP. Within 10 minutes incubation potential mutants can be divided into two classes, *viz.*, those completely expressed and not sensitive to CMP, and those which are not capable of being expressed and are sensitive to CMP.

If we assume that X-ray has the capacity to mutate genetic material at two different levels, *viz.*, (a) in the parental DNA, presumably by direct physical action, and (b) in the daughter DNA, by some copy-error mechanism, these results can be readily accounted for.

Those mutations induced in the parental DNA should require only a short period of protein synthesis following induction. With mutations to prototrophy, expression on M agar would occur as soon as protein synthesis for new enzyme production takes place. Treatment with CMP after this time would not interfere with their expression. On the other hand, mutations induced in daughter DNA, presumably during the process of replication, might be dependent on mutagenic precursors or on chemical mutagens produced by the X-ray and acting at some stage in the replicative process. These would correspond to the so-called "delayed" or "indirect" mutations produced by X-ray. In Figure 7 these correspond to that fraction susceptible to MFD after CMP addition at 10 minutes and revealed by the broth-supplemented M agar platings.

This hypothesis is readily testable when CMP challenge is run on X-irradiated cells, and the cells then plated on both broth-supplemented M agar and on M agar (Figure 8). From the supplemented M agar plates it is evident that during the first 10 minutes of postirradiation incubation in complete medium about half of the mutations become susceptible to CMP challenge. This sensitivity remains constant for approximately 20 minutes and then is lost with further incubation. All mutations are insensitive to CMP after 30 to 40 minutes incubation. Similar results have been obtained using the uracil analogue 6-AU, and are additional evidence that the CMP-sensitive fraction of X-ray-induced mutants is induced in daughter DNA during the replicative process and that RNA and protein are involved in replication.

Other experiments designed to determine whether the CMP-resistant fraction of mutants can be expressed in the absence of DNA synthesis were carried out using E. coli strain $15_{T-Me-Tyr-}$, following reversion of the tyrosine requirement in this strain. It was found that immediately after irradiation few mutants were expressed on M plus thymine plus methionine agar but large numbers were expressed on broth-supplemented M agar. In further experiments with this strain thymine was withheld from tyrosine and methionine-supplemented postirradiation incubation medium for increasing periods of time and then added. All platings were on thymine and methionine supplemented M medium after a total postirradiation incubation period of 80 minutes. In these experiments approximately

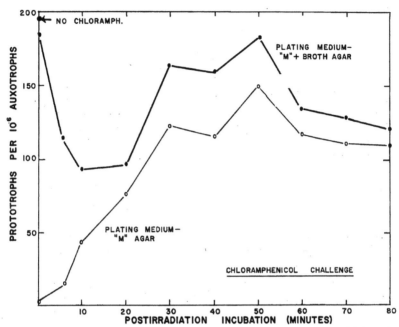

FIGURE 8.—*Separation of X-ray-induced mutations in Escherichia coli strain WP2 into two classes by CMP challenge followed by plating or M agar and broth-supplemented M agar. Following X-ray irradiation (10 kr), the bacterial suspension was diluted 1:4 into M medium plus casein hydrolysate (2 mg/ml) and dl–tryptophan (0.2 mg/ml) and incubated on a reciprocal shaker at 37°C. At the indicated times CMP (20 micrograms/ml) was added to aliquots of the culture and incubation continued in the presence of CMP for 40 additional minutes. All samples were then plated on M agar and on M plus 2.5 per cent nutrient broth agar.*

half of the tyro[t] mutations were able to be expressed on this medium after 20 minutes incubation. During the 20 to 40 minute incubation period no more mutations were expressed. At this time thymine was added to the incubation tubes and the incubation continued, and during the next 20 minutes the remaining half of the mutations were expressed. Therefore, after 60 minutes the mutation frequency obtained was the same as that of the control, which had been incubated in complete medium. No DNA synthesis was observed in the test culture until thymine addition, and then DNA synthesis started immediately and was correlated with expression of the second mutant fraction.

Summary and Conclusions

Several years ago Witkin (26) expressed the opinion that "the time interval between absorption of radiant energy and production of stable genetic changes can no longer be regarded as infinitesimal." She pointed out that many postirradiation treatments altered the ultimate fate of a potential genetic change. Our experiments have indicated that the basis for the delay between application of inducing agents and production of stable genetic changes is due to the necessity for replication of the genetic DNA.

Our experiments with X-ray-induced mutations indicate that certainly with this inducing agent mutations are produced through at least two mechanisms, or at two different levels of genetic structure development. Recent studies by Witkin and Thiel (28) and by Kada, Brun, and Marxovich (11) indicate that the same is probably true for UV-induced mutations. In one type of mutation induction with either agent the initial damage is unstable and capable of modification by postirradiation treatments. The second type is stable, and is not influenced significantly by postirradiation conditions. Both types are analyzed in our X-ray experiments, but we have dealt only with the former type in the UV experiments.

In regard to the unstable type of UV-induced mutation, while our studies suggest that the potential mutation is initially in the form of a mutagenic nucleic acid precursor formed by action of the radiation on a purine- or pyrimidine-containing monomer (4), we have no conclusive evidence for this; and the possibility remains that the initial irradiation damage is to the cellular DNA. Sinsheimer (22, 23) has shown that exposure of uridylic or cytidylic acids to UV results in stable and unstable byproducts, and our experiments indicate that cytidine and uridine are effective compounds in increasing UV-induced mutation frequency.

Studies on the postirradiation events involved in mutation induction by UV have led to delineation of several biochemical interrelations influencing mutation induction and probably involved in DNA replication. Those biochemical interactions occurring prior to any detectable macromolecular syntheses and tending to increase the mutation frequency are said to be involved in *mutation stabilization*. The frequency of mutation is therefore determined by the effectiveness of the MS processes. The nature of this process is not known; however, the fact that 5-HU will lower the induced mutation frequency (presumably through incorporation into RNA) in a process always correlated with MS suggests that stabilization involves "templating" preparatory to macromolecular synthesis.

If MS does not take place an antagonistic process occurs which removes the potential mutation from pathways leading to mutation induction. This process, which we have termed *mutation frequency decline,* is enzymatically mediated . (27, 5) and requires energy, but not RNA or protein synthesis. Also DNA synthesis is not required since the process occurs in the absence of thymine with a thymine-requiring auxotrophic strain (5). Moreover, MFD can be carried to completion without influencing either the timing or the rate of subsequent syntheses definitely involved in mutation induction. Therefore, the processes involved in either MFD or MS must take place prior to the RNA and protein synthesis made necessary by UV irradiation to further DNA replication (2, 5, 6, 10). These results would seem to eliminate hypotheses for MFD based on modification of timing of DNA synthesis so as to give more time for "decay of a premutational state" at a constant rate (13), or disappearance of "mutagens" at a rate independent of the treatment (14).

Ultimately, the potential mutation is fixed in the cell and no longer subject to conditions causing MFD. The striking correlation between RNA synthesis and MF (measured by challenge procedures specifically blocking RNA or protein synthesis) leads us to believe that MF is mediated by RNA synthesis. There appears to be a relation between the *amount* of postirradiation RNA synthesized at the time of CMP addition and the *rate* of DNA synthesis in the presence of the inhibitor, as well as the level of induced mutation obtained. If the RNA and protein necessary for genetic replication have not been formed prior to CMP addition, there is no DNA produced and the potential mutation is lost through MFD. It is probable that the RNA and protein involved in MF are identical with that involved in postirradiation recovery of the DNA-synthesizing system. The experiments make quite clear that acquisition of *capacity* to synthesize DNA rather than DNA synthesis *per se* is necessary for mutation fixation, since DNA synthesis during CMP-challenge is not necessary.

A number of hypotheses have been advanced in regard to the mechanism of mutation induction by UV. However, it is quite difficult to explain most of the postirradiation data on the basis of "direct-induction" theories. In our opinion, the hypothesis for the unstable portion of UV-induced mutation which best accommodates the experimental data is that which we have previously advanced (4). According to this hypothesis, DNA synthesis involves transfer of information from the parental DNA to the daughter DNA through an RNA-protein intermediate. Incorporation of a UV-modified precursor into the RNA leads to a "copy error" involving substitution of a different nucleotide pair or perhaps deletion of the usual nucleotide pair in the subsequently formed daughter DNA. While little experimental support at the molecular level exists for this indirect mechanism of DNA replication, the same situation also exists for theories of a more direct mechanism. There are also several indications that, at least during phage DNA replication, the genetic information is at certain stages carried in some structure or molecule other than the DNA (16, 24).

Our findings thus far for X-ray-induced mutations can be briefly summarized as follows: Approximately half of the X-ray-induced reversions of certain amino acid-requiring auxotrophs of *E. Coli* are sensitive to CMP, or to 6-AU. A short period of incubation involving protein synthesis is required for this sensitivity to become manifested. These mutations require RNA, DNA, and protein syntheses for induction and are apparently induced in a manner comparable to that for the major portion of UV-induced reversions. It is probable that these mutations are induced during replication of the daughter DNA. The other fraction of X-ray-induced mutations are not lost by MFD processes when the cells are incubated in the presence of CMP or 6-AU. These mutations are expressed when the culture is plated on M agar after a short period of incubation in complete medium, and during this incubation no detectable DNA synthesis occurs. This suggests that the second fraction of mutations is induced by X-ray in the gene, and that DNA replication is not necessary for induction. However, RNA and protein syntheses subsequent to their induction are required for phenotypic expression of the reverted character.

References

1. Burton, K. 1956. A study of the conditions and mechanisms of diphenylamine reaction for the colorimetric estimation of deoxyribonucleic acid. *Biochem Jour.*, **62**: *315–323*.

2. Doudney, C. O. 1959. Macromolecular synthesis in bacterial recovery from ultraviolet light. *Nature*, **184**: *189–190*.

3. ———— and Haas, F. L. 1958. Postirradiation modification of ultraviolet-induced mutation frequency and survival in bacteria. *Proc. Nat. Acad. Sci.*, **44**: *390–398*.

4. ————, ————. 1959. Mutation induction and macromolecular synthesis in bacteria. *Proc. Nat. Acad. Sci.*, **45**: *709–722*.

5. ————, ————. 1960. Some biochemical aspects of the postirradiation modification of ultraviolet-induced mutation frequency in bacteria. *Genetics*, **45**: *1481–1502*.

6. Draculic, M., and Errera, M. 1959. Chloramphenicol sensitive DNA synthesis in normal and irradiated bacteria. *Biochim. Biophys. Acta*, **31**: *459–463*.

7. Haas, F. L., and Doudney, C. O. 1957. A relation of nucleic acid synthesis to radiation-induced mutation frequency in bacteria. *Proc. Nat. Acad. Sci.*, **43**: *871–883*.

8. ————, ————. 1958. Interrelations of nucleic acid and protein synthesis in radiation-induced mutation in bacteria. *United Nat. Proc. Second Intern. Conf. Peaceful Uses of Atomic Energy*, **22**: *336–341*.

9. ————, ————. 1959. Mutation induction and expression in bacteria. *Proc. Nat. Acad. Sci.*, **45**: *1620–1624*.

10. Harold, F. M., and Ziporin, Z. Z. 1958. Synthesis of protein and of DNA in *Escherichia coli* irradiated with ultraviolet light. *Biochim. Biophys. Acta*, **29**: *439–440*.

11. Kada, T., Brun, E., and Marxovich, H. 1960. Etude comparee de l'induction de mutants chez *Escherichia coli*, par les rayons U.V. et les rayons X. *Ann. Inst. Pasteur.* In press.

12. Kelner, A. 1953. Growth, respiration, and nucleic acid synthesis in ultraviolet-irradiated, and in photoreactivated *Escherichia coli*. *Jour. Bact.*, **65**: *252–262*.

13. Kimball, R. F., Gaither, N., and Wilson, S. M. 1959. Reduction of mutation by postirradiation treatment after ultraviolet and various kinds of ionizing radiations. *Radiation Res.*, **10**: *490–497*.

14. Lieb, M. 1960. Deoxyribonucleic acid synthesis and ultraviolet-induced mutation. *Biochim. Biophys. Acta*, **37**: *155–157*.

15. Lowry, O. H., Rosebrough, A. C. and Randall, R. J. 1951. Protein measurement with the Folin phenol reagent. *Jour. Biol. Chem.*, **193**: *265–275*.

16. Luria, S. E., and Latarjet, R. 1947. Ultraviolet irradiation of bacteriophage during intracellular growth. *Jour. Bact.*, **53**: *149–163*.

17. McNair-Scott, D. B., and Chu, E. 1957. A consideration of conditions necessary for the production of synchronized division of cultures of *Escherichia coli*. *Bact. Proc.*, **57**: *114*.

18. Newcombe, H. B. 1955. Mechanisms of mutation production in microorganisms. *Atomic Energy of Canada Ltd., Publ. No. 144: 326–338*.

19. ———— and McGregor, J. F. 1954. Dose-response relationships in radiation-induced mutation saturation effects in *Streptomyces*. *Genetics*, **39**: *619–627*.

20. Agur, M., and Rosen, G. 1950. The nucleic acids of plant tissue: I. The extraction and estimation of desoxypentose nucleic acid and pentose nucleic acid. *Arch. Bochem.*, **25**: *262–276*.

21. Schwartz, N. M., and Strauss, B. S. 1958. Effect of tryptophan analogues on reversion of a tryptophan-requiring strain of *Escherichia coli*. *Nature*, **182**: *888*.

22. Sinsheimer, R. L. 1954. The photochemistry of uridylic acid. *Radiation Res.*, **1**: *505–513*.

23. ————. 1957. The photochemistry of cytidylic acid. *Radiation Res.*, **6**: *121–125*.

24. Stent, G. S. 1955. Decay of incorporated radioactive phosphorus during reproduction of bacteriophage T-2. *Jour. Physiol.*, **38**: *853–865*.

25. Visser, E., and Chargaff, E. 1948. The separation and quantitative estimation of purines and pyrimidines in minute amounts. *Jour. Biol. Chem.*, **176**: *703–714*.

26. Witkin, E. M. 1956. Time, temperature, and protein synthesis: A study of ultraviolet-induced mutation in bacteria. *Cold Spr. Harb. Symp. Quant. Biol.*, **21**: *123–140*.

27. ————. 1958. Post-irradiation metabolism and the timing of ultraviolet-induced mutations in bacteria. *Proc. 10th Intern. Cong. Genetics*, **1**: *280–299*.

28. ———— and Thiel, E. C. 1960. The effect of posttreatments with chloramphenicol on various ultraviolet-induced mutations in *Escherichia coli*. *Proc. Nat. Acad. Sci.*, **46**: *226–232*.

29. Zelle, M. R. 1955. Effects of radiation on bacteria. *Chapter X in "Radiation Biology"*, A. Hollaender, Ed. Book II, 365–430. New York: McGraw-Hill Book Co, Inc.

Comments

AUERBACH: Is the separation into mutation stabilization and mutation frequency decline not an artifact, caused by your having used for baseline mutation frequency on plates with an intermediate amount of broth? If you had used mutation frequency on minimal medium as base line, would not both phenomena have turned out to be the same—or rather reverse sides of the same phenomenon?

HAAS: It may very well be that mutation stabilization and mutation frequency decline are reverse sides of the same phenomenon, but they most certainly are not the same thing; nor is the separation an artifact caused by using mutation frequencies obtained on plates supplemented with an intermediate amount of broth for the base line. We use the intermediate level of broth supplementation so that we can demonstrate and study both processes at the same time on identical samples, but the phenomena of mutation stabilization and of mutation fixation, as well as the relation between the two, can be demonstrated just as well with plating on minimal medium. Similarly, the phenomena of mutation stabilization and mutation frequency decline can be demonstrated on broth-agar plates containing a much higher degree of broth supplementation.

It is quite probable that both mutation stabilization and mutation frequency decline are affecting the same biosynthetic step in DNA replication, but one process (stabilization) leads to a *maximum* frequency of mutations (provided mutation fixation takes place before any decline process occurs), while the other (mutation frequency decline) leads to a *less* number of mutations. Therefore, the processes certainly must be different. When one incubates irradiated cells in buffer containing a high level of amino acids, no increase in the mutation frequency will be obtained with incubation, and neither will there be any mutation frequency decline. However, these incubated cells are still fully susceptible to either phenomenon and if, after a period of say 30 minutes buffer-amino acid incubation, conditions suitable for measuring mutation fixation (addition of minimal medium to the culture) or conditions suitable for mutation frequency decline (addition of chloramphenicol) are initiated, we find that both phenomena will take place to the same extent as when they are initiated immediately after irradiation. Thus the potential mutations have been *stabilized* by the amino acids and they have neither been fixed nor have they disappeared. Furthermore, if one interrupts the process of mutation fixation at some intermediate level and then subjects the cells to conditions causing mutation frequency decline, no decline is obtained but neither is any further mutation fixation. Similarly, if at some intermediate level of mutation frequency decline one interrupts this process and introduces conditions favorable for mutation fixation, no increase in the frequency of mutants fixed above this level is obtained and neither is there any further mutation frequency decline. It is quite apparent that in the former case, where mutation fixation was being measured, the mutations already fixed were not subject to processes causing mutation frequency decline, while the fraction remaining unfixed was equally susceptible to either mutation fixation or mutation frequency decline. In the latter case, when mutation frequency decline was being measured, the potential mutants that have disappeared cannot be made to reappear by introducing conditions favorable to fixation, and the fraction which has not disappeared is as susceptible to mutation frequency decline as to mutation fixation.

Discussion of Session II

BERNARD S. STRAUSS

The University of Chicago, Chicago, Ill.

·IT IS EVIDENT now that mutation is a complex process; the intermediate steps depending on the particular mutagen used to initiate the process. As Doctor Auerbach points out, we interpret the action of mutagenic agents in terms of what we know about the structure of the genetic material rather than learning about the genetic material as a result of studies of mutagenic agents. This process of interpretation may lead to useful results even though its presuppositions are incorrect.

In 1939, Thom and Steinberg (16)[1] reported that nitrous acid induced genetic changes in fungi. Their experiments were based on the idea that the genetic material was protein in nature and their interpretation was based on the well-known reaction of nitrous acid with the amino groups—of the amino acids. Nitrous acid is one of the more useful of the mutagens and its action is interpreted today on the basis of its reaction with the amino groups—of the purines and pyrimidines—since we now know, or feel sure that we know, that genetic material is polynucleotide in nature. I think Thom and Steinberg should be given more credit for having first studied the mutagenic action of nitrous acid and I think also that it is instructive to note the pitfalls a favorite hypothesis can create.

Nitrous acid-induced mutation represents a direct type of mutagenesis. The agent produces a base analogue *in situ* which can then yield a slight increase in the rare frequency of pairing mistakes during DNA duplication. Ultraviolet light-induced mutation as described by Doctor Haas probably represents the other extreme, a complex of photochemical and biochemical events being required for the complete mutation process. (In the discussion, Doctor Atwood suggested that ultraviolet light-induced mutation might be due to an *in situ* photocatalytic hydroxylation of a pyrimidine DNA constituent. Since ultraviolet light does not cause muta-

[1]See References, page 178.

173

tion of an isolated transforming principle (6) in contrast to the action of nitrous acid, I believe ultraviolet-induced mutation must be a more complex process.) Doctor Haas, and others (17), have demonstrated that synthetic processes intervene between the primary ultraviolet absorption product and the final mutant state.

The process of ultraviolet-induced mutation is in fact a very special case. A portion of the ultraviolet-induced changes can be reversed by visible light and this photoreactivation is unique to ultraviolet-induced mutation. (The photoreactivation process is itself only partially understood. It is known to be enzymatically mediated.) While it is established that protein synthesis, or some closely related process, is required for the fixation of ultraviolet-induced mutations, it is equally evident, both from the work reported by Doctor Haas at this symposium and from other work, that protein synthesis is not a *general* requirement for the fixation of mutations. Bromouracil is mutagenic for bacteriophage even when introduced along with chloramphenicol as an inhibitor of protein synthesis (7). Protein synthetic processes do not play an important role in the fixation of those mutations induced by the alkylating agent ethyl sulfate, although they may be required for the expression of the mutations (14). The frequency of ethyl sulfate-induced mutation is not reduced when *E. coli* is treated in the presence of any one of the inhibitors methyltryptophan, ethionine, or 6-mercaptopurine nor does post-treatment incubation in buffer, or in buffer containing these inhibitors, seem to alter the mutation frequency significantly.

According to the ideas presented by Doctor Atwood earlier in this symposium, the genetic material is a linear sequence of deoxyribonucleotides. The sequence forms a code which is deciphered by the cell and the process of deciphering represents gene action. Base analogues incorporated into the DNA lead to mutation only by increasing the probability that a mistake will be made on duplication leading to the substitution of one nucleotide pair for another. DNA containing a base analogue is not of itself mutant—bacteriophage with bromouracil completely substituted for thymine can produce normal progeny (8). The mutation process can only be considered complete when a changed DNA (itself able to replicate) containing only normal bases is formed as a result of a replication.

Consider the behavior of bacteriophage treated with ethyl methanesulfonate. Although phage treated *in vitro* produce mutants when used to infect bacteria, such treated phage are not yet mutant. Single bursts yield groups with a few mutants and others with many, indicating that the "mutation" may occur at any time during the replication of the treated DNA (9).

Our present ideas require that a mutation represent a change in base order within the genetic DNA. All mutagens must lead to such a changed base order. If this hypothesis is accepted, the problem becomes a determination of how each particular mutagen may bring about a change. It is not required, for example, that a chemical mutagen react directly with DNA. As Szybalski (15) has shown, the alkylating agent triethylene melamine may react with pyrimidines and pyrimidine derivatives to give products which are in turn incorporated into the DNA where they behave as analogues leading to an increased mutation frequency. It must not be supposed that mutations need represent changes in only single nucleotide pairs. Freese (1) points out that the class of mutations he calls transversions may represent merely the change of a purine for a pyrimidine base. A transversion may also result in a more extensive change in the genetic material since incorporation of a purine at a pyrimidine site could result in sufficient deformation of the double helix of DNA to allow for the substitution of other bases. Fresco and Alberts (2) have shown how noncomplementary bases can be accomodated in a helix by the formation of loops coming out of the helix. Such loops might, on DNA duplication, result in the addition or deletion of portions of the nucleotide chain. It is important to realize that current ideas of mutation as a result of base substitution are not so restrictive that they require *all* mutations to be simple changes in single base pairs.

It is also essential to emphasize that the mutagenic agents, particularly the reactive mutagenic alkylating agents, may have several different reactive sites within the organism and that their reaction, even at a single site, may have both mutagenic and nonmutagenic consequences. The alkylating agent diethyl sulfate (ethyl sulfate) not only is mutagenic but also produces a permeability change in the bacterial cell which makes magnesium ion more available to chelation and removal by citrate (13). Many alkylating agents combine with bacteriophage in a two stage process; the first step is a true combination, while the second is a decay event leading to phage inactivation (10). Only certain of these alkylating agents produce bacteriophage mutations *in vitro* (9).

There is a post-treatment decay reaction in phage plaque-forming ability and in the activity of transforming principle from *B. subtilis* following treatment with either the mutagenic ethyl methanesulfonate or the nonmutagenic (for virus) methyl methanesulfonate. Incubation of bacteriophage T2 following treatment with ethyl methanesulfonate does not greatly alter the frequency of *r* mutants recovered. It seems likely that there is more than one site of reaction of these alkylating agents with genetic material and it is still difficult to determine the chemical difference between the mutagenic and nonmutagenic reactions. Certainly it is established that the reaction of alkylating agents with cellular constituents is a complex process.

The hypothesis of mutation due to base substitution is logically satisfying but there are cases in which the mechanism by which this substitution occurs is not yet apparent. Thymine starvation is mutagenic and Kanazir (3) has shown that this mutagenic action occurs during the first 30 minutes of thymine starvation and before there is any lethality due to the thymine starvation. DNA duplication does not occur. To my knowledge there is no way at present to account for this mutagenicity. Caffeine and other methylated purines are mutagenic (11), but these compounds are not incorporated or metabolized (5) and presumably act by inhibiting some enzyme involved in DNA metabolism. We do not know how this inhibition results in a change in base order.

Breaking a single one of the chains of the DNA double helix can be mutagenic. Radioactive decay of P-32 leads to a transmutation event in which the phosphorus changes to S-32. In most of the decay events the daughter nucleus recoils from the molecule leaving a "hole". Kaudewitz, *et al.* (4) have shown that this decay event is mutagenic and that the production of mutation is independent of the β radiation accompanying the radioactive decay. Kaudewitz reports (personal communication) that about 30 to 40 per cent of the mutants obtained were able to revert to the wild-type, indicating that a proportion (at least) of the strains obtained were true mutants as distinguished from deletions. The removal of phosphorus from the DNA chain constitutes a definite break in that chain. How then does a break in the backbone chain cause a change in the base order? Kaudewitz (personal communication) suggests that a break in *one* of the two (replicating) chains of the DNA molecule leaves the distance between these chains unfixed permitting a mispairing, the actual mutation (change in base order) in this case being a secondary effect. Somewhat similar is the observation of Freese that treatment of bacteriophage at 45° C and pH 5 results in the production of mutants. Freese (1) supposes that these mutations are due to the formation of apurinic acid at lower pH. The initial reaction at pH 5 is different from those mentioned above, but once again the DNA is unstabilized in a way that permits unusual base pairings.

In such cases the mutation is not "fixed" and in fact by definition can not be considered established until a new DNA strand has been synthesized with the new base order. In order to have a mutation which is functional before DNA duplication it is necessary to produce, as a result of the action of the mutagen, a strand which is read differently by the deciphering mechanism. It is this requirement of the basic hypothesis that makes Doctor Haas's report of X-ray-induced mutation in the absence of DNA synthesis so interesting. Spontaneous mutation can occur in the absence of DNA synthesis, but an energy source is required before such mutations can occur (12). Doctor Haas reports that reversions can be induced in the absence of DNA synthesis but that these mutations require protein and RNA synthesis for their expression. Unfortunately, it is not possible to exclude the possibility of DNA synthesis from some of the data as presented. The frequency of prototrophs induced is of the order of 100 per 10^6 cells, or 0.01 per cent. Doubling of the DNA of these particular cells would give a change of only 0.01 per cent, three cell divisions of this selected population would still give less than a 0.1 per cent increase. Considering the analytical methods which must be employed, it is doubtful if the results are good to within 1 per cent. But 1 per cent could represent over five divisions of a selected cell population. If one accepts the fact that the thymine-requiring organisms do not make DNA, then the results using these mutants do seem to indicate mutation and mutation expression in the absence of DNA reproduction.

We assume that mutation represents a functional change in base order. These X-ray-induced mutations must therefore have a base order which looks different to the "reading mechanism" of the cell. Perhaps the X-rays produce a truly different base (base pair?) within the DNA or perhaps there is some sort of "transnucleotidation" corresponding to the known transpeptidation induced by agents such as X-rays which cause breaks in DNA chains. It is almost necessary to suppose a mechanism for rare base substitution within intact DNA molecules to account for mutation in the absence of DNA duplication.   ⌐

References

1. Freese, E. 1959. On the molecular explanation of spontaneous and induced mutations. *Brookhaven Symp. in Biol.*, **12**: *63–75*.
2. Fresco, J. R., and Alberts, B. 1960. The accomodation of noncomplementary bases in helical polyribonucleotides and deoxyribonucleic acids. *Proc. Nat. Acad. Sci.*, **46**: *311–321*.
3. Kanazir, D. 1958. The apparent mutagenicity of thymine deficiency. *Biochem. Biophys. Acta,* **30**: *20–33*.
4. Kaudewitz, F., Vielmetter, W., and Friedrich-Freksa, H. 1958. Mutagene Wirkung des Zerfalles von radioaktiven Phosphor nach Einbau in Zellen von *Escherichia coli*. *Zeit. Naturforsch.*, **13b**: *793–802*.
5. Koch, A. L. 1956. The metabolism of methylpurines by *Escherichia coli*: I. Tracer studies. *Jour. Biol. Chem.*, **219**: *181–188*.
6. Litman, R. M., and Ephrussi-Taylor, H. 1959. Inactivation et mutation des facteurs genetiques de l'acide desoxyribonucleique du pneumocoque par l'ultraviolet et par l'acide nitreux. *Compt. Rend.*, **249**: *838–840*.
7. ———— and Pardee, A. B. 1959. Mutations of bacteriophage T2 induced by bromouracil in the presence of chloramphenicol. *Virology*, **8**: *125–127*.
8. ————, ————. 1960. The induction of mutants of bacteriophage T2 by 5–bromouracil: III. Nutritional and structural evidence regarding mutagenic action. *Biochem. Biophys. Acta,* **42**: *117–430*.
9. Loveless. A. 1959. The influence of radiomimetic substances on deoxyribonucleic acid synthesis and function studied in Escherichia coli phage systems: III. Mutation of T2 bacteriophage as

a consequence of alkylation in vitro: the uniqueness of ethylation. *Proc. Roy. Soc. London,* **Ser. B 150:** *497–508.*

10. ———— and Stock, J. 1959. The influence of radiomimetic substances on deoxyribonucleic acid synthesis and function studied in Escherichia coli phage systems: I. The nature of the inactivation of T2 phage in vitro by certain alkylating agents. *Proc. Roy. Soc. London,* **Ser. B 150:** *423–445.*

11. Novick, A., and Szilard, L. 1951. Experiments on spontaneous and chemically induced mutations of bacteria growing in the chemostat. *Cold Spring Harbor Symp. Quant. Biol.,* **16:** *337–343.*

12. Ryan, F. J. 1959. Bacterial mutation in a stationary phase and the question of cell turnover. *Jour. Gen. Microbiol.,* **21:** *530–549.*

13. Strauss, B. 1961. Production of a permeability defect in *Escherichia coli* by the mutagenic alkylating agent, ethyl sulfate. *Jour. Bact.,* **81:** *573–580.*

14. ———— and Okubo, S. 1960. Protein synthesis and the induction of mutations in *Escherichia coli* by alkylating agents. *Jour. Bact.,* **79:** *464–473.*

15. Szybalski, W. 1960. The mechanism of chemical mutagenesis with special reference to triethylene melamine action. In *Developments in Industrial Microbiology (Miller, ed.), New York: Plenum Press, 231–241.*

16. Thom, C., and Steinberg, R. A. 1939. Chemical inductions of genetic changes in fungi. *Proc. Nat. Acad. Sci.,* **25:** *329–335.*

17. Witkin, E. M. 1959. Post-irradiation metabolism and the timing of ultraviolet-induced mutations in bacteria. *Proc. 10th Intern. Congr. Genetics,* **1:** *280–299.*

Comments

AUERBACH: If UV produced mutations only at gene replication, then all lactose mutants on EMB agar should be sectored. As far as I remember Witkin's experiments, she obtained mainly whole-colony mutants when nuclear number in the treated cells had been reduced to one.

STRAUSS: The operational definition of mutation is that it is an inherited change. It is therefore required that any change be subjected to the test of reproduction. One might conceive of an alteration in DNA that produced a change in heterocatalytic activity but which could not be reproduced. We could not call this a gene mutation. At this level the discussion is one of semantics.

Doctor Auerbach has pointed out a very surprising experiment and her memory is correct. In addition, Kaudewitz has shown that nitrite-treated bacteria produce nonsectored mutants. Mutants arising as a result of P–32 decay tend to be sectored. If one assumes that the Watson-Crick structure as confirmed by the experiment of Meselson and Stahl represents the genetic material at all times, and further assumes that a mutagenic agent affects only one of the two DNA strands, then all mutants should be sectored. With chemical mutagens such as alkylating agents, the sectoring should be more complex for reasons indicated in my discussion.

Now Witkin's experiment (for lactose nonfermenters) is quite clear and, yet, Witkin herself at the Montreal Congress reported experiments indicating that *certain* UV-induced mutations were "fixed" only at the time of gene replication. There appears to be a paradox and essentially the same point raised here by Doctor Auerbach was raised by Marshak in the discussion of Witkin's paper on sectoring. If UV acted on a "chromosomal" level and affected both strands, there would be no sectoring. Perhaps the paradox is due to our assumption that the genetic material is always double-stranded.

Witkin irradiated her material to a survival of 10^{-3}. At this survival one expects about seven lethal hits per organism. Since this Symposium, it has been shown by Marmur and Grossman that one of the effects of UV is to tie the two strands of DNA together by covalent bonds. Nitrous acid treatment has a similar effect. DNA treated with either of these two mutagens does *not* consist of two strands which can separate merely by breaking hydrogen bonds. I think these findings will lead to an explanation for the lack of sectoring.

Session III

Evaluation of Mutations in Plant Breeding

G. F. Sprague, *Chairman*
U. S. Department of Agriculture,
Beltsville, Md.

Use of Spontaneous Mutations in Sorghum

J. R. QUINBY

Texas Agricultural Experiment Station, Substation No. 12, Chillicothe, Texas

SORGHUM in the United States is separated by usage into grain sorghum, dual-purpose sorghum, sweet sorghum, sudangrass, and broomcorn. Statistics on world production of sorghum are inadequate, but the total area devoted to the crop for grain production is thought to be more than 80 million acres. Sorghum is grown on all the continents below latitudes of 45 degrees and on many of the islands of the East and West Indies. Sorghum is the chief food grain in parts of Africa, India, and China. As a world food grain, sorghum ranks third, being exceeded only by rice and wheat.

In the United States, sorghum is used for human food only as sorghum sirup, dextrose, starch, and oil. As a grain crop, sorghum is exceeded in production only by wheat and corn and has been grown on over 20 million acres in recent years. The sorghum crop is now about one fifth as large as the corn crop, whereas 10 years ago it was about one tenth as large. Prior to 1942, the acreage harvested for forage exceeded that harvested for grain, but since that time the reverse has been true.

Sorghum, *Sorghum vulgare* Pers., is a large grass of many varieties that probably originated in Africa and has been cultivated since ancient times. An Assyrian ruin dating from 700 B. C. contains a carving depicting sorghum. In the United States, where the species was introduced, as many as 400 varieties have existed, many of which are sweet sorghums or sorgos. The number of varieties grown on farms has shrunk rapidly as horses and mules have disappeared from farms. It has become a cash grain crop and hybrids have appeared. In Africa and Asia, numerous other varieties exist.

These varieties, even including sudangrass [var. *sudanesis* (Piper) (Hitch.)], cross-pollinate readily and produce fertile offspring. Vinall, *et al.* (43)[1] have considered all varieties as belonging to the same species, but Snowden (35) has classified some 3,000

---

[1]See References, page 202.

forms into 31 species. Snowden's collection was worldwide and his classification is useful even though his giving races specific rank may be questionable. Characteristics such as plant color, juicy or pithy stem, pericarp color, presence or absence of testa, presence or absence of awns, starchy and waxy endosperm, white or yellow endosperm, dehiscent or indehiscent spikelets, height and duration of growth, are known to be genetic. But the inheritance of the complex of characters that makes up the groups of varieties or races such as the broomcorns, kafirs, feteritas, milos, kaoliangs, hegaris, shallus, sumacs, etc., is not understood. Laubscher (16) has considered modifying complexes to be important in separating the races of sorghum. Reviews of sorghum genetics literature have been made by Martin (20), Myers (23), and Quinby and Martin (31).

Introduction Into the United States

Sorghum has been grown in the United States for a little over a century. Accounts of the early introductions and early work with sorghum have been presented by Vinall, *et al.* (43), Martin (20), Quinby and Martin (31), and Quinby, *et al.* (30).

The first sweet sorghum that reached the United States came from China by way of France in 1853 and is now known as Black Amber. Before the distinction between sorghum and sugarcane, *Saccharum officinarum,* was well understood, the Indian Service sent seed of this sorghum to the Comanche and Brazos Reserves in Throckmorton and Young Counties in Texas as Chinese Sugarcane. To this day, farmers in West Texas call sweet sorghum "cane". Fifteen sorgo varieties, several of which survive, were introduced in 1857 from South Africa. Two other sweet sorghums were introduced in 1881 and 1891, the first from Natal, South Africa, and the second from Australia where the variety was undoubtedly introduced. The present American sorgo varieties originated from these 17 introductions and a 1951 introduction from Ethiopia.

The first grain sorghums grown in the United States were White and Brown Durras which reached California from Egypt in 1874. These two varieties were never widely grown outside California and have been unimportant as parents of present-day varieties and hybrids. The progenitor of the milo varieties was introduced into South America and reached South Carolina from Columbia in 1879. The variety was called "Millo Maize" (14). How Black-hull Kafir reached the United States is unknown (43), but the variety was growing on farms in Kansas about 1890. A late-maturing kafir was introduced from Columbia in 1880 (14) and was distributed in South Carolina as "White Millo Maize". Perhaps Blackhull Kafir originated as an early-maturing mutation in this late-maturing kafir. White Kafir and Red Kafir from Natal were shown at the Centennial Exposition at Philadelphia in 1876 and seed from the exhibit was later distributed. Pink Kafir, Feterita, Hegari, and Sundangrass were introduced in 1904, 1906, 1908, and 1909, respectively. Broomcorn has been grown in the United States since Benjamin Franklin began its culture with seeds plucked from an imported broom.

Transformation of Sorghum in the United States

The mechanization of agriculture and changes in sorghum that began about 30 years ago have changed sorghum from a feed crop for use on the farm into a cash grain crop. The chief changes in sorghum on the farm since 1930 have been a reduction of about 40 centimeters in height and a shortening of the duration of growth by about a week. The reduction in height and dry-headedness associated with early maturity have made combine harvesting possible. Sorghum production has expanded into higher latitudes and higher elevations as extremely early-maturing varieties have become available. Recently, a method of hybrid sorghum seed production using cytoplasmic male-sterility has been devised and hybrids have resulted in an expansion of acreage in areas outside the "sorghum belt" as well as within it because of higher yields.

These changes have come about from making use of alleles discovered in the species. The transformation of sorghum in the United States within the last half century is actually a further domestication of the species to make it fit the needs of mechanized farming. Most improvement in sorghum has resulted from accumulating desirable alleles in a single variety.

Genes Used in Sorghum Improvement

Many of the contrasting characters present in sorghum were in one or another of the varieties introduced into the United States, but a few alleles resulted from mutations after sorghum was introduced about 100 years ago. All of the alleles must have arisen as spontaneous mutations originally and no effort will be made to distinguish between recent and ancient mutations.

Nevertheless, eight mutations have occurred in milo and have been preserved during 80 years in the United States. Four of these eight that control response to photoperiod occurred at three loci. Two were alleles that reduce height, one affects pericarp color, and one causes resistance to a root-rotting fungus. The Y gene that causes colored pericarp mutated to recessive y in milo before 1910 and White Milo resulted. The recessive mutation causing resistance to periconia root-rot must have occurred many times because a few resistant plants could be found in almost any badly diseased field of any of several milo varieties when the disease first became serious. Probably mutations at some loci occur again and again. Conversely, many other genes have not been seen to mutate in the 50 years that sorghum has been worked with by plant breeders in the United States. For instance, the yellow endosperm character was never seen in the United States until O. J. Webster returned with it from Africa in 1951.

Maturity Genes

The history and evaluation of milo in the United States has been presented by Karper and Quinby (13, 14) and by Vinall, et al. (43). Another early-maturity genotype has appeared in more recent years. When "Millo Maize" reached the United States in 1879 it was a tall, late-maturing variety. When tested at the Louisiana and Kansas Experiment Stations in 1888, the variety matured in about 120 days, indicating a time of blooming about 90 days from planting. By 1900, the variety in use in western Texas was Standard Milo that bloomed in about 65 days and matured in 95 days. Before 1910, farmers in western Texas were growing Early White Milo that bloomed in 50 days and matured in 80 days. In 1938, a farmer in California found a still earlier maturing milo that bloomed in 44 days and matured in 75 days. Commercial production of this early-maturing milo, called Ryer Milo, began about 1948.

The inheritance of duration of growth in milo has been reported by Quinby and Karper (25, 29). Three genes influence plant response to photoperiod and control the time of floral initiation.

The genes have been named Ma_1, Ma_2, and Ma_3. Only one recessive allele is known at each of the Ma_1 and Ma_2 loci, but two recessive alleles are known at the Ma_3 locus. Recessive ma_1 and ma^R_3 act like amorphic alleles since they cause early floral initiation when homozygous. Recessive ma_2 and ma_3 modify the expression of the dominants and are not amorphic. The various milo maturity geno- types (but not the infrequent cross-over genotypes) that emerge from crosses of Early White Milo with Yellow Milo and with Ryer Milo are shown in Table 1. The "Millo Maize" introduced

TABLE 1.—MILO MATURITY AND PERICARP COLOR GENOTYPES AND PHENOTYPES, EXCLUDING CROSSOVER CLASSES, AT CHILLICOTHE, TEXAS, FROM EARLY JUNE PLANTINGS.

Genotype	Days to bloom	Pericarp color
$Ma_1 Ma_2 (Ma_3y)$..	100	White
$Ma_1 Ma_2 (ma_3Y)$..	90	Colored
$Ma_1 Ma_2 (ma^R_3Y)$..	44	Colored
$Ma_1 ma_2 (Ma_3y)$....	80	White
$ma_1 Ma_2 (Ma_3y)$.	50	White
$Ma_1 ma_2 (ma_3Y)$....	60	Colored
$Ma_1 ma_2 (ma^R_3Y)$..	44	Colored
$ma_1 Ma_2 (ma_3Y)$..	50	Colored
$ma_1 Ma_2 (ma^R_3Y)$.	38	Colored
$ma_1 ma_2 (Ma_3y)$.	50	White
$ma_1 ma_2 (ma_3Y)$...	50	Colored
$ma_1 ma_2 (ma^R_3Y)$...	38	Colored

in 1879 was evidently dominant for all three maturity genes and has been reconstituted by selecting the late-maturing genotype from the F_2 generation of appropriate crosses. A genetic linkage exists between the white pericarp color gene y and Ma_3 and between the maturity gene Ma_1 and the dwarfing gene dw_2.

Of the many varieties grown in the United States and tested for photoperiod response, only Yellow Milo (60-day), Hegari, and Early Hegari, and a few of their derivatives, are dominant Ma_1. The remainder, including the parents of the present sorghum hybrids, are recessive ma_1 and have a critical photoperiod high enough to allow relatively early floral initiation and maturity. The dominant or recessive condition of the genes at the Ma_2 and Ma_3 loci is unknown for most varieties other than the milos, but most are apparently dominant for either Ma_2 or Ma_3 or both. Yellow milo (60-day) is genetically Ma_1 ma_2 ma_3. Hybrids of 60-day Milo and Texas Blackhull Kafir, Hegari, Early Hegari, and California White Durra are extremely late in maturity. Hybrids of 60-day Milo and Early Kalo, Kalo, Bonar Durra, Feterita F. C. 811, Manko, and Fargo are later in maturity than either parent but not extremely late. The latter group is probably dominant for either Ma_2 or Ma_3 and the former for both Ma_2 and Ma_3. Obviously, more than three major maturity genes may exist in sorghum and all the milos might be dominant for a fourth gene. The late-maturing kafir introduced into the United States in 1880, since it was so late in maturing, must have been dominant for all three genes. If Blackhull Kafir came from this late-maturing kafir, it arose as a recessive mutation at the Ma_1 locus.

Several varieties are sensitive to photoperiod at the temperatures that prevail in the summer months at Chillicothe, Texas, but many other varieties are insensitive or show varying degrees of sensitivity. Data in Table 2, taken from Quinby and Karper (26), show the response of 12 varieties to photoperiod at Chillicothe,

TABLE 2.—EFFECT OF 10-HOUR PHOTOPERIOD ON TIME OF FLORAL INITIATION, LEAF
NUMBER, AND TIME OF ANTHESIS OF SORGHUM VARIETIES PLANTED JULY 13, 1941,
AT CHILLICOTHE, TEXAS.

| | Number of days from planting to | | | | Number of leaves on mature plant | |
| | Head differentiation | | First anthesis | | | |
Variety	Short day*	Normal day	Short day	Normal day	Short day	Normal day
Sooner Milo	23	32	43	49	11	13
Texas Milo	23	39	47	68	11	18
Hegari	23	48	47	77	13	18
Kalo	23	39	47	64	11	17
Calif. White Durra	23	34	51	55	13	14
Spur Feterita	23	36	56	65	16	19
Freed	23	32	46	51	10	12
Manko	25	47	50	77	12	17
Bishop	28	39	61	71	14	17
Sumac	29	39	60	65	14	16
Blackhull Kafir	29	39	59	69	14	16
Dwf. Broomcorn	39	39	68	68	15	15

*Short day exposure to sunlight was 10 hours. In July at Chillicothe, Texas, the sun is above horizon for about 14 hours.

Texas, in 1941. Several varieties are sensitive to photoperiod, several are less sensitive, and one, Dwarf Broomcorn, is quite insensitive. Although this point has not been established experimentally, it can be inferred from work with other species that the thermal requirements have not been met if a sorghum variety is not sensitive to photoperiod. Likewise, it can be inferred that varying degrees of sensitivity indicate that the thermal requirements have been partially met. Insensitivity to photoperiod is undoubtedly the mechanism that allows late maturity of varieties in the tropics. In some tropical areas, sorghum is planted before a rainy season of 3 or 4 months duration and the crop heads and matures after the rainy season. Such a long duration of growth under 12-hour days would not be possible with varieties sensitive to photoperiod. Three or four maturity genes, allelic series at those loci, and varying degrees of insensitivity to photoperiod might well account for all the maturities found in sorghum.

Early floral initiation, low leaf number, and small plant size are associated, as shown by data presented by Quinby and Karper (25) and reproduced in Table 3. At the time these data were collected,

TABLE 3.—DATA SHOWING THE SIZE OF FOUR MILO MATURITY GENOTYPES GROWN UNDER NORMAL PHOTOPERIODS FROM A PLANTING AT CHILLICOTHE, TEXAS, MADE ON JUNE 20, 1944.

Criteria	Genotype			
	50-day $ma_1 Ma_2 ma_3$	60-day $Ma_1 ma_2 ma_3$	80-day $Ma_1 ma_2 Ma_3$	90-day $Ma_1 Ma_2 ma_3$
No. of days to anthesis	48.6 ± 0.3	68.6 ± 0.4	82.6 ± 0.7	102.4 ± 0.04
No. of leaves	16.4 ± 0.1	22.0 ± 0.1	23.5 ± 0.3	31.5 ± 0.2
Height of plant, cm	85.8 ± 1.2	87.7 ± 1.3	121.6 ± 3.4	151.8 ± 2.7
Length of leaf, cm	54.1 ± 1.5	65.1 ± 0.5	65.3 ± 0.8	78.1 ± 0.5
Diameter of stalk, cm	1.42 ± 0.02	2.30 ± 0.04	2.22 ± 0.05	2.53 ± 0.01
Weight of head, lbs.	0.24 ± 0.02	0.36 ± 0.02	0.33 ± 0.03	0.27 ± 0.02
Weight of plant, lbs.	0.43 ± 0.03	0.62 ± 0.04	0.64 ± 0.04	0.97 ± 0.06

Ryer milo maturity was not known which accounts for the fact that a Ryer population was not measured. The data in Table 3 show that the duration of the vegetative period is positively correlated with plant size, the best measure of which is total dry weight per plant. The figures show that a two-fold difference in size resulted from a difference of 54 days in blooming. This difference in size is apparently the result of the time element alone and not a difference in rate of growth because no appreciable difference occurred in the time that corresponding successive leaves appeared on the four genotypes. The fact that larger leaves appeared on the later-maturing genotypes indicates only that the growing point continued to grow in circumference from day to day in the early life of the plant, and a leaf arising from a higher node would be expected to be larger than one arising from a lower node. Some additional data are presented in Table 3 to show the morphological relationships of the various genotypes. As would be expected, there is a positive correlation between leaf number and number of days to anthesis, height, stalk diameter, and leaf size. The four genotypes measured are similar to one another in appearance, except that each genotype is a larger counterpart of the one immediately preceding it in earliness of maturity. Early maturing varieties are used by farmers because the later maturing varieties exhaust the soil moisture before maturity or need additional irrigations.

A logical explanation of the nature of the maturity genes in milo would be that Ma_1, Ma_2, and Ma_3 are dominant inhibitors that block some essential reaction leading to floral initiation. This assumption would mean that the "wild type" was small and very early in maturing because of early head initiation. Sorghum must have been domesticated by the preservation in the tropics of larger and later maturing plants that carried inhibitors that prevented early floral initiation. When tropical varieties came to the United States and were grown at latitudes above 30 degrees, they were too late in maturity to be of greatest use and man again preserved mutations, but this time those that inactivated the inhibitors to some degree.

Height Genes

In most of the sorghum-producing areas of the world, tall stature is preferred; but, in the United States, farmers have used shorter and shorter varieties as they became available. The dwarfing genes that have been useful in sorghum improvement shorten only the internodes. Four such dwarfing genes are known in sorghum and are inherited independently (28).

Recessive dw_4 was in Standard and Early White Milos and

in Blackhull Kafir that farmers were growing by 1900. Recessive dw_1 was found before 1905 and recessive dw_2 about 1910, both in milo. The combine height varieties that originated in the 1920's and later were recessive for three height genes were obtained by selecting 3-dwarf segregates from crosses between milo $(dw_1\ Dw_2\ Dw_3\ dw_4)$ and kafir $(Dw_1\ Dw_2\ dw_3\ dw_4)$. The dominant gene in all of the 3-dwarf varieties of milo-kafir parentage is Dw_2 and in the 3-dwarf milos Dw_3. Four-dwarf strains have been obtained by crossing these two 3-dwarf genotypes $(dw_1\ Dw_2\ dw_3\ dw_4$ and $dw_1\ dw_2\ Dw_3\ dw_4)$ and selecting 4-dwarf segregates. The genetic height constitution of many varieties and parents of sorghum hybrids and broomcorn is known. The genetic identity of many strains is given in Table 4. At Chillicothe, Texas, on dryland 2-dwarfs were

TABLE 4.—CLASSIFICATION OF SORGHUM VARIETIES ACCORDING TO HEIGHT.

Genotype	Variety
$Dw_1\ Dw_2\ Dw_3\ Dw_4$	None identified
$Dw_1\ Dw_2\ Dw_3\ dw_4$	Tall White Sooner Milo SA 1170, Spur Feterita, Manchu Brown Kaoliang, Shallu SA 401, Sumac
$Dw_1\ Dw_2\ dw_3\ Dw_4$	Standard Broomcorn
$Dw_1\ dw_2\ Dw_3\ Dw_4$	None identified
$dw_1\ Dw_2\ Dw_3\ Dw_4$	None identified
$Dw_1\ Dw_2\ dw_3\ dw_4$	Texas Blackhull Kafir, Kalo, Early Kalo, Chiltex, Shantung Dwarf Kaoliang
$Dw_1\ dw_2\ Dw_3\ dw_4$	Bonita, Early Hegari, Hegari
$dw_1\ Dw_2\ Dw_3\ dw_4$	Dwarf Yellow and White Milos, Sooner Milo
$Dw_1\ dw_2\ dw_3\ Dw_4$	Acme Broomcorn
$dw_1\ Dw_2\ dw_3\ Dw_4$	Japanese Dwarf Broomcorn
$dw_1\ dw_2\ Dw_3\ Dw_4$	None identified
$Dw_1\ dw_2\ dw_3\ dw_4$	None identified
$dw_1\ Dw_2\ dw_3\ dw_4$	Combine Kafir 60, Combine 7078, Martin, Plainsman, Wheatland, Caprock, Combine White Feterita SA 396, the Redbines, Day
$dw_1\ dw_2\ Dw_3\ dw_4$	Double Dwarf Yellow Milo, Double Dwarf Yellow and White Sooner Milos, Ryer Milo
$dw_1\ dw_2\ dw_3\ Dw_4$	None identified
$dw_1\ dw_2\ dw_3\ dw_4$	SA 403, 4–dwarf Martin, 4–dwarf Kafir

usually about 50 cm shorter than 1-dwarfs; 3-dwarfs about 40 cm shorter than 2-dwarfs; and 4-dwarfs about 10 cm shorter than 3-dwarfs. As more recessive genes accumulate in a strain, the dwarfing effect is obviously lessened.

Two of four height genes are unstable. Mutations to tallness in kafir were reported by Karper (11) at a ratio of 1 mutation to 604 zygotes. Such a mutation rate produces almost 100 tall plants per acre when plant population per acre is above 50,000. Tall plants make fields unsightly and farmers are suspicious of seed production practices when such large numbers of off-type appear in their fields.

All varieties unstable for height are recessive dw_3, except for Early Hegari. Some varieties recessive for dw_3, such as Shantung Dwarf Kaoliang and Acme and Japanese Dwarf Broomcorns, are stable. Hegari and Early Hegari have the same height constitution, $Dw_1 \ dw_2 \ Dw_3 \ dw_4$, but Hegari is stable for height and Early Hegari is unstable. The unstable gene in Early Hegari has not been identified. The cause of the instability of height genes in sorghum is not known.

Sorghum hybrids of the future will probably be heterozygous 3-dwarf or homozygous 4-dwarf. The seed-parents will be 4-dwarfs. The pollinators will exist in two versions: the 4-dwarf and the stable 3-dwarf, $dw_1 \ dw_2 \ Dw_3 \ dw_4$. If the 4-dwarf version of the hybrid is too short in stature in some areas, the heterozygous 3-dwarf hybrid would be used. In the original height inheritance paper (28), the possibility of using stable 3-dwarfs was pointed out. Stephens (38) then suggested the conversion of unstable 3-dwarfs to 4-dwarfs by a backcrossing process followed by a recovery of stable 3-dwarfs following mutation to dominant Dw_3. In producing hybrids using 4-dwarf seed-parents and stable 3-dwarf pollinators, mutations to 3-dwarf would still occur in the 4-dwarf female. The mutated gamete would be $dw_1 \ dw_2 \ Dw_3 \ dw_4$ and the plant produced from it would be homozygous 3-dwarf. The rest of the hybrid population would be heterozygous 3-dwarf and only a few centimeters shorter than the plants that carry the mutated gene. In the field, the few homozygous 3-dwarfs would be inconspicuous.

These dwarf genotypes in sorghum are brachytic. Brachysm, by definition, is dwarfness characterized by shortening of the internodes only. Some unpublished data bearing on this point and collected in 1947 are presented in Table 5. Plants of two Sooner milo populations segregating for height were studied. The data from plants heterozygous for height were discarded, leaving only the data from homozygous height genotypes to be summarized.

The mean difference in height between 1-dwarf plants in pop-

TABLE 5.—COMPARISON OF PLANT CHARACTERS IN SOONER MILO GENOTYPES THAT DIFFER BY ONE HEIGHT ALLELE IN POPULATIONS GROWN AT CHILLICOTHE, TEXAS, IN 1947.

Plant character	Height class		Difference	P
SA 1295 population	1–dwarf	2–dwarf		
Height to flag leaf, cm	136.4 ± 1.65	95.8 ± 1.11	40.6	0.01
No. of leaves	21.1 ± 0.17	21.1 ± 0.17	0.1	0.7
Days to bloom	62.6 ± 0.31	62.0 ± 0.23	0.6	0.1
Width of 15th leaf, mm	72.0 ± 1.20	69.8 ± 1.06	2.2	0.2
No. of stalks per plant	2.3 ± 0.18	2.5 ± 0.12	0.2	0.3
Weight of heads, gms.	186 ± 16.2	181 ± 13.1	5.0	0.8
SA 5535 population	2–dwarf	3–dwarf		
Height to flag leaf, cm	102.1 ± 1.37	55.4 ± 0.73	46.7	0.01
No. of leaves	19.4 ± 0.21	19.0 ± 0.17	0.4	0.1
Days to bloom	59.0 ± 0.44	58.7 ± 0.45	0.3	0.6
Width of 15th leaf, mm	68.6 ± 0.73	68.6 ± 0.77	0	0.9
No. of stalks per plant	3.2 ± 0.14	3.2 ± 0.13	0	0.9
Weight of heads, gms.	209 ± 10.3	172 ± 10.0	37.0	0.01

ulation SA 1295 was 41 cm. The difference between 2-dwarf and 3-dwarf plants in population SA 5535 was 47 cm. These differences were highly significant. No significant differences occurred between the height genotypes in number of leaves, days to bloom, width of leaf, or number of stalks per plant. The 5-gram difference in weight of heads in the height classes of 1-dwarf and 2-dwarf plants in population SA 1295 was not significant. However, the 37-gram difference in favor of the 2-dwarf over the 3-dwarf class of population SA 5535 was significant statistically. Why the 3-dwarf plants in population SA 5535 produced lighter heads than 2-dwarf plants is not apparent. Perhaps the smaller head yield of the 3-dwarf class could be due to shading from the taller 2-dwarf plants as the two heights occurred at random in the rows. Regardless of this exception in head weight, the dwarfing genes in sorghum appear to be truly brachytic.

This brachysm, true to its definition, does not influence length of the sheath of the upper leaf, peduncle length, or head length. Fortunately, the length of the peduncle is not shortened, otherwise the head of short plants would not emerge from the boot. Also, if the rachis and seed branches of the head were shortened as much as the internodes, short-statured sorghums would have heads as compact as those of Pig-nosed durras. Data to show the lack of influence of dwarfing genes on leaf sheath, peduncle length, and head length are shown in Table 6. Kafir plants of these height genotypes growing together at Chillicothe, Texas, in 1960 were tagged

TABLE 6.—DAYS FROM PLANTING TO BLOOMING AND SIZE OF VARIOUS PLANT PARTS OF THREE HEIGHT GENOTYPES OF KAFIR AT CHILLICOTHE, TEXAS, IN 1960.

Criteria	Height genotype		
	2–dwarf	3–dwarf	4–dwarf
Days from planting to blooming	55.5 ± 0.6	55.6 ± 0.4	56.8 ± 0.4
Flag leaf sheath length, cm	33.4 ± 0.4	32.5 ± 0.6	30.6 ± 0.4
Height to upper node, cm	79.4 ± 1.0	34.2 ± 0.5	20.9 ± 0.5
Peduncle length, cm	32.5 ± 0.9	31.7 ± 0.8	35.6 ± 0.8
Head length, cm	20.2 ± 0.4	23.1 ± 0.4	21.6 ± 0.3
Total height, cm	132.2 ± 1.3	89.0 ± 1.1	78.1 ± 1.0
Stalk diameter, cm	1.65 ± 0.04	1.83 ± 0.03	1.58 ± 0.04

for days from planting to blooming and were measured for height from the base of the culm to the upper node for peduncle length, head length, and flag leaf sheath length. The 2- and 3-dwarf classes were isogenic lines, but the 4-dwarf class was a derived line that was similar in maturity but not isogenic with the other two. The data show that even the 4-dwarf genotype has a peduncle and head as long as those of taller 3-dwarf and 2-dwarf genotypes.

Rapid internode elongation in sorghum follows floral initiation, indicating some connection between the presence of a floral bud and internode elongation. The mechanism involved in the shortening of the internodes without a shortening of the peduncle and head is as yet unknown. Far-red radiation is now known to induce floral initiation and influence internode elongation (8). Probably dwarfing genes in sorghum slow down the synthesis of the substance that causes elongation.

Genes For Resistance to Disease and Insects

The diseases of sorghum and the organisms that cause them have been described by Leukel, et. al. (19). Resistance to some of the diseases is present in one or more varieties and resistance to some diseases has been bred into some varieties.

All of the different strains of milo grown on farms in the United States prior to 1937 were susceptible to milo disease, later identified as periconia root rot (18) and caused by the fungus *Periconia circinata*. This disease suddenly became widespread during the drought years beginning in 1934. A number of varieties of milo parentage, including Day, Colby, Wheatland, and Beaver, were also susceptible as was Darso and one strain of Sumac, a sorgo variety. Resistant plants in Dwarf Yellow milo were quickly found in Kansas (44) and in Texas (27). Bowman, *et al.* (4) reported susceptibility as being partially dominant and controlled by a single major gene. In Texas (27), resistant plants were invariably found to be homozygous resistant. Resistant plants that occurred occasionally in susceptible strains must have occurred due to mutation. In pure milo, resistant selections were found in one strain in Kansas, two in Texas, and three in California (31). Resistant plants of Beaver were found at both Dalhart, Texas, and Garden City, Kansas. Resistant plants of Day and Colby were found in Kansas. Resistant Darso selections were found at Temple and Chillicothe, Texas, and at Stillwater, Oklahoma. Resistant plants of exact Wheatland type were never found in diseased fields of Wheatland, but resistant plants found in such fields were distributed as Martin, Resistant Wheatland No. 288, and Dalhart Resistant Wheatland.

Yield trials over long periods of years at the Texas Substations at Chillicothe and Lubbock (27) have shown that losses of 50 to 60 per cent can be expected whenever a susceptible variety is grown on infested soil. Fortunately, all varieties grown to any extent and most parents of hybrids now in production are resistant to periconia root rot.

The three smuts of sorghum in the United States are covered kernel smut, *Sphacelotheca sorghi,* loose kernel smut, *S. cruenta,* and head smut, *S. reiliana.* Both kernel smuts can be controlled with chemical seed treatments, but head smut is soil-borne. Resistance to the four races of covered kernel smut is present in Spur Feterita. Casady (5) has recently reported on the inheritance of resistance to *S. sorghi.*

Head smut has recently become a problem in Texas, probably because of the widespread planting of a susceptible variety, Combine 7078, and several susceptible hybrids. The inheritance of resistance has not yet been determined, but heritable resistance exists since selections from resistant strains in a nursery in Refurio County on the Texas Gulf Coast in 1959 were resistant again in

1960. Resistance occurs in a number of varieties, including milo, hegari, and feterita. Among the parents of sorghum hybrids, Combine White Feterita is highly resistant but not immune. Other resistant parents are Wheatland, Plainsman, Caprock, and several of the Redbines. Most of the kafirs are moderately susceptible and Martin slightly so. The head smut resistance of a hybrid cannot be predicted if one parent is resistant and one susceptible to the fungus. Some parents, such as Caprock, are quite resistant themselves, but their hybrids with susceptible parents, such as Combine Kafir Tx 3197, are susceptible. Combine White Feterita, Tx 09, is also resistant, but its hybrid with Tx 3197 is quite resistant. Most hybrids with two resistant parents are resistant.

A destructive stalk rot of sorghum, caused by the fungus *Macrophomina phaseoli* (Maubl.) Ashley, is called charcoal rot. Severe lodging results from the presence of the disease because the culms break over close to the ground level when the pith of the lower internodes rots away. The fungus, *Fusarium moniliforme* Sheldon, produces symptoms that are sometimes quite similar to those produced by *M. phaseoli,* and there is a tendency to attribute the damage to charcoal rot regardless of which fungus is responsible. Tullis (42) reported on work done with *moniliforme* and cited the pertinent literature. Resistance to both these diseases exists in sorghum, but the expression of susceptibility is variable from year to year and it has been impossible to determine the mode of inheritance of resistance to these stalk-rotting fungi. Breeding for resistance to charcoal rot has not been very effective in sorghum up to the present.

Anthracnose of sorghum caused by *Colletotrichum graminicolum* (17) severely damages sorgo varieties grown in Mississippi. Leaves are injured sometimes to the point of defoliation and rotting causes a break-down of the tissues of the stalk. LeBeau and Coleman (17) found resistance to the leaf phase of this disease in recently introduced varieties from Africa. They found resistance to the disease to be a simple dominant.

The chinch bug, *Blissus leucopterus* (Say.), is a serious pest on sorghum in some areas in certain years. In Texas and Oklahoma, the insect migrates from grass pastures or barley fields to sorghum by flying and the usual measures of building barriers to stop the crawling insects is not effective. An effective breeding program to produce adapted chinch bug-resistant varieties was carried on at Lawton, Oklahoma. This breeding work, and studies of resistance that went on at the same time, gave some information on the inheritance of resistance. Snelling, *et al.* (34) reported on work at Manhattan, Kansas, and Lawton, Oklahoma. Resistance was found to be dominant and transgressive segregation for resistance indicated that more than one gene was involved. Chinch bug resistance was not put on a definite basis by this work, but there was ample evidence of its inheritance. Varieties of sorghum have been classified for susceptibility to damage from chinch bugs by a number of workers in several states and this information has been summarized in a book by Painter (24). Dahms (6), working in Oklahoma, found that resistant varieties showed the greatest tolerance to a uniform infestation of chinch bugs, that the insects preferred susceptible varieties as hosts, that female chinch bugs lived longer on susceptible varieties than on resistant ones, that more eggs were laid on susceptible varieties, and that nymphs developed faster on susceptible varieties. This work is significant and shows that there is in plants actual resistance to insects produced by influencing the biology of the insect.

Male-sterile Genes

Hybrid vigor was known to exist in sorghum long before the difficulties inherent in producing crossed seed in a species with perfect flowers were overcome. The story of the advent of sorghum hybrids has been told at least twice (31, 30). Cytoplasmic male-sterility was the final answer to the problem of hybrid seed production, but several genetic male-steriles that were apparently mutations were worked with before cytoplasmic male-sterility was discovered.

Antherless was the first genetic male-sterile worked with and was found in 1929 (15). Male-steriles 1 and 2 were found about 1935, one in India and the other in Texas. It was proposed by Stephens (36) in 1937 and ms_2 might be used for the production of hybrid sorghum seed. By 1946, work had progressed with ms_2 to a point where it appeared that hybrid sorghum might be put into production. However, a still better male-sterile had been found and work with ms_2 was abandoned.

In 1943, Glen H. Kuykendall found in the Day variety a male-sterile that must have arisen as a mutation. This abnormality proved to be a genetic male-sterile that gave male-sterile offspring when crossed to some varieties and fertile offspring when crossed to others. The proposed procedure for the production of hybrid seed by a 3-way cross was published by Stephens, et al. in 1952 (40). The Day male-sterile was distributed to plant breeders by the Texas Agricultural Experiment Station in 1951. Because cytoplasmic male-sterility was found, the Texas Station did not put any hybrids into production using the Day male-sterile. However, one seed company did produce hybrids for a year or two using the Day male-sterile.

Beginning in 1949, work was done that established the existence of cytoplasmic male-sterility. In 1950, F_2 populations of reciprocal crosses involving Sooner Milo and Texas Blackhull Kafir were grown and partial male-sterility was found in the populations whose female parent was milo. Holland (10) reported the early data from this study in a thesis. It soon became apparent that male-sterility in this case was due to the interaction between milo cytoplasm and kafir nuclear factors as reported by Stephens and Holland in 1954 (39). The degree of male-sterility was increased as the proportion of kafir chromosomes in milo cytoplasm was increased. The number of genes involved was not determined by Stephens because of the pressure of work to get sorghum hybrids into production and because unfavorable weather in 1952 and 1953 at Chillicothe confused the expression of genetic sterility. Maunder and Pickett (21) reported cytoplasmic male-sterility to be dependent on a single pair of recessive genes, $ms_c \ ms_c$, interacting with sterile cytoplasm. Plants in their fertile classes in segregating populations varied in seed set from 5 per cent upward. Obviously, important modifying genes must exist. Finding fertility restorers was no problem in sorghum. All milos and milo derivatives with sterile cytoplasm carry the dominant allele, Ms_c; otherwise they would be male-sterile. The dominant restorer also exists in many varieties with normal cytoplasm.

The genetic male-steriles that arose as mutations are not important today in the production of hybrid sorghum seed, but work with them contributed to the discovery of cytoplasmic male-sterility. J. C. Stephens had been working with mutant male-steriles in sorghum for more than 20 years when he and R. F. Holland announced, in 1954, the attainment of cytoplasmic male-sterility.

Other Useful Genes

Seed color is an important economic characteristic of sorghum. The colors are varied and occur in different layers of the caropsis. (7, 41, 32, 37). The contrasting colors, such as white, yellow, red, and brown, occur as a result of the color content of the epicarp, the thickness of the mesocarp, the presence or absence of a brown testa, and the dominant or recessive condition of a gene designated as S that controls the presence or absence of brown pigment in the epicarp when a brown testa is present. Six genes, exclusive of the S gene, have been reported to cause color in the epicarp. The genetics of seed color in sorghum has been summarized by Quinby and Martin (31).

All of the possible pericarp colors are not represented by varieties in the United States. Most of the grain produced in the United States is genetically red or pink. Small amounts of white grain are produced. White grain frequently is used as poultry feed as chickens seem to prefer white grain for some reason. True yellow is not represented by a commercial variety in the United States but may soon be as the yellowest yellow endosperm varieties or hybrids have a yellow pericarp.

Brown seeds result when a brown testa or undercoat is present along with a dominant spreader gene that allows the presence of brown in the epicarp. When the seeds are dominant red or pink also, the seeds are dark reddish brown. The brown color is caused by the presence of tannin and the astringency of developing brown seeds is some protection from birds. In Africa, brown-seeded varieties are used to make beer. In the United States brown-seeded varieties are grown only in the humid areas or where birds are a problem.

Most sorgo varieties have brown seeds and the brown color is associated with high tannin content. Atlas and Tracy, two widely grown sorgo varieties that have no brown testa in their seeds, originated as selections from crosses involving brown-seeded and non-brown-seeded varieties.

Most discolored spots on sorghum grain result from injury caused usually by sucking insects or diseases. When sorghum plants are damaged, any dead tissue becomes pigmented. The spots become blackish brown, reddish brown, or tan in color, depending upon the genetic constitution of the plant with respect to two plant color genes, Q and P. Plants with black glumes *(QP)* have blackish brown spots, those with red glumes *(qP)* have reddish black spots, and those with sienna *(qp)* or mahogany glumes *(Qp)* have tan spots. The pigment in the spots is water-soluble and tan spots are the least objectionable in starch manufacture by the wet milling method. The beautiful sienna glume color of Sweet sudangrass is the result of the variety being recessive for both q and p.

Two kinds of starch are found in varieties of many cereals, including rice, maize, millet, coix, barley, and sorghum. Waxy starch, which stains red with iodine, is amylopectin, a branched glucose polymer. Common starch that stains blue with iodine is a mixture of amylopectin along with an unbranched or linear polymer, amylose. The properties of waxy starch make it of particular interest to industry as it forms a clear paste which has little tendency to gel on standing. Extensive studies of the chemical properties of sorghum grain and of the properties of sorghum starch have been made by Barham, et al. (3).

Waxy endosperm of sorghum was first reported by Meyer (22). Karper (12) found waxy to be a simple recessive to starchy in segregating populations. Waxy varieties and hybrids are in commercial production. The waxy allele is present in Black Amber, the first sorghum to become established after being introduced into the United States. The waxy allele bred into waxy grain sorghums came from the Batad variety that came from the Philippines. Seeds of a number of varieties from India and several kaoliangs are also waxy.

A sugary endosperm character in sorghum resembles sugary in maize. The character occurs in a number of Indian varieties and the sugary allele was found as a mutation at Chillicothe (31). Ayyangar, et al. (1) found sugary to be a simple recessive. Sugary varieties in India have "dimpled" grains. An analysis of sugary grains in India showed the amount of reducing sugars to be three times the quantity found in nondimpled grains, the nonreducing sugars being equal in both. No commercial use has been made of this character in the United States. In India, sugary grain sorghum is parched by placing green heads in beds of coals. The grain is then knocked from the heads and eaten.

A yellow endosperm character was introduced into the United States from Nigeria by O. J. Webster in 1951. The yellow color is caused by carotene and xanthophyll. The yellow endosperm character is being bred into adapted varieties. The amounts of carotene and xanthophyll in the yellowest sorghum varieties obtained thus far are one-quarter to one-half those contained in yellow corn.

Sorghum stalks are pithy or juicy and sweet or nonsweet. According to Hilson (9), pithy is a simple dominant to juicy. According to Ayyangar, et al. (2), 17 to 20 per cent of juice can be extracted from pithy-stalked varieties and 33 to 48 per cent from juicy varieties. Pithy-stalked plants have white leaf midribs and juicy-stalked plants opaque midrids. Sorgos are quite juicy and grain sorghums less juicy or quite dry. Kafirs have juicy stems and feteritas and broomcorn dry stems.

The inheritance of sweet and nonsweet stalks was studied by Ayyangar, et al. (2) and they found nonsweet to be a simple dominant to sweet. These authors also reported genes for sweetness and juiciness to be independent in inheritance. Sorgos have sweet stems, but most grain sorghums have little sweetness. However, feterita, which has dry stems, tastes sweet if the dry pith is chewed. Sweet sudangrass is both juicy and sweet, whereas the introduced variety is pithy and nonsweet.

Summary

Sorghum is an introduced species in the United States and is grown on about 20 million acres. The mechanization of agriculture and changes in sorghum that began about 30 years ago have changed sorghum from a feed crop for use on the farm into a cash grain crop.

The species has been transformed by producing early maturing and shorter statured varieties. Alleles for resistance to several plant diseases and to one insect have been found. Cytoplasmic male-sterility has been found and methods of hybrid sorghum seed production devised. Several other desirable characters have been bred into improved varieties, including improved seed and glume colors, waxy endosperm, yellow endosperm, and either pithy or juicy stems.

References

1. Ayyangar, G. N. R., Ayyar, M. A. S., Rao, V. P., and Nambiar, A. K. 1936. Inheritance of characters in sorghum—the great millet, 9, dimpled grains. *Indian Jour. Agr. Sci.*, **6**: *938–945*.

2. ————, ————, ————, ————. 1936. Mendelian segregations for juiciness and sweetness in sorghum stalks. (Research note.) *Madras Agr. Jour.*, **24**: *247–248*.

3. Barham, H. N., Wagoner, J. A., Campbell, C. L., and Harclerode, E. H. 1946. The chemical composition of some sorghum grains and the properties of their starches. *Kansas Agr. Exp. Sta. Tech. Bul. 61*.

4. Bowman, D. H., Martin, J. H., Melchers, L. E., and Parker, J. H. 1937. Inheritance of resistance to Pythium root-rot in sorghum. *Jour. Agr. Res.*, **55**: *105–115*.

5. Casady, A. J. 1961. Inheritance of resistance to psychotic races 1, 2, and 3 of *Sphacelotheca sorghi* in sorghum. *Crop. Sci.*, **1**: *63–68*.

6. Dahms, R. G. 1948. Effect of different varieties and ages of sorghum on the biology of the chinch bug. *Jour. Agr. Res.*, **76**: *271–288*.

7. Graham, R. J. D. 1916. Pollination and cross-fertilization in the juar plant *(Andropagon sorghum* Brot). *India Dept. Agr. Mem. Bot.*, **Ser. 8**: *201–216*.

8. Hendricks, S. B., and Borthwick, H. A. 1959. Photocontrol of plant development by simultaneous excitations of two interconvertible pigments. *Proc. Nat. Acad. Sci.*, **45**: *344–349*.

9. Hilson, G. R. 1916. A note on the inheritance of certain stem characters in sorghum. *Agr. Jour. India*, **11**: *150–155*.

10. Holland, R. F. 1952. *Thesis, Texas Agricultural and Mechanical College, College Station, Texas*.

11. Karper, R. E. 1932. A dominant mutation of frequent recurrence in sorghum. *Amer. Nat.*, **66**: *511–529*.

12. ————. 1933. Inheritance of waxy endosperm in sorghum, *Jour. Hered.*, **24**: *257–262*.

13. ———— and Quinby, J. R. 1946. The history and evolution of milo in the United States. *Jour. Amer. Soc. Agron.*, **38**: *441–453*.

14. ————, ————. 1947. Additional information concerning the introduction of milo into the United States. *Jour. Amer. Soc. Agron.*, **39**: *937–938*.

15. ———— and Stephens, J. C. 1936. Floral abnormalities in sorghum. *Jour. Hered.*, **27**: *183–194*.

16. Laubscher, F. X. 1945. A genetic study of sorghum relationships. *Union South Afr. Dept. of Agr. and For., Sci. Bul. 242.*
17. LeBeau, F. J., and Coleman, O. H. 1950. The inheritance of resistance in sorghum to leaf anthracnose. *Agron. Jour.,* 42: *33–34.*
18. Leukel, R. W., and Martin, J. H. 1953. Four enemies of sorghum crops. *U. S. Dept. Agr. Yearbook, 368–377.*
19. ———, ———, and Lefebvre, C. L. 1951. Sorghum diseases and their control. *U. S. Dept. Agr. Farmers Bul. 1959.*
20. Martin, J. H. 1936. Sorghum improvement. *U. S. Dept. Agr. Yearbook, 523–560.*
21. Maunder, A. B., and Pickett, R. C. 1959. The genetic inheritance of cytoplasmic-genetic male-sterility in sorghum. *Agron. Jour.,* 51: *47–49.*
22. Meyer, Arthur. 1886. Ueber Stärke Körner, Wekhe sich mit Jod roth Färben. *Bericht der Deutsch. Bot. Gesell.,* 4: *337–363.*
23. Myers, W. M. 1947. Cytology and genetics of forage grasses. *Bot. Rev.,* 13: *319–421.*
24. Painter, R. H. 1951. Insect Resistance in Crop Plants. *New York: The MacMillan Co.*
25. Quinby, J. R., and Karper, R. E. 1945. The inheritance of three genes that influence time of floral initiation and maturity date in milo. *Jour. Amer. Soc. Agron.,* 37: *916–936.*
26. ———, ———. 1947. The effect of short photoperiod on sorghum varieties and first-generation hybrids. *Jour. Agr. Res.,* 75: *295–300.*
27. ———, ———. 1949. The effect of milo disease on grain and forage yields of sorghum. *Agron. Jour.,* 41: *118–122.*
28. ———, ———. 1954. Inheritance of height in sorghum. *Agron. Jour.,* 46: *211–216.*
29. ———, ———. 1961. Inheritance of duration of growth in the milo group of sorghum. *Crops Sci.,* 1: *8–10.*
30. ———, Kramer, N. W., Stephens, J. C., Lahr, K. A., and Karper, R. E. 1958. Grain sorghum production in Texas. *Texas Agr. Exp. Sta. Bul. 912.*
31. ——— and Martin, J. H. 1954. Sorghum improvement. *Advances in Agronomy,* 6: *305–359. New York: Academic Press, Inc.*
32. Sieglinger, J. B. 1933. Inheritance of seed color in crosses of brown-seeded and white-seeded sorghums. *Jour. Agr. Res.,* 47: *663–667.*
33. ———. 1940. A sorghum seed color chimera. *Jour. Hered.,* 31: *363–364.*
34. Snelling, R. O., Painter, R. H., Parker, J. H., and Osborn, W. M.

1937. Resistance of sorghums to the chinch bug. *U. S. Dept. Agr. Tech. Bul. 585.*

35. Snowden, J. D. 1936. The Cultivated Races of Sorghum. *London: Adlard and Sons.*

36. Stephens, J. C. 1937. Male-sterility in sorghum: Its possible utilization in production of hybrid seed. *Jour. Amer. Soc. Agron.,* **29:** *690–696.*

37. ————. 1946. A second factor for subcoat in sorghum seed. *Jour. Amer. Soc. Agron.,* **38:** *340–342.*

38. ————. 1956. A breeding method to eliminate tall mutations in combine grain sorghum. *Texas Agr. Exp. Sta. Tech. Article 2522.*

39. ———— and Holland, R. F. 1954. Cytoplasmic male-sterility for hybrid sorghum seed production. *Agron. Jour.,* **46:** *20–23.*

40. ————, Kuykendall, G. H., and George, W. W. 1952. Experimental production of hybrid sorghum seed with a three-way cross. *Agron. Jour.,* **44:** *369–373.*

41. ————, and Quinby, J. R. 1938. Linkage of the Q B Gs group in sorghum. *Jour. Agr. Res.,* **57:** *747–757.*

42. Tullis, E. C. 1951. *Fusarium moniliforme,* the cause of a stalk-rot of sorghum in Texas. *Phytopathology,* **61:** *529–535.*

43. Vinall, H. N., Stephens, J. C., and Martin, J. H. 1936. Identification, history, and distribution of common sorghum varieties. *U. S. Dept. Agr. Tech. Bul. 506.*

44. Wagner, F. A. 1936. Reaction of sorghums to the root, crown, and shoot rot of milo. *Jour. Amer. Soc. Agron.,* **28:** *643–654.*

Comments

GABELMAN: You assumed the wild-type genes for maturity were probably all recessive and that the selection in the tropics was in favor of the dominant genes. Under what photoperiods would this natural selection have taken place? Do you assume that this is selection brought about by domestication, or was it natural selection?

QUINBY: It seems reasonable to suppose that late maturity and large size resulted from domestication. Photoperiodic response is influenced by temperature and floral initiation would be delayed even with 12-hour days at some temperatures. Snowden, in his book "The Races of Sorghum", presents evidence that leads him to believe that present-day varieties evolved from grassy types. However, my opinion in regard to the "wild type" being early maturing grows out of the fact that the alleles for late maturity act like dominant inhibitors.

SINGLETON: The dwarf$_3$ (dw$_3$) appears like the effect of pseudo alleles similar to the case of Star-asteroid eye characters in Drosophila analyzed by E. B. Lewis who found crossing over within the locus. When two wild-type alleles were located on a single chromosomal strand (*cis* position), the phenotype was wild. If a similar situation exists for dwarf$_3$ in sorghum, a tall plant would be expected following a rare crossover within the locus. If this is the proper interpretation, it might be corrected by radiating the recovered tall plants, deleting the + allele so that crossing over would not give rise to a tall (+) plant.

NUFFER: I would like to suggest a method of solving the problem of instability of the dw_3 locus. This situation seems especially well-suited for the application of ionizing radiation. Since the reversion of dw_3 to Dw_3 prevents the production of true breeding dw_1 dw_2 dw_3 dw_4 lines, a likely procedure would be to induce a deficiency at the Dw_3 locus. This could best be done by subjecting seed of a reverted dw_1 dw_2 Dw_3 dw_4 line to X-radiation. Plants from the treated seed would then be selfed and a search made in the progenies for 4–dwarf plants. Since X-ray-induced changes are usually irreversible, the new dw_3 should be stable.

QUINBY: Dr. W. R. Singleton has also suggested this same procedure and I shall send him seed of several reverted dw_1 dw_2 Dw_3 dw_4 strains to be X-rayed and returned to me.

MEHLQUIST: Relative to your suggestion that the mutation for disease resistance has occurred many times, is it not reasonable to assume that this gene has existed for a long time but did not show until the nonresistant plants were exposed to the disease?

QUINBY: Perhaps the genes for resistance might have been carried along in the populations, but resistance was not always found in a variety the first time it was grown in a diseased field. Also, resistance was found in many different varieties, or the same variety at different locations, and the incidence of resistant plants was quite low, sometimes as low as two or three plants to 5 or 10 acres.

Use of Induced Mutants in Seed-propagated Species

HORST GAUL[1]

Max-Planck-Institut für Züchtungsforschung
Köln-Vogelsang, Germany

P LANT breeding is controlled evolution. Two of the major factors of evolution, recombination and selection, are extensively used by breeders and refined methods were developed during the first half of this century to exploit them. In the last 30 years research has shown that mutations, the third major factor in evolution, offer an additional tool, which is potentially able to modify and improve cultivated plants in a way similar to the conventional breeding methods. These results, however, do not imply that the efficiency of the new method is equal to or greater than that of the older. If the efficiency can be increased, the mutation technique could be used more often in conjunction with the other breeding methods. These questions will be discussed throughout this paper, particularly in the last part.

The interest in the induction of mutations for plant breeding has increased considerably all over the world in the last 10 years. However, the enhanced interest is not only a consequence of pure objectivity, but also results from the fact that funds for mutation research are often easily obtained. Although theoretical work on the action of mutagenic agents, on the nature of mutations, etc., has a long history, strict and intensive investigations concerning the practical application in breeding has not been conducted until recently. This should be kept in mind when considering the objection often made to the new method, *viz.*, that up to date so few varieties which have their origin in induced mutants have been released.

In the last few years a series of meetings and symposia has taken place dealing with general radiobiology and with physically and chemically induced mutations from both a theoretical and practical point of view. Among the reviews of the last 5 to 6 years relevant to the present paper, the following may be men-

---
[1]The author is grateful to Doctors R. S. Caldecott and R. A. Nilan for critical reading of the manuscript and for linguistic corrections.

tioned (9, 14, 25, 28, 34, 48, 59, 65, 66, 71, 74, 75, 80, 82, 83, 94, 99, 100, 101, 104, 115, 123).[2] In the review by Prakken (83), there is an appendix with a bibliography covering 789 titles and the symposia, proceedings, handbooks, etc., are quoted separately. In this connection the bibliography of Sparrow, *et al.* (102) also deserves mention, since it covers nearly 2,600 titles, published between 1896 and 1955, on the effects of ionizing radiations on plants.

In these numerous reviews on mutations and plant breeding there is naturally much repetition. There is, therefore, no need to present again a detailed picture and to quote every relevant publication. Instead, I shall try to give a critical evaluation of the present stage and future possibilities of breeding with mutations. Certainly this review will sometimes be colored by my personal point of view.

First, the nature of induced mutations will be discussed, including genetics and types of mutants. Then the methods of breeding with mutations will be outlined. Thereafter, the problem of induction and selection will be reviewed and some special hints will be given. Finally, an attempt will be made to evaluate the use of mutations in plant breeding.

Nature of Induced Mutants

In plants, the most comprehensive information about characters which can be modified by induced mutations has been obtained in barley and snapdragon. Recently, however, the number of vital mutations is rapidly increasing in many other species, e.g., tomatoes (113, 114, 116), flax (60), soybeans (129), peanuts (41), peas (35, 38, 68, 69, 126), bush beans (85), potatoes (52), and millet (107, 108). There are also numerous reports on wheat and rice (cf. symposium on the effects of ionizing radiations on seeds, Karlsruhe, Germany, 1960, in press). Concerning barley there are large collections of induced mutants in Sweden, Germany (Gatersleben, Halle, Köln-Vogelsang), and Belgium (Gembleaux), each of them consisting of several hundred forms.

The induced variability in the species mentioned is striking. Practically all morphological and physiological characters can be

---

[2]See References, page 240.

changed by means of induced mutations within the framework of the species or even beyond it.

Genetics of Mutants

The great majority of these mutations are recessive and segregate in a 3:1 ratio; sometimes, however, with a deficiency of recessives and occasionally with an excess. In diploid organisms completely dominant mutations have scarcely been found, except some types which are lethal or semilethal in the homozygous condition (cf. 79, 110). Therefore, whenever a dominant deviation is met with in mutation experiments with diploids, extreme caution is advisable. For instance, a great deal of the fungi-resistant variants, claimed to be dominant mutants, may probably be a consequence of contamination only.

There are, however, a few reports of true dominant mutations. Nötzel (78), for instance, investigated 40 different barley mutations, including types of *erectoides, intermedium, macrolepis,* and earliness. One of the earliness mutants was dominant, while the other 39 were recessive. Another dominant mutation of earliness has been reported by Scholz (93). Out of 70 *erectoides* mutants investigated by Hagberg (50), 2 were almost completely dominant for ear density and 1 was partially dominant for the same character. However, other characters affected by the same mutation may behave as recessive or intermediate. For example, *erectoides* mutants may be recessive for ear density, but dominant for length of the upper internode of the culm (48, 49).

Along with the pronounced pleiotropic action of all (or nearly all) mutations, the manifestation of the relative degree of recessivity in the different characters concerned can vary considerably. Moreover, superdominance and superrecessivity have been repeatedly observed for various characters in lethal, sublethal, and vital mutations (47, 49, 112). The degree of dominance may also be greatly influenced by the genetic background (e.g. 45, page 624; 51, 72) and/or the environment. The dominant tomato mutant *subsistens,* for instance, which was studied by Endlich (21), is lethal in the homozygous condition. In the heterozygot the mutant allel showed a varying degree of expressivity in different environments. Under low light intensity or high temperature, the mutated character appeared to be recessive or nearly recessive.

In contrast to diploids, dominance or semidominance seems to be a fairly common phenomenon for many characters of vital mutations in polyploids like wheat (70). Here, the phenotypic change is probably more often caused by chromosome mutations instead of point mutations and polyploids are more tolerant to chromosomal aberrations as compared with diploids.

There is no essential difference between the alleles available from the world collections of natural forms and those from induced mutations. This has been adequately demonstrated for some characters in barley. Thus, some *erectoides* mutants (50, 78), *macrolepis* mutants (78), naked-kernel mutants (92), smooth-awned mutants, and a mutant with waxy stalks and leaves (Scholz as cited by Stubbe, 117), have been shown in crosses to be allelic with corresponding forms of the world collections. In tomatoes several loci of induced mutants of *Lycopersicon pimpinellifolium* have been proved identical with those of the corresponding *esculentum* mutants (117).

Identity of the locus does not mean that the alleles are identical. The occurrence of multiple alleles or of pseudoalleles can usually be studied only with mutants which do not differ otherwise in their gene content. From a careful morphological study of 70 *erectoides* mutants in barley, including the various pleiotropic effects, Hagberg (50) arrived at the conclusion that it seems to be impossible to copy an individual mutation. These 70 mutants were scattered over 22 separate loci. Concerning the phenotype belonging to an individual *erectoides* allele there are important differences between the loci, but there is also a remarkable variation within a locus. This finding is in general accord with the evidence accumulated from microorganisms and Drosophila in the last 10 to 15 years which has shown that genes have a complex nature. So far as plant breeding is concerned, these results imply that the induction of mutations does not simply reproduce the natural variability, but may expand it to a large extent. Through mutations of a given character, not only may loci which are not yet known be discovered, but also new alleles within the loci may be created.

Phenotypic Classification of Mutants

The alteration of characters induced by mutations may be large or small. There are transitions all the way from macro-mutations to micro-mutations. In Table 1 vital mutations are classified in four types according to degree of change of the phenotype.

TABLE 1.—PHENOTYPIC CLASSIFICATION OF VITAL-MUTATIONS.

1. **Macro-mutations**	Large mutations, drastic mutations (Grossmutationen)
	Detectable in a single plant
a. Transspecific	Systemic mutations, organization mutations
b. Intraspecific	
2. **Micro-mutations**	Small mutations (Kleinmutationen)
	Detectable in a group of plants
a. Manifest	
b. Cryptic	Detectable if the environment and/or genetic background is changed

Those deviations, which are easy to recognize in a single plant, may be called macro-, or large, mutations. Macro-mutations usually affect characters which are already known in the species, but sometimes the induced character is unknown in the species, in the genus, or even in the family. (Particularly in cultivated plants the previous separation into different species is sometimes not justified, e.g., in wheat (61).)

In contrast to these are micro-, or small, mutations (Kleinmutationen) which cannot be detected with certainty in a single plant but only in a group of at least, say, 30 or more individuals. These micro-mutations may be classified into manifest and cryptic micro-mutations, depending upon their degree of detection. Manifest micro-mutations are detectable if, in the unchanged environment and genetic background, appropriate methods of screening and observations are applied. Cryptic micro-mutations cannot be recognized even in a large group of plants under "normal" growing conditions. A drastic change of the environment and/or the genetic background may enable them to become manifest, whereas under normal conditions they behave in a neutral manner. The existence of cryptic mutations is more or less hypothetical and there is scarcely any experimental evidence for them until now (see, however, 72). Nevertheless, they may have significance in evolution and they could be of value for certain breeding purposes. The existence of the other types of mutations has been shown repeatedly. Further use of the term micro-mutation in this paper refers to manifest ones only.

Naturally, the grouping as suggested in Table 1, is arbitrary and has no meaning relative to the nature of changes of the genetic material. Whether in individual cases the phenotypic change is considered to be small or large will often depend on the methods and personal view of the observer. Though it is not possible to construct sharply distinguished categories of mutations, the classifications suggested may be of value for the communication of evolutionary and plant breeding problems.

Transspecific Macro-mutations

The occurrence of transspecific mutations is very rare. Nevertheless, a number of more or less clear cases of both induced and spontaneous mutations have been reported (see reviews in 44, 45, 79, 111, 117, 119). Recently, in barley, another induced macro-mutation (57) has been described in more detail (91). The culms of that mutant have no differentiated nodes. All the nodes are located in close succession immediately above the root, and the culm is formed only by one internode. Such a character is common in the Cyperaceae. Another interesting example of a spontaneous macro-mutation was recently described by Staudt (105) in Fragaria.

Macro-mutations that are transspecific usually have a stronger reduced vitality and/or fertility. They are of no immediate value for plant breeding. However, if a definite character has to be transferred from a foreign species, it might be easier to use transspecific mutants instead of interspecific hybridization. In agricultural plants transspecific mutations which could serve for practical breeding are scarcely known. Yet, it might be possible that they will be found in the future, owing to the increased production of mutations. Such mutations might also be of considerable interest in ornamentals.

Sometimes the manifestation of the mutant characters is not constant but variable. This holds true, e.g. for some mutants of *Antirrhinum majus* which were investigated by Stubbe, (111, 117). Thus, the mutant *transcendens* has a tendency to reduce the number of stamens per flower from four to two. Flowers with two stamens are characteristic in adjacent genera of the Scrophulariaceae to which Antirrhinum belongs. However, in the mutant, only 40 per cent of the flowers possess two stamens and the others form three and four. Selection of stable types with two stamens

proved impossible. When crossing this mutant with wild species
of the genus Antirrhinum, Stubbe (111, 117) had success in the
selection of forms which possess two stamens in every flower. From
Figure 1 it can be seen that already in the F_5 generation, 98 per

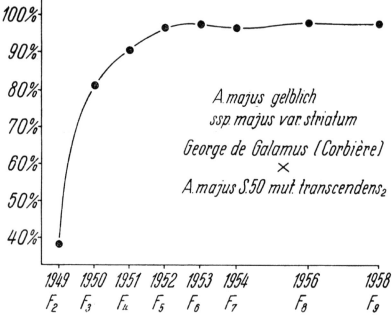

FIGURE 1.—*Curve of selection of F_2 to F_9 for two stamens, after crosses
of the mutant trancendens$_2$ of* Antirrhinum majus to A. majus gelblich
ssp. majus var. strictum. *From Stubbe (117).*

cent of the flowers had two stamens. The introduction of numerous
other genes evidently increased the penetration of the mutated
character and led to its stabilization in the new "genetic environment". With other mutants forming variable numbers of stamens
up to eight, similar success in stabilizing was possible (111, 117).

These examples have been broached here because they offer
an interesting model for plant breeding. Macro-mutations may not
only be stabilized in a new genetic background, but it also seems to
be possible to modify their action and particularly to eliminate
undesired pleiotropic effects, as discussed later in this paper.

Intraspecific Macro-mutations

Intraspecific macro-mutations are the most common type of all mutations that have been selected. Numerous characters of value for plant breeding have been found, but usually these mutants have a reduced vitality. Reduced general performance may, however, not be unexpected on *a priori* reasoning. A gross change in the genetic system, though it may mean the induction of a valuable character, will usually result in a disturbance of the delicate gene balance, which had been built up by the breeder with great efforts over a long time. Perhaps the frequency of productive large mutants might be higher in those plants in which greater breeding work has not yet been done (Cf. 25). However, a small portion of these large mutations represent a valuable source for further recombination work as discussed later.

Micro-mutations

Small mutations generally deserve more attention from plant breeders than macro-mutations. Like the large mutations, the small ones may affect all morphological and physiological characters. The significance of small mutations in evolution was first recognized and emphasized by Baur (3). Since then it has been repeatedly discussed (119). From the fact that most character differences in race and species hybrids show complex segregation, the majority of research workers (see, e.g., 106) agree today with East (18, page 450) that "the deviations forming the fundamental material of evolution are the small variations of Darwin". Though small mutations were early described in Antirrhinum (109) and plenty are known in Drosophila (63, 120), it is surprising that relatively little attention has been paid to them. Also, in the extensive work with barley, small deviations of various characters, such as kernel size, leaf size, straw height, protein content, etc., have been known for a long time (43, 57, 73, 79, 93, 95, 96). Yet, the suggestion to put the emphasis in breeding programs on small mutations was not made until recently by Nybom (79) and Gregory (39, 40), and meets perhaps with even stronger emphasis by myself (25, 28).

Most plant attributes of interest to the breeder are quantitative characters which are controlled by many genes. These are called polymeric genes, multiple genes (or factors), or polygenes. When compared with genes controlling so-called qualitative characters, there are no convincing reasons to assume that polymeric genes are a different sort of genes so far as their nature is concerned. A mutation step in one of these multiple genes may affect a quantitative character in a measurable way. For instance, increasing yielding potential of, say, 5 per cent in a top variety of barley, wheat, or oats is usually considered to be a desired goal of breeding. It seems possible that such progress can be obtained by mutation of only one or a few polymeric genes without a pronounced morphological effect. However, the experimental evidence for such a speculation is as yet meagre, except for the extensive evidence with peanuts from the careful investigations of Gregory (39, 40, 41, 42). This author found a striking increase in the genetic variance for yield in progenies of normal-appearing M_2 plants which were selected at random. He also succeeded in the selection of mutants with higher yielding capacity.

In addition, Cooper and Gregory (15) presented evidence of small mutations exerting a quantitative effect on leaf-spot resistance. Starting with progenies of normal-appearing M_4 plants in rice, Oka, et al. (81) found a considerable increase in the genetic variance for plant height and heading date. For both these characters, the authors succeeded in selecting plus and minus mutants. In barley, a study of induced variability for quantitative characters has recently been presented by Moës (73). He describes positive and negative alterations of a great number of characters, such as number of tillers per plant, number of seeds per spike, kernel size, lodging-resistance, leaf size, straw height, etc. Also at our laboratory in Köln-Vogelsang, following a program to develop selection methods for small mutations in barley, we succeeded in obtaining a greater number of small variants in spring barley and winter barley, and some of the results will be presented later in this paper.

Despite these studies, the field of small mutations is largely unexplored as compared with that of large mutations. More information is needed about their features and their frequency for an evaluation of their significance in plant breeding. A priori it may be expected, that the more genes that are involved in a character, the higher the probability of obtaining an alteration by a mutation of one of the multiple genes concerned. Indeed the results obtained in Drosophila (63, 97, 120) and peanuts (39, 41, 42) suggest that the frequency of small mutations is considerably higher than that of large mutations.

Methods of Breeding with Mutations

Mutants can be utilized in various ways in plant breeding and the most important methods are reviewed in Table 2. The term "mutation breeding" is often loosely used. In the classification of Table 2, it is suggested that the term mutation breeding be used only when a new variety results from the direct propagation of the mutant. Extensive use of mutants can be made in the various methods of cross-breeding. Besides, there are a number of special aims for which mutagens may prove valuable, and these are outlined in Table 2.

TABLE 2.—METHODS OF BREEDING WITH MUTATIONS AND USE OF MUTAGENIC AGENT IN SEED-PROPAGATED SPECIES.

A. Self-pollinating species mainly
  1. Immediate use of mutants: Mutation breeding proper
  2. Use of mutants in cross-breeding
    a. Within the same variety
    b. With foreign varieties
    c. In hybrid populations, as an additional source of variability
    d. In heterosis-breeding (including cross-pollinating species)
B. Cross-pollinating species
  3. Induction of mutations in cross-pollinating populations
C. Self- and cross-pollinating species
  4. Mutagenic agents as a tool for special purposes
    a. Induction of translocations for interspecific or intergeneric transfer of desired characters
    b. Diploidization of artificially produced polyploids
    c. Use of translocations (with localized breakage points) for "directed" duplications
    d. Use of radiation to induce transitory sexuality in apomicts
    e. Use of radiation to produce haploids
    f. Use of radiation to break incompatibility in distant crosses

Mutation Breeding

The direct use of mutants offers theoretically the greatest advantage in utilizing mutations in plant breeding. As compared with cross-breeding, it may save nearly half of the time necessary to create a new variety. Evidently here lies the field in which the significance of small mutations should be tested. The experience with large mutations has shown that, besides their relative rarity, only a very small fraction exerts a positive influence on yield.

The chance of producing more or less large mutations with higher yield is, however, a matter of fact which cannot be overlooked. In Table 3 some examples are given in barley. There is no

TABLE 3.—SOME EXAMPLES OF BARLEY MUTANTS WITH WELL-ESTABLISHED HIGHER YIELDING CAPACITY. *

Character and Serial No.	Number of years tested	Relative yield	Author
	Spring Barley		
Early, M$_s$ 90	4	108	HOFFMANN (57)
Large-kernelled, 44/7	10	107	FROIER (24)
Early, less straw-stiff, 44/4	9	109	FROIER (24)
Large-kernelled, early, 3978	6	115	SCHOLZ, pers. com., also (93, 96)
Erectoides, 2660	5	105	SCHOLZ, pers. com., also (93, 96)
Erectoides, 2654	8	103	SCHOLZ, pers. com., also (93, 96)
Smooth-awned, 4033	7	107	SCHOLZ, pers. com., also (93, 96)
Semi-smooth-awned, 3945	6	113	SCHOLZ, pers. com., also (93, 96)
Early, W 3	4	107	GAUL
	Winter Barley		
Early, 506	5	114	SCHOLZ pers. com., also (93)
Early, 481	8	107	SCHOLZ pers. com., also (93)

*Kernel yield of the mother line is taken as 100.

doubt that the higher yield is well established because these figures are based on drill tests with several replicates over a period of at least 4 years. It should be noticed, however, that in these examples, the higher yielding power is proved only for one location.

The objection often raised as to why such mutants have not been released as varieties is not valid. Under the conditions in which they were produced, the total expenditure in cross-breeding as compared with "mutation-breeding", may have been of the order of 50:1, or even 500:1. It is not surprising that by the time the superiority of these mutants was established, through conventional breeding methods, new lines were developed which were superior even to these "yield-mutants". Moreover, a great deal of those progressive mutants reported in the past were just derived as a by-product of theoretical mutation experiments. This also holds true for our mutant of Haisa II (Table 3), which was selected from a theoretical experiment described elsewhere (26). Haisa II was marketed in 1950 and used to be one of the most extensively grown varieties in Western Germany. However, in recent years, it has continuously lost acreage. The mutants investigated by Scholz (93 and personal communication) belong to "fairly old" varieties, and most of those recorded in Table 3 have now reached the same level of yield as the present top varieties in that area of Germany.

Scholz (96) also gave an example showing that the yield of mutants can be raised in a second radiation cycle. A barley mutant with naked kernels (2) proved to possess a yielding capacity of 90 to 91 per cent as compared with the mother variety Haisa, which has been established in yield trials over a period of 7 years. Because the glumes contribute about 10 per cent of the total kernel weight, the mutant reached the level of the mother variety in terms of "net yield". Irradiation of this mutant resulted in a new mutant which yielded 98 per cent of Haisa on an average of 4 years of testing and which, in addition, is 7 to 8 days earlier. Thus, the net yield of the new mutant has been raised beyond that of Haisa.

A few years ago the first varieties derived by propagation of a mutant were marketed. In Middle Germany the winter barley Jutta was released, which is derived from an induced mutant in the variety Kleinwanzlebener Mittelfrühe. In this mutant winterhardiness, strawstiffness, and yield are improved. At present this mutant covers about 10 per cent of the total acreage of winter barley (125 and personal communication). In Sweden a barley *erectoides* mutant, named Pallas, and a pea mutant, called Weibulls original Strålärt, are grown (8). Earlier a new variety of oil rape and another of white mustard were marketed, which were selected from X-rayed material. However, both these species are cross-pollinating and there is no clear evidence that the new varieties originated by mutations.

Use of Mutants in Cross-breeding

Information about the use of mutants in cross-breeding is still more limited than that of mutation breeding proper. If characters of two mutants belonging to the same variety are to be combined, a small number of F_2 plants is sufficient. The breeding procedure is very simple, because the two mutants differ (practically) only by two genes. This is in contrast to the great expenditure necessary in combination breeding on the basis of variety crosses. Occasionally, two or even more genes may be changed in one mutant at the same time, but this does not complicate seriously the breeding procedure. Usually, however, crossing of two mutants results in normal bifactoriel segregation (78, 93).

In extensive crosses among different *erectoides* mutants of barley, Hagberg (50) studied the additive effect of the combined loci as expressed by the internode length in the ear. He succeeded in the addition of up to four homozygous mutations in one plant. By this, the average internode length was reduced from 33 mm of the mother line (no mutation) to 24 mm in those containing a single mutation, to 16 mm for those containing two mutations, to 13 mm where three genes were involved, and to 12 mm in the line containing all four mutated genes. This result indicates that the addition of each further gene had a smaller effect than the previous one. Hoffmann (discussion of 50) reported that in another polyfactorial character, namely, earliness, the combination of two mutations sometimes results in a multiplication effect. Thus, crossing a 3-day earlier with a 6-day earlier mutant resulted in a double recessive being about 18 days earlier. Crosses between various intermedium mutants of barley (Hoffmann, *loc. cit.*) led sometimes to double recessives which were dwarfs. These examples indicate that certainly not all mutations can be combined without reducing the vitality. On the other hand, crossing of small mutations *inter se* has led in peanuts to heterosis (41, 42) and the use of micro-mutations should also be explored in combination breeding.

Also crosses of mutants with foreign varieties offer the chance to obtain rapid and simple results. Their advantage is particularly to be expected when a "rare" character, which is known only in primitive or non-adapted forms, is to be transferred. If that character is available from a mutant collection of an adapted variety, it may be much easier to use the mutant instead of the non-adapted variety for a cross-breeding program. Characters from mutant collections which may be useful are for instance: resistance against fungi, variability in relation to different soil and climatic conditions, strawstiffness, high protein content of seeds (95, 95), earliness, and numerous other properties. In barley, for example, the smooth-awned and naked-kernel mutants of Bandlow (2) are certainly a valuable source for an easy production of top varieties (96).

An example along this line has already been given by Down and Anderson (17) in bush beans. These authors crossed an earliness mutant, induced by X-rays from the variety Michelite, with other strains resistant to *Colletotrichum lindemuthianum*. The new variety Sanilac, derived from these crosses, combines earliness with resistance and is higher yielding than Michelite.

A similar case which theoretically deserves still more interest is reported by Stubbe (118) in Antirrhinum. In breeding this ornamental plant, an erect growth has been desired for a long time. The mutant *eramosa* approaches that aim and forms usually one culm only. It has almost completely lost the ability of branching and, in addition, it possesses a number of other pathological characters, like inhibition and deformation of the flowers. By crossing *eramosa* with other varieties, Vogel (according to Stubbe, 118) succeeded in breeding a new nice-looking and vigorous snapdragon which forms only one culm and has normal flowers. Evidently, the undesired pleiotropic by-effects of the mutation have been "dissolved" in a new genetic background.

This model demonstrates how macro-mutations may be of use in cross-breeding agricultural plants. Little has been done in this field. There are some reports from practical breeders; and though the evidence is not conclusive, two examples may be mentioned. In barley breeding, the *erectoides* mutants deserve great interest because of their strawstiffness. However, the dense spike of the mutants is often considered a disadvantage because generally the seed quality is lowered. Thus, the German breeder v. Rosenstiel (84 and personal communication) crossed one of the Swedish *erectoides* mutants *(ert* 12) with other strains of his material carrying the mildew-resistance of *Hordeum spontaneum*, H 204 (86). He selected lines with extremely stiff straw and, in some of them, the dense spike had disappeared. Because these high-yielding lines with a "normal" spike carry also the mildew-resistance of *H. spontaneum*, the procedure of v. Rosenstiel is an example of how a character from a wild species and a mutant may be combined. Two of these lines are now in the stage of official yield trials of the Federal

Variety-Board in Germany. From the experience he has in his cross-material, v. Rosenstiel is fairly sure that the unusual strawstiffness obtained was derived from the *erectoides* parent.

Another example that may be given is the mildew-resistant mutant of barley found by Freisleben and Lein (22). This mutant is resistant against all race groups isolated in Germany. However, the older plants have strongly chlorotic spots and, in turn, a reduced yield. Using this mutant as cross parent, Vettel (personal communication) was able to "separate" the resistance from the leaf spots. He selected high-yielding lines carrying the mutant resistance.

There are further possibilities for a combination of cross-breeding with mutations. Thus, induction of mutations in F_2 seeds of a hybrid population may be useful because it means an additional increase of variability. Up to date there is little experimental evidence of such an approach. Likewise, the possibilities of using mutants for heterosis breeding are almost completely unexplored. In the latter case there is, however, the interesting instance of the commercial use of a tobacco mutant (121, 122) which has been overlooked in the older literature concerning mutations and plant breeding. This (lightgreen) *chlorina* mutant of Vorstenland tobacco was produced in Java. The leaves had an attractive color and quality. Since the homozygous *chlorina* mutant produced too few leaves, its F_1 hybrid, with the ancestral type, was used, and had to be produced again every year. In the second half of the 1930's these hybrid plants were grown extensively in the Netherlands East India, but the mutant was lost during the war as far as is known. This seems to be the first case of an induced mutant that has ever been used in practice (8, 83).

Induction of Mutations in Cross-pollinating Populations

There is no experiment known to me which proves crucially the use of mutations in cross-pollinating species. However, the new varieties of oil rape and white mustard in Sweden already mentioned were selected from X-rayed material. There are also a number of other reports where variants with valuable characters have been selected after application of mutagens, either with or without selfing of the M_1 generation. This is, for example, the case in *Trifolium pratense* (10, 89), in *Phalaris arundinacea* (53), in *Alopecurus pratensis* (128), and in *Melilotus albus* (90). It may be expected that mutagenic treatment of cross-pollinating populations will result in additional and possibly new genetic variability.

A conclusive experiment of recurrent irradiation with successful selection for small mutations was reported in *Drosophila melanogaster* (11, 97). This may serve here as a model for plant populations. The subject of selection was a quantitative character, the sternopleural bristle number of the flies. In this experiment, two strains were treated with X-rays at every second generation. Two further strains, serving as controls, were nonirradiated but subjected to the same selection procedure as the irradiated lines. As illustrated in Figure 2, the two irradiated strains (Ap and Bp) responded strikingly to the selection. The hair number could be

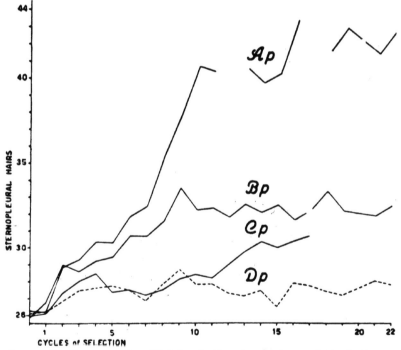

FIGURE 2.—*Progress of artificial selection for high number of sternopleural hairs of* Drosophila melanogaster *in two X-ray-treated lines (Ap and Bp) and in two nontreated lines (Cp and Dp). From Buzzati-Traverso and Scossiroli (11).*

raised from 26 to 41 after 10 selection cycles. In contrast, the control strains (Cp and Dp) did not undergo any appreciable change.

Mutagenic Agents as a Tool for Special Purposes

Physical and chemical mutagens may be exploited in plant breeding in various other ways than those discussed above. A number of other possibilities are outlined in Table 2 and will be briefly discussed here. The use of translocations in transferring a valuable character from one species or genus to another has recently been reviewed (28), including the typical and fully analysed example of Sears (98). In the same review (28), the problem of diploidization of artificially produced tetraploids is outlined.

Use of duplications to produce "directed mutations" has been suggested by Hagberg (50, see there also older references). For this project, it is necessary to have a very large set of translocations in the genotype in which the breeding work shall be done and to have these translocations localized. If a greater gene content of the chromosomes is also mapped, then it should be fairly easy to produce definitely localized duplications (and deletions) of short segments carrying desired genes through crosses of adequate translocation types.

Irradiation may also be useful in apomictic species to produce transitory sexuality for one or a few generations, as has been shown by Julen (62) for *Poa pratensis*. This enables the breeder to make crosses and to select in later generations new apomictic forms with improved characters.

That irradiation can be exploited for the production of haploids has been known for a long time and is outside the scope of the present paper. Attempts have also been made to overcome the interspecific incompatibility by exposing male or female gametes or tissues to gamma radiation prior to crossing. In one instance, that of the cross *Brassica oleracea* × *B. nigra*, the technique appeared to be successful and many interspecific hybrids were produced (16).

Methods of Induction and Selection of Mutants

The significance of mutations in plant breeding will largely depend on the progress of their production methods. It is important to obtain a greater yield of total mutations and particularly of progressive mutants. The final output of mutants for plant breeding depends on both the methods of original induction and of selection. According to the arrangement of Table 4, I have recently

TABLE 4.—POSSIBILITIES OF OBTAINING A HIGHER YIELD OF MUTANTS FOR USE IN PLANT BREEDING.

A. Control of induction
 1. Through raising of mutation rate per surviving M_1 plants
 2. Through alteration of the proportion of chromosome mutations vs. factor mutations
 3. Through alteration of the mutation spectrum
B. Control of selection
 1. Through knowledge of chimera formation and diplontic selection in M_1 plants
 2. Through improved screening methods of mutants

reviewed the progress in obtaining more efficient production methods of mutants (28). The arrangement of Table 4 will also be the basis of the following considerations.

Induction of Mutations

There is no need to review the subject of a control of mutation induction again in detail, though in the meantime some remarkable results have been obtained (13, 46, 67, 75, 76, 77). Such a limited control appears to be possible through the different actions of the great number of known physical and chemical mutagens that can be used either alone or in combination with various secondary factors. Also the stage of the plant development and the parts of the plant being treated are important.

Mutation research is in a stage of rapid development, and the theoretical progress is fascinating in the fields outlined under A1 to A3 of Table 4. Yet, its practical application relative to "recipes" for useful techniques in plant breeding is still limited. With regard to the theoretical progress, I am, however, inclined to suppose that methods to obtain a greater total mutation frequency (Table 4, A1) will be available for everybody in the near future. I also would like to speculate that techniques for a certain control of the relative frequency of chromosome mutations vs. factor mutations (Table 4, A2) will be developed in, say, the next 5 to 15 years. Alteration of the mutation spectrum (Table 4, A3) is most apart from being used in practical plant breeding. Up to date, conclusive evidence in this field has only been obtained with chlorophyll mutations (cf. 28).

Exciting progress has been made with chemical mutagens. At present, two of the most powerful substances for higher plants are ethylene imine (20) and especially ethyl methane sulfonate or EMS (54, 55, 56). The high mutagenic efficiency of EMS is demonstrated in Table 5 for chlorophyll mutations in barley. After X-irradiation of barley seeds the highest mutation rates obtained are of

TABLE 5.—SOME RESULTS OF THE MUTAGENIC EFFICIENCY OF EMS ($CH_3SO_2OC_2H_5$) IN BARLEY SEEDS OBTAINED BY HESLOT.[*]

Concentration	Treatment duration	Temperature	Number of spikes	Number of M_2 plants	Mutants, %
1/250	24 hours	24° C	210	1,760	14.0
1/400	24 hours	24° C	500	7,160	6.2
1/100	3 days	3° C	260	3,690	12.3

[*]Personal communication. LD–50 is approximately for 24 hours, 24° C:1/300, and for 3 days, 3° C:1/100.

the order of 3 to 4 per cent mutants per 100 M_2 plants (12, 33), while in Table 5 with EMS 14 per cent are recorded. According to Heslot (Table 5), around 12 per cent were obtained with an EMS treatment that led to approximately 50 per cent M_1 survivals. The values mentioned for X-ray treatment were only reached with doses exerting a greater M_1 lethality, except in one experiment (33, treatment 30,000 rp). Consequently, the mutagenic efficiency of EMS appears to be somewhere between 3 to 8 times higher than that of X-rays, if these results are reproducible. Similar high mutation rates with EMS have now also been obtained by Ehrenberg (19). Apparently EMS has a relatively low toxic effect and a high genetic effect as compared with X-rays. Moreover, Heslot (56) found many more morphological and physiological mutations per chlorophyll mutation after treatment with EMS than after X-raying. EMS treatment results in much lower chromosome breakage (Heslot, personal communication, also 67), however, it causes high M_1 sterility. This discrepancy needs to be further investigated. The mutagenic activity of EMS has already been proved in other species (56, 87). Apparently, EMS is the most efficient mutagen for plant breeding known at present and it deserves greatest interest in further theoretical and practical applications.

Most of the applied mutation work has been done with radiation of seeds. Certainly seed treatment offers a great number of advantages (77), but this does not necessarily mean that it is the best method for the induction of mutations in plant breeding. In Table 6 some other stages of plant development are shown which

TABLE 6.—PLANT STAGES FOR MUTAGENIC TREATMENT.

1. Seeds, dormant, presoaked, germinating
2. Pollen
3. Flowers, prior to, during, and after meiosis
4. Zygotes, immediately after fertilization
5. Buds in various stages
6. Any other stage

can be used. Particularly, flowers which have passed meiosis and also zygotes deserve attention. In the last case, the whole organism is represented by one cell only; consequently, no chimera formation will take place and high mutation rates might be expected because there is no intercellular competition. After treatment of seeds or buds diplontic selection results in a great loss of mutations (34). With gamma-radiation sources, like cobalt 60 and caesium 137, acute irradiation of various plant-stages can easily be done, and there are also useful methods for the application of chemical mutagens (88).

The experience with chronic irradiation has shown that this treatment apparently offers no advantage as compared with acute irradiation. The "maximal" mutation rate seems to be rather lower with chronic irradiation which may be easily explained by the phenomenon of diplontic selection (27, 28, 30, 33). However, these results are based mainly on chlorophyll mutations and it is unknown if the same holds true for vital mutations. Theoretically, diplontic selection could *eventually* be utilized as a screening procedure for progressive mutations, but experimental evidence is lacking for this speculation.

In relation to radiation of seeds, we have recently presented a number of detailed suggestions on how a breeder may best proceed (34). These suggestions were based mainly on our experience with barley, and they include the description of an early test to find the most efficient dose on the base of the seedling length (31). Furthermore, suggestions are given as to the most suitable sowing technique, the procedure in recurrent irradiation programs, and various other topics. Previously it was suggested to use a dose which leads to about 50 per cent surviving M_1 plants (23, 43). In contrast to these earlier findings we have recommended applying higher doses which result in 10 to 20 per cent (or even less) survival. At 90 per cent lethality of M_1 plants the mutation rate may be nearly doubled as compared with 50 per cent lethality. The discrepancy with the older results is mainly a consequence of the fact that the mutation frequency was not measured correctly (27, 32).

There are large differences in the radiosensitivity between various species, necessitating quite different doses. In an extensive study, Sparrow, *et al.* (103) considered nuclear volume (of meristematic cells), polyploidy, and chromosome number as the major factors determining radiosensitivity. A table of doses leading to approximately 50 per cent survival in various agricultural plants can be found in Gustafsson and v. Wettstein (48).

Diplontic Selection

Mutagenic treatment of seeds or other parts of the plant results in the formation of chimeras. The efficiency of mutant selection depends, therefore, on the problem as to which parts of the plant the progenies should be grown and whether the elimination of mutations within the plant (diplontic selection) can be inhibited. Unfortunately, little is known in this field. After treatment of barley seeds, we were able to show that about the first five tillers per plant possess a considerably higher mutation frequency than tillers formed later (27, 30, 33). It has therefore been recommended to space the M_1 seeds extremely close in order to obtain high mutation rates (34). We also have demonstrated that the size of chimeras varies with the dose. With very low X-ray dose the average size of a mutated sector may comprise about one quarter of the generative tissue of a single spike. With increasing dose this size increases correspondingly and finally all the florets of a spike contain the same mutation. With greater tillering, the mutated sector may then even include several spikes of a plant (30, 33). The decrease of mutation frequencies in later formed tillers has recently also been demonstrated in rice by Bekendam (4, 5). In addition, the same author was able to show that in rice, as in barley, after irradiation of seeds the generative tissue of a single head may derive from one to four embryo cells. The chimeric structure has also been investigated in several other plants, e.g. in maize (1), sorghum (64), pea (6), and blue lupin (36).

In practical breeding work, selection for mutations may perhaps alreardy be started with seedlings of the M_1 generation. We have shown that raising X-rayed barley seeds in the greenhouse before transplanting them to the field may result in 4 to 5 times better survival than when seeds are sown directly in the field (33). The mutation frequency of the "greenhouse seedlings" was not essentially lower than that of the "field seedlings". This finding complements interesting results of Caldecott (13), who demonstrated that heavily damaged seedlings (shortest height class of the length of the first leaf) have a 2 to 4 times higher mutation rate than those which belong to the tallest height class. Thus, with the technique of raising seedlings in the greenhouse, it might be expected (a) that higher doses can be applied, resulting in higher mutation rates, and (b) that the selection of seedlings with short leaves (but which are still viable) will result in an additional increase in mutation rates.

Similarly, Blixt, et al. (7) found recently in peas, after treatment with ethylen imine, a correlation between the M_2 mutation rate and the frequency of leaf-spots on the first leaves of M_1 plants. In order to obtain a high rate of M_2 mutations, they suggest, therefore, taking offsprings from those plants only which show the highest frequency of leaf-spots at the M_1 seedling stage.

Selection of Mutants—General Results

In practical breeding work it will often be advisable to commence with the selection of mutants only after mutagenic treatments have been applied for several years or generations to the same material (23, 34). Recurrent induction of mutations leads to an accumulation of mutants in the treated population and increases the efficiency in breeding with mutations.

Selection for fertility can already start in the M_1 generation or in case of recurrent induction of mutations, it may be repeated in each succeeding generation. After treatment of barley seeds, we have shown repeatedly that the mutation frequency of fertile M_1 spikes is no smaller than that of the M_1 spikes with partial sterility *if* the tillering of the M_1 plants is reduced (25, 28, 29, 34). Recent results in rice (4, 5) agree basically with barley. Selection of fertile (or nearly fertile)

M_1 spikes results therefore in elimination of chromosome mutations without reducing the frequency of factor mutations. We have also shown in barley that the M_2 sterility in progenies of partially sterile M_1 spikes is actually several times greater than in the progenies of fertile M_1 spikes. However, in one experiment, even a twofold selection for fertility, namely, in the M_1 and M_2 generation, still resulted in the M_3 generation in 3 per cent M_2 spike progenies with partial sterility (34). Usually the breeder is interested in point mutations only and not in chromosomal aberrations. This is valid at least for diploid plants. Selection of fertile M_1 spikes means, therefore, an increase of efficiency in breeding with mutations.

Selection of mutants has commonly been done in the M_2 generation. In that generation only macro-mutations can be recognized. Moreover, owing to the chimeric structure of the M_1 plants and their sterility, a great part of mutations are only represented in the heterozygous condition in the M_2 generation. These become manifest for the first time in the M_3 generation. Thus, in peas, Gottschalk (37) recognized only 60 per cent of the mutants in the M_2 generation and 40 per cent in the M_3. Similar results have been obtained also in barley (73). If selection of mutants is started in the M_2 generation, it should be continued in the M_3 and following generations.

The procedure of selection depends on the intention of the breeding program. The selection may be directed towards a special goal, like yield, fungi resistance, baking quality in wheat (58), high protein content in barley (95, 96), germination under low temperature in soybeans (129), strawstiffness, earliness, etc., without any interest for other mutants. Or selection may be nondirected and every mutant which might be of value for plant breeding and which becomes recognized is picked up for further investigations. Of course, directed and nondirected selection may be combined.

Screening for small mutations can start for the first time in progenies of normal-appearing M_2 plants, and it should be commenced on a large scale. Selection has to be continued in the following generations, and the material will become continuously smaller by the elimination of nonmutant and undesired lines. As compared with the pedigree method, for example, the breeder needs a different approach. Breeders using the pedigree method are used to observe large differences among the various progenies and desired types are selected, at least in the beginning, by eye inspection. With small mutations, the basic phenotype is often not changed; and under these conditions it is hard, if not impossible, to detect, for instance, an increase in the yield potential of 5 to 10 per cent by simple eye inspection. There is a need to develop mass selection methods, and these will be different for different purposes and plants. This demand is shared with conventional breeding methods, and indeed many of the known selection methods can be applied to mutants in a more or less modified way. There are, however, some basic differences between conventional breeding and breeding with mutations, and these will often require a different approach and special selection techniques. It is not possible to go into more detail here on this problem.

For the selection of useful mutants, particularly of higher yielding mutants, it might be possible to utilize the pleiotropic gene action which seems to be connected with every (or nearly every) mutation. I have suggested that one looks for "indicator characters" (25, 28). These should be mutations which are relatively easy to detect, either in a "normal" environment or under extremely changed (laboratory) conditions. They are characterized by morphological or physiological deviations which do not necessarily have breeding value. As a consequence of the pleiotropy, a certain proportion of these mutants may possess progressive features, i.e., they may have a greater yielding potential.

Selection of Mutants—Special Results

We have started preliminary studies along this line in our laboratory. Among various such indicator mutations we have considered, the earliness character appears to be suggestive. Because earliness is a polyfactorial character, the probability of a mutation-event is relatively great; and earliness mutations have been frequently met with in all species. Moreover, it is possible to recognize even small differences with regard to the beginning of flowering. In cereals, the date of heading is a character that can be recorded fairly precisely and small differences become evident. In order to recognize small genetic differences, it is however necessary (a) to have a great number of plants, (b) to work with replications, (c) to observe the material at least once a day during the heading time, and (d) to observe the material for several years.

TABLE 7.—HEADING DATA OF 56 EARLINESS MUTANTS OF SPRING BARLEY AS COMPARED
WITH THE MOTHER LINE HAISA II AND TENTATIVE RESULTS ABOUT
THE YIELD POTENTIAL. *

Serial No (fr. No)	Days earlier than Haisa II		Number of obser- vations	% kernel yield of Haisa II		Number of years tested
	Mean value	Range		Mean value	Range	
1	10.1	8–13	8	–	–	–
13	9.3	5–13	3			
5	9.0	8–10	2	–	–	–
2	8.4	4–13	6	83	78–87	2
3	7.0	5–8	3	–	–	–
14	7.0	5–9	2			
4	6.0	3.5–8	8	–	–	–
6	4.5	2–6	10	92	81–102	3
10	4.3	3–6	6	108	99–117	2
7	4.2	2–5	5	82	–	1
18	4.0	2–6	4	72	71–72	2
21	4.0	–	3	84	79–88	2
19	4.0	–	2	–	–	–
12	3.6	2–5.5	9	105	100–111	5
9	3.6	2–5	7	103	85–123	3
22	3.5	3–4	2	–	–	–
25	3.5	3–4	2	132	–	1
8	3.4	2–5	8	90	79–102	3
11	3.3	2–5	3	62	53–70	2
20	3.0	2–4	4	89	81–97	2
32	3.0	2–4	4	99	94–104	2
15	2.5	1–4	9	101	95–107	5
52	2.5	2–3	2	91	87–95	2
16	2.2	1–4	5	73	69–78	2
29	2.1	0–3.5	5	81	–	1
38	2.0	1–3	3	103	98–107	2
28	2.0	1–3	2	–	–	–
17	1.8	0–4	5	100	88–112	2
33	1.8	1–3	4	113	112–114	2
30	1.7	0–3	3	–	–	–
23	1.6	0.5–3	9	100	93–107	5
47	1.6	1–4	6	86	82–100	3
48	1.5	1–2	2	87	85–89	2
35	1.5	1–2	2	119	–	1
27	1.5	0–3	4	72		1
41	1.5	1–2	2	–		–
34	1.3	0–2	4	93	–	1
24	1.3	0–3	4	86	81–90	2
37	1.1	0–2	8	96	93–100	4
59	1.0	0.5–2.5	3	98	91–105	3

TABLE 7.—CONTINUED.

Serial No (fr. No)	Days earlier than Haisa II		Number of obser- vations	% kernel yield of Haisa II		Number of years tested
	Mean value	Range		Mean value	Range	
40	1.0	0–2	4	80	77–84	2
39	1.0	0–2	3	92	86–97	2
42	0.9	0–2	4	98	90–105	2
50	0.8	0.5–1	3	91	88–95	2
26	0.8	0–3	4	94	92–96	2
55	0.8	0.5–1	2	100	–	1
58	0.7	0–2	7	94	85–110	3
57	0.6	0–1	4	107	105–110	2
62	0.5	0–1	3	93	89–95	3
46	0.5	0–1	4	103	94–103	2
60	0.5	0–1.5	3	100	96–104	3
53	0.5	0–1	2	82	75–88	2
61	0.3	0–1	3	99	95–103	3
63	0.3	0–0.5	3	97	96–99	3
51	0.3	0–1	3	87	85–90	2
64	0.2	0–0.5	2	95	90–98	3

*Where no data are recorded the yield is lower than 80 per cent of Haisa II. All mutants originated by X-rays and were selected in M_2 and M_3 between 1954 and 1958 from various experiments.

In Table 7, heading data of 56 earliness mutants are recorded which are all derived from the barley variety Haisa II. These data are based on observations of at least 2 years, but in most cases the results from 3 to 7 years are recorded. Occasionally, the mutants were grown in the same year in two different trials and therefore the number of observations recorded in Table 7 is sometimes greater than the number of years tested. The most extreme mutant we have is 10.1 days earlier than the mother line on the average of eight observations in 7 years, and there are all transitions to mutants only 0.2 day earlier. In each mutant the degree of earliness varies somewhat from year to year, indicating the influence of environmental factors on the expressivity of that character. It seems that the variability is usually greater with more extreme earliness.

The mutant character of 28 of these earliness mutants could be firmly recognized in the M_2 generation. They may be considered, therefore, as large mutations. The other 28 were either questionable in M_2 or were found for the first time in the M_3 generation. Most of these last-mentioned 28 may be considered as small mutations, and generally they are not earlier than 2 days (on the average of several observations). The results reported in Table 7 are not representative with regard to the relative frequencies of small and large earliness mutations because they are derived from various experiments and the selection methods have been inconsistent. There is no doubt that the frequency of small earliness mutants is considerably higher than indicated. These mutants have not been considered enough in the earlier experiments and perhaps not in some later ones too.

The distribution of these earliness mutants on yield classes is indicated in Table 8. The various yield data are derived from different experimental arrangements. Most results are based on drilled

TABLE 8.—DISTRIBUTION OF THE EARLINESS MUTANTS OF BARLEY ON YIELD CLASSES. *

Yield class	< 80	− 90	− 100	− 110	<
Number of mutants	15	11	18	9	3
Mean value of days of earlier heading than Haisa II	4.8	2.5	1.4	2.2	2.3

*Mutants from Table 7. Kernel yield of Haisa II is taken as 100. Tentative results.

trials, some on plots sown by hand, and some on hills. The lack of yield data in Table 7 indicates poor-looking mutants which have not been tested for yield. These mutants yield definitely less than 80 per cent of the control. Particularly, the results of those mutants having more than 95 per cent yield of Haisa II were based mainly on drilled trials, which had often a plot size of about 10 m² and three replications. The yield of two of the three mutants recorded in the class > 110 per cent of Table 8 is based on one year only (1960), and the plot size was smaller.

The results shown in Table 8 seem to indicate that 12 out of the 56 earliness mutants possess a greater yielding potential than the mother line. These tentative investigations will be continued to obtain more reliable results. However, even if the proportion of high-yielding earliness mutants goes down to $1/10$, earliness seems to be an indicator character of practical interest. Yet it has to be considered that these results are valid only for one variety (Haisa II) and for one location (Köln-Vogelsang). It may be expected that with other varieties and growing conditions, the proportion of high-yielding earliness mutants will be lower or higher.

From Tables 7 and 8 it is further evident that, on the average, more extreme earliness mutants have lower yields. The higher yielding mutants are all $\frac{1}{2}$ to 4 days earlier than Haisa II.

Another possible indicator character under investigation in our laboratory is seed size. As with earliness, kernel size is certainly controlled by polymeric genes and the great heritability of 1,000-kernel weight is well known. In one of these investigations winter barley, variety Breustedts Atlas, was used. The procedure of selection is outlined in Table 9. We started with nearly 1,500 normal-appearing M_2

TABLE 9.—PROCEDURE OF RECURRENT SELECTION FOR 1,000-KERNEL WEIGHT IN X-RAYED PROGENIES OF WINTER BARLEY, VARIETY BREUSTEDTS ATLAS.

Year	Generation	Number of plants or progenies investigated	Grown as	1,000-kernel weight determined with
1957	M_2	1,494	Single plants (drilled bulk)	200 and 300 kernels
1958	M_3	135	Hills	2 × 300 kernels (from 2 field replications)
1959	M_4	31	Micro-drill test	2 × 1,000 kernels (from 2 field replications)
1960	M_5	31	Micro-drill test	2 × 1,000 kernels (from 2 field replications)

plants which were taken at random from an M_2 bulk. From these, 135 with the highest 1,000-kernel weight were grown further. By a twofold selection we ended with 31 M_2 progeny lines which were investigated more carefully in micro-drill tests for 2 years.

The results of the first drill trial in 1959 are shown graphically in Figure 3. From that histogram it is evident that only 4 out of 31 lines had a lower 1,000-kernel weight than the mother variety. Most of the differences are only in the range of 6 per cent as compared with the mother variety, but they go up to around 16 per cent. Thus, it appears that the relatively simple selection procedure has been surprisingly effective. According to an analysis of variance, differences between the lines are highly significant (F value lines/error:17.80, F value of the table at 0.1 per cent level:2.97). Also in 1960, differences between the lines were highly significant which is not demonstrated here (simple lattice, F value lines/intra-block error:6.73, F value of the table at 0.1 per cent level:3.58).

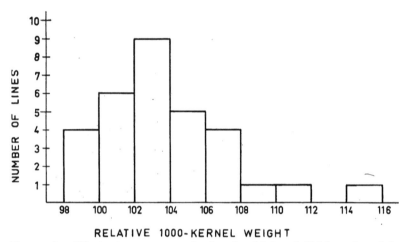

RELATIVE 1000-KERNEL WEIGHT

FIGURE 3.—*Effect of recurrent selection for higher 1,000-kernel weight in progenies of X-rayed winter barley (variety Breustedts Atlas), according to the procedure of Table 9, results of 1959. Control is taken as 100. Only 4 lines out of 31 have a lower 1,000-kernel weight than the control.*

In Figure 4 a scatter diagram of the results in 1958 and 1959 is represented. There is a highly significant correlation between both years ($r = 0.745$, $P > 0.1$ per cent). A similar correlation was obtained between 1959 and 1960, as is shown in Figure 5 ($r = 0.797$, $P > 0.1$ per cent) and also between 1958 and 1960 ($r = 0.790$, $P > 0.1$ per cent). Because of severe lodging in 1960, only marginal plants could be sampled from the drilled plots. Thus, the sampling has been done in 1958 from hills, in 1959 from drilled plots, and in 1960 from marginal plants of drilled plots. Because of the high correlation between the different years and growing conditions, there can be no doubt that the differences between most of the lines are genetically controlled.

At present we are not completely sure, however, whether all these variants of kernel size are true mutants though this is very probable for the majority at least. The seeds, which were irradiated in 1955 were "Zuchtgarten-Elitegemisch" and not a pure line obtained by propagation of a single plant for only a few generations. According to the breeding history, the propagation methods, and his experience, the breeder of this barley believes it is not possible that the variants selected in the irradiated material were already present in the starting material (Breustedt, personal communication).

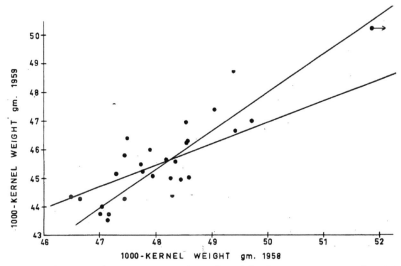

FIGURE 4.—*Correlation of 1,000-kernel weight between 1958 and 1959 in 31 selected lines from X-rayed winter barley with regression lines. One barley line, marked by a dot with arrow, lies outside of the diagram (1958, 55.2; 1959, 50.2). r = 0.745.*

We also investigated this question in more detail by statistical means. The irradiated seeds were in the F_{21} generation and practically all genes were, therefore, in a homozygous condition. The irradiated material was a mixture of 15 "nursery-lines", which were phenotypically very similar, if not alike. In 1960, progenies of eight of these (untreated) lines were available for us to determine the 1,000-kernel weight. In the meantime, 23 lines had been developed from the eight progenies. The variance of the 1,000-kernel weight of the 23 lines was $s^2 = 0.571$ and the corresponding variance of our 31 selection lines was $s^2 = 4.92$ (determined as mean square deviation in relation to the mean value of our 31 selection lines). According to that, the variance of the selection lines is 8 to 9 times greater. The difference is highly significant (P > 0.1 per cent).

A similar selection program as with winter barley was conducted with spring barley, variety Haisa II (Hochzucht). The results of the first drill test in 1959 are demonstrated in Figure 6. It is obvious that a great number of small variants of kernel size have been

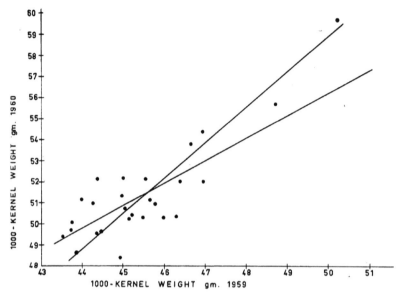

FIGURE 5.—*Correlation of 1,000-kernel weight between 1959 and 1960 in 31 selected lines from X-rayed winter barley with regression lines. r = 0.797.*

screened. Again, differences between the lines are highly significant (simple lattice, F value lines/intrablock error:16.90; F value of the table at 0.1 per cent level:2.79). Also, the correlation between 1959 and 1960 is highly significant (r = 0.652, P > 0.1 per cent). The breeders of Haisa II consider this variety to be a pure line within reasonable limits (Vettel and Lein, personal communication).

The selection experiments described are of a preliminary character. They were conducted to determine the feasibility and value of initiating more exact and intensive experiments, and to gather technical experience. The results have been broached here because they indicate at least that screening for increased kernel size is very simple, if there is a corresponding genetic variability in the starting population. Experiments are underway in our laboratory to obtain conclusive information on the nature of these variants, as well as on their general significance in breeding.

Besides the experiments described, we started, several years ago, a program of recurrent selection for kernel yield in irradiated proge-

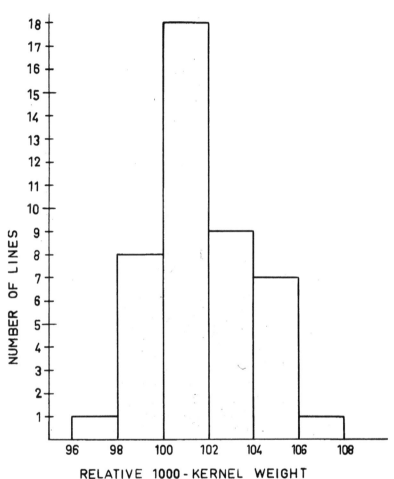

FIGURE 6.—*Effect of selection for higher 1,000-kernel weight in progenies of X-rayed spring barley. Control is taken as 100. Results of 1959. Only 9 lines out of 44 have a lower 1,000-kernel weight than the control.*

nies of spring barley. This program includes already an untreated control. It seems that it is relatively easy to screen directly for small mutations affecting yield and the results obtained so far will be published elsewhere.

Conclusions

It has been shown that the induction of mutations offers a new tool which is potentially able to make progress in plant breeding similar to that obtained with conventional methods. Mutants can be used directly to establish a new variety ("mutation breeding"). They can also be used in cross-breeding and for special aims. Generally, the use of mutants is time-saving as compared with the traditional methods.

However, the rate of progressive mutations and, hence, the efficiency of mutation breeding has been low until now. At present, I doubt, therefore, if generally this method is as useful and more economical than cross-breeding when a given aim is to be reached.

At present, the uses of mutants in cross-breeding seems to be more promising than mutation breeding. In cross-breeding, mutants may complement the natural gene resources. Particular interest attaches to the possibility of removing the undesired pleiotropic effects of a mutation in a new genetic background.

To summarize, the significance and importance of mutations in practical breeding now lies in their complementary use to the well-established breeding methods. Increasing the efficiency of the mutation technique can lead to a greater use of mutations in conjunction with the conventional methods.

It should be emphasized that any evaluation of mutation breeding is speculative at the present time. Crucial evidence on comparative efficiency of traditional cross-breeding versus breeding with mutations is almost completely lacking. Strict utilization of mutations in practical breeding programs has scarcely been attempted until very recently.

In appraising the efficiency of and in defense of breeding with mutations one could challenge the followers of the traditional methods with the view that the efficacy of cross-breeding is not extremely great either. Thus, Vettel (125), one of the most successful practical breeders in Germany, recently reviewed 30 years of his experience in cross-breeding with cereals. He gave a statistic of all crosses he has made in wheat, barley, and oats, analyzing 5,045 different cross-progenies. Only 0.3 per cent from these 5,045 F_2 populations resulted in commercially used varieties. In an evaluation of this figure, one has to realize that very much work of further selection and breeding has to be done in the generations succeeding the F_2. The structure of total work to be done by a breeder may be compared with a pyramid having the top in the F_1 generation and becoming broader with succeeding generations. Another example of efficiency in cross-breeding was recently given by Williams (127) in tomatoes. The frequency of successful hybrid combinations appears to be around 2 per cent in that crop. Two such F_1 hybrids which are used commercially in England were analysed in order to fix heterosis and Williams isolated 1 desirable recombinant in every 1,000 to 1,500 F_2 individuals.

Gregory (42), on the other hand, in his extensive work with small mutations in X-rayed peanuts, indicates that the frequency of mutants which are superior in yield may be of the order of 1 among 500 to 5,000 M_2 population plants. Gustafsson (45, cf. also 43, 44, 48) stated that a higher productive mutant is formed "once in 500 to 1,000 genotypical changes". It should be emphasized that this estimate is based mainly on the selection of superior large mutations. From our preliminary studies in barley it appears that out of 5 to 10 small or fairly small earliness mutations, 1 may outyield the mother line. Earliness mutations are frequent and are easy to detect; the small ones, however, only in a group of plants.

Consequently, if one critically tries to compare both methods, breeding with mutations and traditional cross-breeding, there is no evidence that the first is inferior. Much more experimental evidence and practical experience are necessary for a sound evaluation of the significance of breeding with mutations.

The significance of mutations in the future depends largely on the question of whether a higher total mutation rate can be obtained, and particularly, whether the output of useful mutants can be increased. The final output of mutants depends on the technique of original induction as well as of selection. In both these fields theoretical progress has been made, and there is hope that the efficiency of mutation production will be considerably increased.

It would be useful, if, in various parts of the world, mutant collections of the main crops could be established which are based on adapted varieties with high performance. Collections of large mutations are easy to establish even with present methods. The expenditure is not too great as compared, e.g., with collection trips to centers of genetic diversity. These large mutations imply a valuable source in cross-breeding programs. Selection of small mutations involves more work than that of large mutations. Yet, mutation breeding proper might offer the greatest advantage as compared with conventional cross-breeding. Further theoretical and practical work will serve a final evaluation of mutation breeding.

References

1. Anderson, E. G., Longley, A. E., Li, C. H., and Retherford, K. L. 1949. Hereditary effects produced in maize by radiation from the Bikini atomic bomb: I. Studies on seedlings and pollen of the exposed generation. *Genetics*, 34: *639–646.*

2. Bandlow, G. 1951. Mutationsversuche an Kulturpflanzen: II. Züchterisch wertvolle Mutanten bei Sommer- und Wintergersten. *Züchter,* 21: *357–363.*

3. Baur, E. 1924. Untersuchungen über das Wesen, die Entstehung und die Vererbung von Rassenunterschieden bei *Antirrhinum majus. Bibliotheca Genetica,* 4: *1–170.*

4. Bekendam, J. 1961. Inductie van mutaties bij rijst door röntgenbestraling (X-ray-induced mutations in rice). *Meded. Landbouwhogeschool, Wageningen,* 61 *(1): 1–68.*

5. ————. 1961. X-ray induced mutations in rice. *Symposium on the Effects of Ionizing Radiation on Seeds and their Significance for Crop Improvement, Karlsruhe,* 1960. In press.

6. Blixt, St., Ehrenberg, L., and Gelin, O. 1958. Quantitative studies of induced mutations in peas: I. Methodological investigations. *Agr. Hort. Genetica,* 16: *238–250.*

7. ————, ————, ————. 1960. Quantitative studies of induced mutations in peas: III. Mutagenic effect of ethyleneimine. *Agr. Hort. Genetica,* 18: *109–123.*

8. Borg, G., Fröier, K., and Gustafsson, Å. 1958. Pallas barley, a variety produced by ionizing radiation: Its significance for plant breeding and evolution. *2nd. Intern. Conf. on the Peaceful Uses of Atomic Energy, Geneva. New York: United Nations.*

9. Brock, R. D. 1957. Mutation plant breeding. *Jour. Aust. Inst. Agr. Sci.,* 23: *39–50.*

10. Bruns, A. 1954. Die Auslösung von Mutationen durch Röntgenbestrahlung ruhender Samen von *Trifolium pratense. Angew. Botanik,* 28: *120–155.*

11. Buzzati-Traverso, A. A., and Scossiroli, R. E. 1959. X-ray induced mutations in polygenic systems. *Comitato Nazionale per le Ricerche Nucleari, II. Conferenza di Ginevra, 1–13.*

12. Caldecott, R. S. 1958. The experimental control of the mutation process. *Proc. First Int. Wheat Genetics Symp., Winnipeg, Canada, 68–87.*

13. ————. 1961. Seedling height, oxygen availability, storage and temperature: Their relation to radiation-induced genetic and seedling injury in barley. *Symposium on the Effects of Ionizing Radiation on Seeds and their Significance for Crop Improvement, Karlsruhe, 1960.* In press.

14. Carpenter, J. A., and Gladstones, J. S. 1958. Plant breeding with X-ray-induced mutants in Western Australia. *Australian Atomic Energy Symp., 648–650.*

15. Cooper, W. E., and Gregory, W. C. 1960. Radiation-induced leaf spot resistant mutants in the peanut. *(Arachis hypogaea L.). Agron. Jour., 52: 1–4.*

16. Davies, D. R., and Wall, E. T. 1961. Gamma radiation and interspecific incompatibility in plants. *Symposium on the Effects of Ionizing Radiation on Seeds and their Significance for Crop Improvement, Karlsruhe, 1960.* In press.

17. Down, E. E., and Andersen, A. L. 1956. Agronomic use of an X-ray-induced mutant. *Science, 124: 223–224.*

18. East, E. M. 1935. Genetic reactions in Nicotiana: III. Dominance. *Genetics, 20: 443–451.*

19. Ehrenberg, L. 1960. Chemical mutagenesis: Biochemical and chemical points of view on mechanisms of action. *Abhandlungen d. Deutsch. Akad. d. Wiss. zu Berlin, Kl. f. Medizin, Jahrg. 1960. Chemische Mutagenese. Erwin-Baur-Gedächtnisvorlesungen,* I: *124–136. 1959.*

20. ————, Gustafsson, Å., and Lundquist, U. 1959. The mutagenic effects of ionizing radiations and reactive ethylene derivatives in barley. *Hereditas, 45: 352–368.*

21. Endlich, J. 1959. Genetische und cytologische Analyse einer dominant wirkenden, im Dominanzgrad verschiebbaren Mutation bei *Lycopersicon esculentum* Mill. *Die Kulturpflanze, 7: 131–160.*

22. Freisleben R., and Lein, A. 1942. Über die Auffindung einer mehltauresistenten Mutante nach Röntgenbestrahlung einer anfälligen reinen Linie von Sommergerste. *Naturwiss., 30: 608.*

23. ————, ————. 1943. Möglichkeiten und praktische Durchführung der Mutationszüchtung. *Kühn-Archiv, 60: 211–225.*

24. Fröier, K. 1954. Aspects of the agricultural value of certain barley X-ray mutations produced and tested at the Swedish Seed Asso-

ciation, Svalöf, and its branch stations. *Acta Agr. Scand.*, 4: *515–542.*

25. Gaul, H. 1956. Stand der Mutationsforschung und ihre Bedeutung für die praktische Pflanzenzüchtung. *Arbeiten der DLG*, 44: *54–71.*

26. ————. 1957. Die Wirkung von Röntgenstrahlen in Verbindung mit CO_2, Colchicin, und Hitze auf Gerste. *Zeit. f. Pflanzenzüchtg.*, 38: *397–429.*

27. ————. 1957. Die verschiedenen Bezugssysteme der Mutationshäufigkeit bei Pflanzen, angewendet auf Dosis-Effektkurven. *Zeit. f. Pflanzenzüchtg.*, 38: *63–76.*

28. ————. 1958. Present aspects of induced mutations in plant breeding. *Euphytica*, 7: *275–289.*

29. ————. 1958. Über die gegenseitige Unabhängigkeit der Chromosomen- und Punktmutationen. *Zeit. f. Pflanzenzüchtg.*, 40: *151–188.*

30. ————. 1959. Über die Chimärenbildung in Gerstenpflanzen nach Röntgenbestrahlung von Samen. *Flora*, 147: *207–241.*

31. ————. 1959. Determination of the suitable radiation dose in mutation experiments. *Berichte des II. EUCARPIA-Kongresses, Köln, 65–69.*

32. ————. 1960. Critical analysis of the methods for determining the mutation frequency after seed treatment with mutagens. *Genetica Agraria*, 12: *297–318.*

33. ————. 1961. Studies on diplontic selection after X-irradiation of barley seeds. *Symposium on the Effects of Ionizing Radiation on Seeds and their Significance for Crop Improvement, Karlsruhe, 1960.* In press.

34. ———— and Mittelstenscheid, L. 1960. Hinweise zur Herstellung von Mutationen durch ionisierende Strahlen in der Pflanzenzüchtung. *Zeit. f. Pflanzenzüchtg.*, 43: *404–422.*

35. Gelin, O. 1960. Experimental mutation in Pisum. *Genetica Agraria*, 12. In press.

36. Gladstones, J. S. 1958. Induction of mutation in the West Australian blue lupin *(Lupinus digitatus Forsk.)* by X-irradiation. *Australian Jour. Agr. Res.*, 9: *473–482.*

37. Gottschalk, W. 1961. Genetic problems of mutation breeding in peas *(Pisum sativum). Symposium on the Effects of Ionizing Radiation on Seeds and their Significance for Crop Improvement, Karlsruhe, 1960.* In press.

38. ———— und Scheibe, A. 1960. Untersuchungen an röntgenin-

duzierten Mutanten von *Pisum sativum. Zeit. f. Pflanzenzüchtg.,* **42:** *313–338.*

39. Gregory, W. C. 1955. X-ray breeding of peanuts *(Arachis hypogaea* L.). *Agron. Jour.,* **47:** *396–399.*
40. ————. 1955. The comparative effects of radiation and hybridization in plant breeding. *Proc. First Int. Conf. Peaceful Uses Atomic Energy,* **12:** *48–51.*
41. ————. 1956. Induction of useful mutations in the peanut. *Brookhaven Symposium on Biology and Genetics in Plant Breeding,* **9:** *177–190.*
42. ————. 1957. Progress in establishing the effectiveness of radiation in breeding peanuts. *Proc. 9th Oak Ridge Regional Symposium on Radiation in Plant Breeding, 36–48.*
43. Gustafsson, Å. 1947. Mutations in agricultural plants. *Hereditas,* **33:** *1–100.*
44. ————. 1951. Induction of changes in genes and chromosomes: II. Mutations, environment, and evolution. *Cold Spr. Harb. Symp. Quant. Biol.,* **16:** *263–281.*
45. ————. 1954. Mutations, viability, and population structure. *Acta Agr. Scand.,* **4:** *601–632.*
46. ————. 1960. Chemical mutagenesis in higher plants. *Abhandlungen d. Deutsch. Akad. d. Wiss. zu Berlin, Kl. f. Medizin, Jahrg. 1960. Chemische Mutagenese. Erwin-Baur-Gedächtnisvorlesungen,* **I:** *14–29. 1959.*
47. ————, Nybom, N., and v. Wettstein, U. 1950. Chlorophyll factors and heterosis in barley. *Hereditas,* **36:** *383–392.*
48. ———— and v. Wettstein, D. 1958. Mutationen und Mutationszüchtung. 2. *Auflage Berlin, Paul Parey. Handbuch der Pflanzenzüchtung,* **1:** *612–699.*
49. Hagberg, A. 1953. Heterozygosity in erectoides mutations in barley. *Hereditas,* **39:** *161–178.*
50. ————. 1959. Barley mutations used as a model for the application of cytogenetics and other sciences in plant breeding. *Berichte des II. EUCARPIA-Kongresses, Köln, 235–248.*
51. Hallquist, C. 1953. Change of dominance in different genic environments. *Hereditas,* **39:** *236–240.*
52. Heiken, A. 1960. Spontaneous and X-ray-induced somatic aberrations in *Solanum tuberosum* I.. *Acta Acad. Reg. Sci. Upsaliensis,* **7:** *5–125.*
53. Hertzsch, W. 1957. Mutationsversuch mit Rohrglanzgras. *Zeit. f. Pflanzenzüchtg.,* **37:** *263–279.*
54. Heslot, H. 1960. Induction de mutations chez les plantes cultivées:

Recherches effectuées par les agronomes français. *Genetica Agraria*, **12**. In press.

55. ———, Ferrary, R., Lévy, R., and Monard, C. 1959. Recherches sur les substances mutagenes (halogéno –2 éthyle) amines, dérivés oxygénés du sulfure de bis– (chloro –2 éthyle) esters sulfoniques et sulfuriques. *C. r. séances Acad. Sci.*, **248**: 729–732.

56. ———, ———, ———, ———. 1961. Induction de mutations chez l'orge: Efficacite relative des rayons, du sulfate d'éthyle, du méthane sulfonate d'éthyle et de quelques autres substances. *Symposium on Effects of Ionizing Radiation on Seeds and their Significance for Crop Improvement, Karlsruhe, 1960*. In press.

57. Hoffmann, W. 1951. Ergebnisse der Mutationszüchtung. *Vorträge über Pflanzenzüchtg., Land– und Forstwirtschtl. Forschungsrat e.V., Bonn, 36–53.*

58. ———. 1954. Aussichten der Mutationszüchtung zur Verbesserung der Qualität des Weizens. In *Die Qualitätszüchtung von Brotgetreide. Bericht über die Tagung der Arbeitsgemeinschaft für Getreideforschung e.V., Detmold, 46–54.*

59. ———. 1959. Neuere Möglichkeiten der Mutationszüchtung. *Zeit. f. Pflanzenzüchtg.*, **41**: 371–394.

60. ——— and Zoschke, U. 1955. Röntgenmutationen beim Flachs *(Linum usitatissimum L.). Züchter*, **25**: 199–206.

61. Jakubziner, M. M. 1958. New wheat species. *Proc. First Intern. Wheat Genetics Symposium, Winnipeg, Canada*, 207–220.

62. Julén, G. 1961. The effect of X-raying on the apomixis in *Poa pratensis. Symposium on the Effects of Ionizing Radiation on Seeds and their Significance for Crop Improvement, Karlsruhe, 1960*. In press.

63. Käfer, E. 1952. Vitalitätsmutationen, ausgelöst durch Röntgenstrahlen bei *Drosophila melanogaster. Zeit. indukt. Abstamm. -u-Vererb. Lehre*, **84**: 508–535.

64. Kaukis, K., and Reitz, L. P. 1955. Ontogeny of the sorghum inflorescence as revealed by seedling mutants. *Amer. Jour. Bot.*, **42**: 660–663.

65. Konzak, C. F. 1956. Some procedural problems associated with mutation: Studies in crop plants. *Work Conf. on Radiation-induced Mutations, Brookhaven, Upton, N. Y.* 132–151.

66. ———. 1957. Genetic effects of radiation on higher plants. *Quart. Rev. Biol.*, **32**: 27–45.

67. ———, Nilan, R. A., Legault, R. R., and Heiner, R. E. 1961. Modification of induced genetic damage in seeds. *Symposium on the Effects of Ionizing Radiation on Seeds and their Signifi-*

cance for Crop Improvement, Karlsruhe, 1960. In press.

68. Lamprecht, H. 1957. Durch Röntgenbestrahlung von Pisum-Samen erhaltene neue und bekannte Genmutationen. *Agr. Hort. Genetica,* 15: *142–154.*

69. ————. 1958 Über grundlegende Gene für die Gestaltung höherer Pflanzen sowie über neue und bekannte Röntgen-Mutanten. *Agr. Hort. Genetica,* 16: *145–195.*

. 70. MacKey, J. 1954. Neutron and X-ray experiments in wheat and a revision of the speltoid problem. *Hereditas,* 40: *65–180.*

71. ————. 1956. Mutation breeding in Europe. *Brookhaven Symposium on Biology and Genetics, Brookhaven Nat. Lab., Upton, N. Y.,* 9: *141–156.*

72. Michaelis, P., and Kaplan, R. W. 1950. Manifestationswechsel beim Zusammenwirken chromosomaler und plasmatischer Erbfaktoren. *Naturwiss.,* 37: *494.*

73. Moës, A. 1959. Les mutations induite par les rayons X chez l'orge distique. *Bul. Inst. Agr. et Stat. Rech. de Gembloux,* 27: *1–82.*

74. Nilan, R. A. 1956. Factors governing plant radiosensitivity. *Proc. Atomic Energy Comm. Report TID– 7512, 151–162.*

75. ————. 1959. Radiation-induced mutation research in the United States of America. *Proc. II. Congr. Eur. Assoc. Res. Plant Breed. EUCARPIA, 36–47.*

76. ————. 1960. Factors modifying radiosensibility in plants. *Genetica Agraria,* 12. In press.

77. ————, Konzak, C. F., Legault, R. R., and Harle, J. R. 1961. The oxygen effect in barley seeds. *Symposium on the Effects of Ionizing Radiation on Seeds and their Significance for Crop Improvement, Karlsruhe, 1960.* In press.

78. Nötzel, H. 1952. Genetische Untersuchungen an röntgeninduzierten Gerstenmutanten. *Kühn-Archiv,* 66: *72–132.*

79. Nybom, N. 1954. Mutation types in barley. *Acta Agr. Scand.,* 4: *430–456.*

80. ————. 1957. Växtförädling med hjälp av inducerade mutationer. *Sver. Utsädesför. Tidskr.,* 67: *34–55.*

81. Oka, H. I., Hayashi, J., and Shiojiri, I. 1958. Induced mutations of polygenes for quantitative characters in rice. *Jour. Heredity,* 49: *11–14.*

82. Osborne, T. S. 1957. Mutation production by ionizing radiation. *Proc. Soil and Crop Sci. Soc. Florida, 91–107.*

83. Prakken, R. 1959. Induced mutation. *Euphytica,* 8: *270–322.*

84. v. Rosenstiel, K. 1960. Vom Züchten. *Schriftenreihe der Landw. Fakultät der Universität Kiel, Verl. P. Parey.* In press.

85. Rudorf, W. 1961. Über die mutagene Wirkung von S³⁵ bei *Phaseolus vulgaris*. *Zeit. f. Pflanzenzüchtg.*, **45**: *69–90.*

86. Rudorf, W., and Wienhues, F. 1951. Die Züchtung mehltauresistenter Gersten mit Hilfe einer resistenten Wildform *(Hordeum spontaneum nigrum H 204)*. *Zeit. f. Pflanzenzüchtg.*, **30**: *445–463.*

87. Scarascia, G. T., Avanzi, S., Bozzini, A., Cervigni, T., D'Amato, F., Donini, B., and Giacomelli, M. 1961. Effects of radiations and chemical mutagens in durum and bread wheats. *Symposium on the Effects of Ionizing Radiation on Seeds and their Significance for Crop Improvement, Karlsruhe, 1960.* In press.

88. Scheibe, A. 1959. Mutationsauslösung durch Chemikalien bei Gerste und Weizen. *Berichte des II. EUCARPIA-Kongresses, Köln, 69–75.*

89. ———— and Bruns, A. 1953. Eine kurzröhrige weissblühende Mutante bei *Trifolium pratense* nach Röntgenbestrahlung. *Vorl. Mittl. Angew. Bot.*, **27**: *70–74.*

90. ———— and Hülsmann, G. 1958. Mutationsauslösung durch Chemikalien beim Steinklee *(Melilotus albus)*. *Zeit. f. Pflanzenzüchtg.*, **39**: *299–324.*

91. Schmalz, H. 1960. Der Einfluss von Gibberellin auf eine "knotenlose" Sommergersten-Mutante. *Züchter*, **30**: *81–83.*

92. Scholz, F. 1955. Mutationsversuche an Kulturpflanzen: IV. Über den züchterischen Wert zweier röntgeninduzierter nacktkörniger Gerstenmutationen. *Kulturpflanze*, **3**: *69–89.*

93. ————. 1957. Mutationsversuche an Kulturpflanzen: VII. Untersuchungen über den züchterischen Wert röntgeninduzierter Mutanten verschiedener Merkmalsgruppen bei Sommer– und Wintergerste. *Zeit. f. Pflanzenzüchtg.*, **38**: *181–220; 225–274.*

94. ————. 1959. Strahleninduzierte Mutationen und Pflanzenzüchtung. *Atompraxis*, **5**: *475–481.*

95. ————. 1960. Versuche zur züchterischen Steigerung des Eiweissgehaltes der Gerste mit Hilfe der experimentellen Mutationsauslösung. *Qualitas Plant. et Materiae Vegetabiles*, **6**: *276–292.*

96. ————. 1960. Qualitätsprobleme in der Futtergerstenzüchtung, dargestellt an Ergebnissen von Mutationsversuchen. *Zeit. f. Pflanzenzüchtg.*, **44**: *105–128.*

97. Scossiroli, R. E. 1959. Selezione artificiale per un carattere quantitativo in popolazioni di *Drosophila melanogaster* irradiate con raggi X. *Istituto Genet., Universita di Pavia, 1–132.*

98. Sears, E. R. 1956. The transfer of leaf-rust resistance from *Aegilops*

umbellulata to wheat. *Brookhaven Symposium on Biology and Genetics, Brookhaven Nat. Lab., Upton, N. Y., 9: 1–22.*

99. Singleton, W. R. 1956. The use of radiation in plant breeding. *Atomic Energy and Agr., 183–194.*

100. Smith, H. H. 1958. Radiation in the production of useful mutations. *Bot. Rev., 24: 1–24.*

101. Sparrow, A. H. 1956. Cytological changes induced by ionizing radiations and their possible relation to the production of useful mutations in plants. *Work Conf. on Radiation Induced Mutations, Brookhaven, Upton, N. Y., 76–113.*

102. ————, Binnington, J. P., and Pond, V. 1958. Bibliography on the effects of ionizing radiations on plants, 1896–1955. *Brookhaven Nat. Lab., Upton, N. Y.*

103. ————, Cuany, R. L., Miksche, J. P., and Schairer, L. A. 1961. Some factors affecting the responses of plants to acute and chronic radiation exposures. *Symposium on the Effects of Ionizing Radiation on Seeds and their Significance for Crop Improvement, Karlsruhe, 1960.* In press.

104. ———— and Konzak, C. F. 1958. The use of ionizing radiations in plant breeding accomplishments and prospects. *Camelia Culture, 425–452.*

105. Staudt, G. 1959. Eine spontan aufgetretene Grossmutation bei *Fragaria vesca* L. *Naturwiss, 46: 23–24.*

106. Stebbins, G. L. 1950. Variation and evolution in plants. *New York: Columbia Univ. Press, 1–643.*

107. Steuckardt, R. 1960. Untersuchungen über die Wirkung von Röntgenstrahlen auf Rispenhirse *(Panicum miliaceum* L.*)* nach einmaliger und mehrfacher Bestrahlung: I. Die Mutabilität nach einmaliger Bestrahlung sowie die Beziehung zwischen Auslesebeginn und Selektionserfolg. *Zeit. f. Pflanzenzüchtg., 43: 85–105.*

108. ————. 1960. Untersuchungen über die Wirkung von Röntgenstrahlen auf Rispenhirse *(Panicum miliaceum* L.*)* nach einmaliger und mehrfacher Bestrahlung: II. Die Chlorophyllmutationsrate und das Mutationsspektrum nach mehrfach wiederholter Bestrahlung. *Zeit. f. Pflanzenzüchtg., 43: 297–322.*

109. Stubbe, H. 1934. Einige Kleinmutationen von *Antirrhinum majus* L. *Züchter, 6: 299–303.*

110. ————. 1938. Genmutation: I. Allgemeiner Teil. *Berlin: Gebrüder Borntraeger, Handbuch d. Vererbungswiss., 1–429.*

111. ————. 1952. Über einige theoretische und praktische Fragen der Mutationsforschung. *Abhandlg. d. Sächs. Akad. d. Wiss. Math.–Naturw. Kl., 44: 3.*

112. ———. 1953. Über mono– und digen bedingte Heterosis bei *Antirrhinum majus*. L. *Zeit. indukt. Abstamm. -u-Vererb. Lehre*, **85:** *450–478*.

113. ———. 1957. Mutanten der Kulturtomate *Lycopersicon esculentum* Mill.: I. *Die Kulturpflanze*, **5:** *190–220*.

114. ———. 1958. Mutanten der Kulturtomate *Lycopersicon esculentum* Mill.: II. *Die Kulturpflanze*, **6:** *89–115*.

115. ———. 1958. Advances and problems of research in mutations in the applied field. *Proc. 10th Intern. Congr. Genetics, Montreal*, **1:** *247–260*.

116. ———. 1959. Mutanten der Kulturtomate *Lycopersicon esculentum* Mill.: III. *Die Kulturpflanzen*, **7:** *82–112*.

117. ———. 1959. Considerations on the genetical and evolutionary aspects of some mutants of Hordeum, Glycine, Lycopersicon, and Antirrhinum. *Cold Spr. Harb. Symp. Quant. Biol.*, **24:** *31–40*.

118. ———. 1959. Einige Ergebnisse der Mutationsforschung an Kulturpflanzen. *Berichte d. Deutsch. Acad. d. Wiss. zu Berlin*, *1–39*.

119. ——— and v. Wettstein, F. 1941. Über die Bedeutung von Klein– und Grossmutationen in der Evolution. *Biol. Zbl.*, **61:** *265–297*.

120. Timoféeff-Ressovsky, N. W. 1940. Allgemeine Erscheinungen der Genmanifestierung. *Berlin: S. Springer. Handbuch der Erbbiol. d. Menschen*, **1:** *32–72*.

121. Tollenaar, D. 1934. Untersuchungen über Mutation bei Tabak: I. Entstehungsweise und Wesen künstlich erzeugter Gen-Mutanten. *Genetica*, **16:** *111–152*.

122. ———. 1938. Untersuchungen über Mutation bei Tabak: II. Einige künstlich erzeugte Chromosom-Mutanten. *Genetica*, **20:** *285–294*.

123. Torssell, R., Ed. 1954. Mutation research in plants. *Acta Agr. Scand.*, **4:** *359–642*.

124. Verkerk, K. 1959. Neutronic mutations in tomatoes. *Euphytica*, **8:** *216–222*.

125. Vettel, F. 1957. Züchtungsmethodische Fragen der Neu– und Erhaltungszüchtung bei Weizen, Gerste, und Hafer. *Berichte d. Deutsch. Akad. d. Landwirtschaftswiss. zu Berlin*, **6:** *1–39*.

126. Wellensiek, S. J. 1959. Neutronic mutations in peas. *Euphytica*, **8:** *209–215*.

127. Williams, W. 1959. The isolation of "pure lines" from F_1 hybrids of tomato, and the problem of heterosis in inbreeding crop species. *Jour. Agr. Sci.*, **53:** *347–353*.

128. Wöhrmann, K. 1958. Untersuchungen über den Ährchensitz bei *Alopecurus pratensis* L. *Angew. Bot.*, **32**: *45–51*.
129. Zacharias, M. 1956. Mutationsversuche an Kulturpflanzen: VI. Röntgenbestrahlungen der Sojabohne. *Züchter*, **11**: *321–338*.

Comments

DERMEN: I am still unclear about the difference between macromutations and micromutations.

GAUL: The grouping has practical purposes only and may serve for the communication of evolutionary and breeding problems. It considers the phenotype only and has nothing to do with the change of the genetic structure. The classification is simply based on the method of detection, i.e., whether the mutation can be recognized with certainty in a single plant or in a large group of plants only. It is arbitrary because there are all transitions from large alterations to small ones. However, in plant breeding, the selection procedure entirely depends on whether one is dealing with macro- or micromutations.

STRAUSS: It seems that the terms micromutation and macromutation have been used by you in an operational or field sense only. Should not these terms be defined more carefully in terms of actual gene changes?

GAUL: This would be good if it is possible. *A priori*, I would not expect that a "small change of the gene structure" would necessarily correspond to a small change of the phenotype, and *vice versa*.

MACKEY: As to Doctor Gaul's definition of macromutations versus micromutations, I must say that the plant breeder urgently needs a separation of two such groups in his discussion, since they relate to the selection methods applied. Personally, I do not hesitate to enlarge the earlier concept of macromutation, since such shifts in meanings are common in biological denomination rules. Doctor Gaul's "transpecific macromutation" will be the old concept now in a strict sense.

FREY: You have suggested improving the efficacy of mutation by (a) better control of mutation induction, and (b) better control of selection. I would like to see you add another category of great importance, namely, analysis of the elements that constitute the characteristic to be modified and then an intelligent and logical selection of parent material (the genotype) to be irradiated.

GAUL: I agree completely with you. For instance, it might be expected that it will be easier to find high-yielding earliness mutants in a late-ripening variety than in an early one. I think however, that a thorough selection of the genotype of the starting material is an objective not only valid for the use of mutations but is common for all breeding methods.

SARVELLA: Are the first five tillers referred to in the theory of diplontic selection formed from the five buds present in the seed after irradiation (primary axillary spikes), or from the first five tillers emerging from the ground? Would you expect the mutation rate at a node to be lower in the tillers which are formed after the primary axillary tiller is formed?

There is evidence that the mutation rate per spike and per seedling in the plant appear to be the same for the whole plant, for the primary axillaries, and for the whole plant excluding the apical spikes and the primary axillary spikes (secondaries, etc.) at some nodes. The primary axillary spikes were tagged when the X–1 seedlings were 3 weeks old.

GAUL: In reference to your first question. The hypothesis of diplontic selection considers the ontogenetically first five tillers, which do not always correspond with the first five tillers emerged from the ground. In reference to your second question. The first second-order tillers derived from axillary bud 1, 2, and perhaps also 3 presumably contain fewer mutations. With excessive tillering, the mutation rate of the later second-order tillers depends, I suppose, on the number of·surviving L II-initial cells of the corresponding axillary bud. If there are several surviving initial cells in these buds or primordias, the mutation rate of the later second-order tillers will usually be smaller than that of the tillers derived from primary axillary buds. This is a consequence of inter-cellular competition. But if there is only one surviving L II-initial cell and this carries a mutation, no drop of the mutation frequency is expected because the generative tissue of the later secondary tillers is supposed to originate from this initial cell. This question is fully discussed in the paper "Studies on diplontic selection" which is now in press (Symposium on the Effects of Ionizing Radiation on Seeds and their Significance for Crop Improvement, Karlsruhe, 1960).

We have now evidence from four different experiments that the mutation rate of the later formed tillers is smaller than that of the first by about five.

NILAN: From the literature and some discussions at this symposium, I find that frequently the induced mutation technique and the cross breeding method are compared as methods in plant breeding. I feel that this comparison is not fair since for the most part induced mutations will be just a complement but not a supplement to a cross-breeding program. As plant breeders, we realize that genetic diversity is the basic ingredient of any plant improvement program. This diversity, which has arisen by mutation, can be introduced into a cross-breeding program through varieties and related species and genera. This diversity can also be obtained through induced mutations. Thus, induced mutations comprise an important source of genetic diversity for a cross-breeding program.

The Use of Induced Mutations for the Improvement of Vegetatively Propagated Plants

NILS NYBOM

Balsgård Fruit Breeding Institute,
Fjälkestad, Sweden

I F WE LOOK at lists of new fruit varieties, a great many will be found to constitute spontaneous mutations. A glance at a rose nursery catalog will tell us the same thing, and we also know that much of the variation in form and color among flower bulbs and other ornamental plants goes back to spontaneous mutations that have been taken care of and have been cultivated as new and more attractive varieties. In orange trees, potatoes, strawberries, and many other clonal plants there are often said to occur different "races", some of which may be claimed to be better keeping, better yielding, or deviating in other respects from the original types (65, 72, 77, 92, 100).[1]

For several reasons, these sports seem to have been of special importance among the vegetatively propagated plants. Their greater constancy, preserved through clonal propagation, permits the detection of even slight phenotypical changes. Thanks to the vegetative mode of reproduction, practically all such changes may be propagated, even such that would lead to complications and perhaps rapid elimination in a seed-propagated plant.

The vegetatively propagated plants often have a long period of sexual reproduction. They usually turn out to be highly heterozygous, and our knowledge of their genetical relationships is still very imperfect. All this has made improvements by means of the classical methods, involving crossing, recombination and selection, time-consuming and dearly bought. To a considerable extent this may also have favored the use of spontaneous mutations in these plants.

Towards the end of the twenties, means were found for the artificial induction of mutations, as it seemed of the most varying kinds and in unlimited amounts. No wonder that plant breeders tried to apply this possibility in their work. I am not going to relate this story in detail here as it has been dealt with elsewhere (46, 49, 50, 86, 104).

---

[1]See References, page 285.

We all know that the experiences and the opinions based on these studies in many cases turned out to be quite negative. Either one did not find any greater amounts of mutations at all, or those obtained did not appear very promising as a basis for the production of new and better varieties.

The interest in induced mutations was revived when it indeed turned out to be possible to produce undeniable, progressive mutations of practical interest for the breeder (17, 39, 46). The increasing interest in the application of atomic sciences also has helped to focus the interest of the plant breeder on this "new" path or tool in plant breeding, the mutation method (47, 107).

Although the inclination of many plant breeders has changed "from fairly sceptic to moderately optimistic" (86), one still meets fairly conflicting statements concerning the prospects of the mutation method. It is one of the chief aims of the present conference to collect our experiences, to focus the interest on the most promising areas, and, as far as possible, to present a realistic evaluation of the mutation method in plant breeding.

A Survey of Mutation Experiments in Vegetatively Propagated Plants

In order to give an idea of the achievements arrived at in various plants and to provide a basis for the discussion to follow, I shall begin by summarizing experiences from the more important mutation experiments in the vegetatively propagated plants that have been published.

Potatoes

Some of the more extensive and most promising work on induced mutations has been reported with the potato by my compatriot, A. Heiken. As I happen to be rather familiar with his results, I shall take the liberty to deal with them at some length. This may be the more justified as his work is published in a journal that may not be generally available.

However, for the sake of justice, I should like to begin by referring to the classical work of Asseyeva. She published during the twenties and early thirties a series of papers on the chimaeric structure of different spontaneous potato mutants (2). She found that most of the spontaneous mutants were periclinal chimaeras; and with the

eye excision method, she was able to unveil the genotypical constitution of the deeper lying tissues. She also published some work on
radiation-induced changes (3, 4). X-ray treatment was found to be
"a powerful and reliable means" for bringing about tissue recombinations similar to those obtained by eye excision. Not all changes
were such tissue recombinations, however. Some behaved like original mutations. Unfortunately, her promising experiments were dis-
continued. It was 20 years before they were taken up again (51, 54,
55, 58).

Heiken's own results were supplemented by a survey of the
reported spontaneous mutations in the potato. His list of "tuber skin
color aberrations" contains more than 50 cases, and to this list may
be added changes in skin structure, shape and color of tuber, stem,
foliage and flowers, as also the very characteristic potato mutations
"bolters" and "wildings".

The author also presents the results of his own extensive studies
on the occurrence of similar aberrations in untreated material. In
addition to 1 per cent bolters, he found 25 cases of such spontaneous
aberrations in a material of several million plants, i.e., somewhat
more than 1 in 200,000 plants.

For irradiation, Heiken divided each tuber into two parts. One
half was irradiated and the other was sown as a control. In this way
it was possible to detect and remove all cases of contamination and
tuber-carried diseases. The irradiations were carried out at different
times. With regard both to the lower lethality and the considerably
higher mutation frequency, the best results were obtained just at the
beginning of germination, during February to March. A certain
number of primary tubers then yielded six times more mutations
than if irradiated during dormancy, i.e., during November to December. After germination had started, April to May, a drop in mutation
frequency was noted. The most suitable X-ray dose was around
4,000 r.

All tubers from the surviving X_1 plants (plants growing from the
irradiated tubers) were harvested and sown the next year to form
the X_2 families, the plants in which then carried the X_3 tubers.

In the first year, irradiation gave rise to typical "primary effects",
viz., increased fleshiness of leaves, disturbances in vein branching,
deformation of leaves, etc. However, only a very small portion of

these effects persisted into the next generation. The frequency of transmitted aberrations was very much the same in those families that had passed the X_1 germination without primary effects. Observations on the X_3 plants, on the other hand, showed that most of the induced changes had already been found in X_2.

As far as distinct, easily observed changes are concerned, selection work may therefore be concentrated in the second year only. However, a breeder looking for less drastic changes might prefer to postpone selection to X_3, when it can be based on more or less pure tuber lines.

In the second year, changes turned up in certain families. Only changes in the above-ground plant parts were looked for. The dose response was evidently linear. Thus, 2,000 r gave 4.9 per cent mutated X_2 families, 4,000 r 10.5 per cent, and 8,000 r 20.8 per cent. However, due to the differences in lethality, the absolute number of mutated families always was higher after 4,000 r. The 640 tuber-halves irradiated with 4,000 r gave 409 surviving X_1 plants and the same number of X_2 families. Of these, 364 were normal while 45 contained mutations. The "segregation" ratio in X_2 was about 2 aberrant plants to 8 normal plants per family.

By means of eye excisions Heiken demonstrated that most, though evidently not all, e.g., none of the "wildings", of the isolated changes were periclinal chimaeras, having normal tissue below the mutated. All isolated types were also grafted on virus-free stocks in order to test the presence of any sap-transmissible principles, but no such indications were found.

Even though these studies were intended as a "preliminary exercise" for mutation breeding work in the potato, like most other studies on induced mutation hitherto they were not planned as a mutation breeding project. The author wanted, in the first round, to determine the range of variability induced and concentrated mainly on distinct and easily identifiable mutations. Thus not less than 55 of the isolated 109 aberrations consisted of various malformations, 17 of them were dwarf types, and 27 were different flower color variations. Strange enough, no bolters were found after irradiation.

The average yield of these isolated, drastic changes was, as one could expect, considerably reduced. Most of them gave between 70 to 80 per cent of the original clone, whereas some were rather similar to the mother variety in vigor.

The dormancy period of the tubers was also studied in special experiments. Some of the isolated mutants had increased periods of dormancy while others started germination earlier than the normal clones.

After the last, main irradiation experiment a series of less marked and probably more interesting changes were isolated, but as they required another year for verification and testing, they were not included in the report. Summarizing his results, Heiken states that, "the somatic aberration frequencies obtained have been high enough to make continued research in this field very desirable". He suggests that a systematic selection of induced mutations could be combined with the present virus-testing routines.

One of the keys to Heiken's success was the fact that he had the privilege of working at an institute specializing in virus diseases of potatoes and in growing virus-controlled stocks of the varieties used. I should like to stress this fact, as it is of general importance for muta- tion studies in vegetatively propagated crops. Virus diseases often simulate genetical changes and may, thereby, cause serious trouble.

Chrysanthemum

A very interesting and promising piece of work on induced muta- tions in *Chrysanthemum indicum* was published some years ago by Jank (59) in Germany. Chrysanthemum is also a plant where spon- taneous mutations, especially concerning flower color, have often been observed and also have been of great practical importance in the breeding work.

Jank also combined his work with an extensive survey of the reported spontaneous color variations. Not less than 318 such "sports" have been described in 170 different varieties. It turned out that rose-colored varieties especially often give rise to mutations with varying new colors. White varieties usually are somewhat less mutable, and the same is true of the bronze-colored ones. There are also violet, red, orange, yellow, brown, and other colors which seem to be more stable.

Three varieties were selected for the main experiment, Day Dream, Vogue, and Berta Talbot, all with different shades of rose. Cuttings were taken on February 2. On March 18 the rooted and potted plants were decapitated in order to give rise to side shoots, and a few weeks later they were X-rayed. The suitable doses had been found to lie between 1,000 and 2,000 r. The roots were also irradiated in order to avoid excessive root shoot formation.

In addition to reduced growth rate, the first shoots to appear showed the expected primary effects, thick succulent leaves with irregular margine and uneven surface. The author was well aware that the mutations should be expected to form aberrant sectors in the shoots formed, and that repeated bud and branch formation would be necessary in order to isolate the changes in a pure condition.

Therefore, the shoots formed were again decapitated and the tips so obtained planted in order to form new roots. This process was repeated as often as possible; in all, seven times between April and June. The original 144 irradiated plants thus gave rise to 1,144 new plants, which were allowed to grow undisturbed and which flowered during October. Some changes in foliage were then noticed, as also some types with possibly deviating flowering periods, late flowering being a desirable character.

In addition, a great many variations in flower color were recorded, in all 281 different cases. Practically all of the originally irradiated plants yielded such color changes, most of them a whole series. The following new nuances were found: intensive rose, flesh-colored, copper-rose, copper-red, cream-yellow to cream-rose, cream, yellow, bronze, brown, red, and violet. White was obviously not found.

The new colors were either found as sectors in the flowers or as single deviating flowers on a plant. Whole, changed plants were also observed, mostly after the later decapitations; the early ones usually giving narrow sectors.

The percentage of recorded changes was highest after the highest dose, 2,000 r. The author points out that it is important to use doses that permit an extensive clonal propagation of the irradiated individual in order to isolate the numerous changes induced at the moment of irradiation.

On the whole, Jank's results appear very impressive and promising, and he also summarized his work by stating that, "the experimental induction of mutations by means of X-rays may be regarded as an effective way of creating new color variations in *Chrysanthemum indicum*".

Flower Bulbs

Another category of vegetatively propagated plants where spontaneous mutations have had, and indeed still have, a great commercial importance is the ornamental bulb plants, particularly tulips and hyacinths, but also iris, daffodils, lilies, freesias, and others. The variety collections are continuously enriched by new, and at least sometimes improved variations of the old standard varieties. Let us take one example of such a "mutation family", that of the Bartigon tulip. This old and popular variety is now found in the catalogs as a double-flowered form, "Double Bartigon"; as a parrot form, "Red Champion"; as a giant or "maximum" mutation, "Bartigon Max"; as a white-variegated type, "Cordell Hull"; and as a series of new color sports, including salmon red "Queen of the Bartigons", scarlet red, "All Bright", deep rose "Philip Snowden", etc.

The flower bulbs have also been the object of some relatively successful experiments with induced mutations, primarily by de Mol in Holland who described his results in a long series of papers. He summarizes the results of the first 13 years of X-ray experiments with tulips and hyacinths (74, 75). More than 70 different cases of induced changes are described, found in about 30 different varieties. The mutations mostly concerned flower color, but flower shape was also involved, ranging from highly irregular monstrosities to rather slight changes, e.g., from rounded to more angular flowers.

During 1933–34 he irradiated in all 1,472 bulbs belonging to 60 different varieties, using 1,200 r units of X-rays. In this material he later isolated 41 different mutations out of 25 of the varieties. Most of these mutations seem to have been cultivated further on in pure condition.

De Mol was aware that the older spontaneous sports may be periclinal chimaeras, e.g., the parrot-tulip Gemma, and that a radiation-induced reversion to the mother type, La Reine, was not a true mutation from the genetical point of view but as a "modification", or rather as a tissue recombination. But he also induced changed color in parrot tulips without loss of the parrot character, and also new parrot types from normal varieties. Except for a few possible tissue recombinations, most of de Mol's changes obviously were original mutations.

In a special chapter, at the end of his 1944 paper (74), de Mol presents practical hints for the mutation breeder. The best time for irradiation seems to be September because the bulbs are then easier to handle than earlier in the summer when the new side-bulbs are formed. A suitable dose is said to be 800 r units. Like Jank, de Mol advocates the use of doses low enough to permit a rich formation of new buds and shoots. The material should be propagated by side-bulbs for 4 years in order to obtain all changes in a pure state. The growing conditions should also be modified so as to stimulate the production of new side-bulbs.

There was a pronounced similarity between the induced mutations and those known to occur spontaneously. There was also, in the main, parallel variations in induced and spontaneous mutability among the varieties. Many of the induced changes occurred, however, in varieties in which they were not known before.

More recently a Swedish plant breeder, Carlsson, at the plant breeding station at Gullåker, Hammenhög, has taken up irradiation experiments with flower bulbs on a comparatively large scale. I believe we have irradiated some 25,000 flower bulbs for him at our cobalt 60-source at Balsgård since 1955. The doses used have been considerably higher than those mentioned by de Mol, namely, between 2,000 and 5,000 r. This difference may to some extent depend on the deviating irradiation conditions, acute X-rays compared with semi-chronic gamma during 5 to 6 days in the open air.

Carlsson (22) has not yet published any results from his works, most of his material still being in an early stage. He has, however, already isolated a series of different changes and seems convinced that he shall be able to select types among them that ought to become of future value. In addition to the more common color changes, he has found several cases of white-variegated flowers, similar to the "Cordell Hull" sport of Bartigon. He also believes he has obtained giant sports in several varieties, corresponding to "Bartigon Max" out of Bartigon.

Other Ornamental Plants

Preliminary results have also been reported from mutation experiments with some other ornamental plants, the best known examples being those with carnations described by Richter and Singleton (91) and by Sagawa and Mehlquist (94, 95).

Some of these changes, such as reversions from the spontaneous sports White Sim and Pink Sim back to the original red William Sim, obviously are tissue recombinations. However, some other new colors were also obtained as well as changes in flower morphology. These studies are being continued by Mehlquist (71).

In Saintpaulia, Sparrow and his co-workers have reported very interesting irradiation experiments (107 and 108). In the white-flowered and semi-double variety Dwight's White mutations with violet and lavender-colored flowers were induced as also types with single and double flowers and a series of leaf changes.

After irradiating leaf-petioles with 2,000 and 3,000 r X-rays, 14.2 and 25.2 per cent, respectively, mutated plants were recorded, in all 154 cases. A feature of special interest concerning these mutants is that they are mostly homogeneous, i.e., not chimaeric. This is also to be expected as the new plants are derived from single cells of the primary, irradiated leaf-petiole.

In this connection there are some unpublished Swedish results with roses that I have been allowed to mention here. Doctor Gelin (39) at Weibullsholm Plant Breeding Institute irradiated 1957 summer buds of five varieties of roses with 2,500 to 10,000 r gamma-rays from cobalt 60. After irradiation these were budded into common rootstocks. About half of them took and developed into new shoots next year. In order to isolate the possible induced changes, buds were again taken from these new shoots in 1958.

In addition to the primary effects the first year, the isolation the next year also gave rise to persistent changes. Besides changes in thorniness (increased thorniness), leaf color (increased anthocyanin content), and leaf shape, there was also found, after 5,000 r, a darker colored mutation in the variety Peace, somewhat similar to the spontaneous "Pink Peace," which is now under propagation for further tests (40).

Thus, these first experiments with roses did give at least one mutation that might become of direct commercial importance if further trials show it to deviate in a positive way from the other types.

It might be added here that similar-looking mutations should always be worth further trial and comparison, as varying pleiotropic changes seem to be a rule rather than an exception. Some of the red Delicious mutations in apples, for example, at the same time show a distinctly different and, from some points of view, an improved mode of growth and fruit production (1, 64).

Fruit Trees and Small Fruits

Most of the mutation experiments with vegetatively propagated plants fall within this group. For the sake of surveyability, the various mutation projects known through publications have been collected in Table 1.

A special kind of mutation of great practical interest is represented by the self-fertility mutations of Lewis, who has given detailed descriptions of their induction (66, 67). Irradiation was done during the resting stage before meiosis. Irradiation of mature pollen grains had no effect as the substances responsible for the incompatibility reaction are obviously by then already formed. In cherries, 800 r on an average gave one seed after self-pollination (or incompatible cross-pollination) of 130 flowers. Ten per cent of the seedlings so formed turned out to be completely self-fertile and to give rise to self-fertile offspring in their turn.

The artificial induction of self-fertile types must be said to be of very great importance, not only from the point of view of fruit production but also with regard to the possibility of obtaining individuals homozygous for rare recessive genes, leading, e.g., to extreme earliness (67).

The method most commonly used for the induction of somatic mutations in fruit trees has been to irradiate dormant scion wood during winter or early spring and then to graft these scions into other trees. In order to ensure a better union between stock and scion, Gröber (45) and Zwintzscher (112) recommend that only the upper part of the scions be irradiated. Granhall (42) and Zwintzscher (112) also overcame the same difficulty by irradiating complete young trees already growing on an understock. Granhall used 1-year-old "yearlings", while Zwinzscher used somewhat older trees, the crowns of which had been pruned back in order to form new shoots.

Other methods of irradiation described, e.g., by Hough and Weaver (57), involve continuous exposure of growing trees at a stationary cobalt 60-source. This method of irradiation has also been practiced in Sweden, although we are not prepared to judge whether it is more efficient than acute irradiation. We have also, like Gröber (45), irradiated bud sticks of apples and pears during August and inserted the irradiated buds into suitable rootstocks. As far as we can judge, this method seems to offer certain advantages.

TABLE 1.—MUTATION BREEDING PROJECTS IN FRUIT FROM WHICH PUBLISHED INFORMATION IS AVAILABLE.

Ref. No.	Leader	Place	Plant	Mutations recorded	Technical remarks
6	Bauer	Cologne, Germany	Black currants	Mutations in various characters	X-irradiation followed by back-pruning
9, 10, 11, 12, 13	Bishop	Kentville, Canada	Apples	Various fruit characters, e.g., color and shape	Neutron- and X-irradiation of dormant scions
18, 19, 87	Breider	Würzburg, Germany	Grapes	Mutations affecting yield and other characters	Dormant buds irradiated
37, 41	Einset	Geneva, U.S.A.	Apples	Fruit shape	Neutron irradiation of dormant scions
33, 42, 43, 44	Granhall, Nybom	Balsgård, Sweden	Apples, pears, and other fruits	Various fruit characters, e.g., color and shape	Mostly gamma-irradiation of young trees
45	Gröber	Gatersleben, Germany	Apples, peaches, and other fruits	Fruit color mutations and other changes	Mostly X-irradiation of dormant scions
57	Hough	New Brunswick, U.S.A.	Peaches	Mutations affecting period of ripening and other characters	Chronic gamma-irradiation
66, 67	Lewis	John Innes, England	Cherries and other top fruits	Self-fertility mutations	X-irradiation of flower buds
85	Olmo	Davis, U.S.A.	Grapes	"Self-thinning" mutants	Dormant buds X-rayed
110, 111	Williams	John Innes, England	Apples and pears	Fruit color and skin properties	Internal beta-irradiation from P 32.
112, 113	Zwintzscher	Cologne, Germany	Cherries, apples, and other top fruits	Changes in period of ripening, defoliation, and other characters	X-irradiation of dormant scions and young trees

On the branches formed from the irradiated scions various effects may be observed, primary changes like shoot bifurcations, disturbances in the leaf spiral, deformed leaves, and so on, but also fruit changes that may be genetical in nature, like sectors with deviating over-color. Such sectors may be rather common at first, but it is ·a general experience that if the shoots are allowed to grow undisturbed for several years these sectors eventually disappear. They may get broader later on but at the same time more rare, and only in a very few cases thus far have "stable" or "pure" changes been observed affecting whole branches (10, 12, 45).

German workers especially have paid much attention to the problem of recovery and isolation of induced changes. A piece of work of great interest in this connection is that of Bauer (6) on black currants. According to his experiences, the first primary shoots coming from the irradiated buds only very rarely reveal any mutational changes. But if these shoots are pruned back to the originally irradiated stock and new shoots thereby forced to develop, a certain proportion of these will show such changes. These second-year shoots may then be removed, transplanted, and new shoots forced to develop again. This was repeated for several years, and after five years of selection in this way not less than 324 aberrant plants had been isolated from the originally irradiated 343 shoots.

The reason why the unpruned primary shoot (like the X_1 potato plant) only exceptionally gives rise to any changes may be explained by intrasomatic elimination which is supposed to take place in an irradiated multicellular organism and leads to an elimination of the mutated cells and a "normalization" of the plant (34, 63, 83). In barley, e.g., the primary roots may contain up to 100 per cent cytologically disturbed cells immediately after irradiation; but still, after completed ontogenesis, most of the cells in the spikes and the roots may look quite normal again. When mature buds are irradiated, the changes will form narrow sectors along the primary shoot. When the shoots grow, these sectors run a great risk of being eliminated due to intrasomatic competition. This must be part of the reason for the meagre results of many mutation breeding projects in fruit trees.

Zwintzscher (112, 113) has adopted a special system in order to unravel the irradiated tissue and to isolate the induced changes in pure condition. The shoot developed from an irradiated bud is cut back close to the originally irradiated scion so that adventitious buds are forced to develop. The basal buds on the detached shoot may be expected to contain induced, sectorial changes and are, therefore, budded over on new rootstocks, either a certain number beginning from the base, e.g., five, or only those that sit in the axis of a leaf showing primary effects. The primary effects can certainly not be called "mutations" but may perhaps be taken as an indication that the tissue in question was in a sensitive condition during irradiation and that it has not had the chance of getting rid of the changes to the same extent as the buds sitting with normal leaves further to the tip of the growing shoot. The originally irradiated wood may be kept for several seasons as a reservoir of induced changes.

Gröber (45) is using a similar pruning method. After the end of the first growing season, all shoots coming from irradiated buds are cut back on three or four buds, which are then allowed to break next year.

The types of mutations induced are also briefly indicated in Table 1. On the whole, these mutations are very similar to those known to occur spontaneously in fruit trees, as listed, e.g., by Shamel and Pomeroy (102). From the genetically better explored organisms we know that most mutations, whether induced or spontaneous, consist of destructive changes, deformations with more or less reduced vigor, etc. The same is, of course, true of mutations found among the fruit trees.

From almost all mutation projects, however, there are also reported mutations that may be of practical importance, e.g., the solid red sports of Cortland apple (11); the high-quality, late-ripening Elberta peach (57); mutations with later flowering period in the black currant (6); with earlier ripening in grapes (18); the "self-thinning" changes in the seedless Perlette Grape (85); and certainly the self-fertile cherry mutations (67).

Most of the mutation breeding experiments in these time-consuming plants have been going on for a relatively short time. We are undoubtedly still in the beginning of the exploration of the mutation method, and even if the positive achievements are far from definite, this is of course still less true of failures. Preliminary results are also reported from other mutation experiments with vegetatively propagated plants (16, 23, 69, 70, 73, 88). Of special interest is the still unpublished work of Shapiro (103) carried out at the Brookhaven National Laboratory with roses, chrysanthemums, geraniums, and gerbera.

Mutation Experiments at Balsgard

As briefly as possible, I should like to account for the mutation experiments on fruit trees and small fruits carried out at my own station at Balsgård. In cooperation with The Swedish group for Theoretical and Applied Mutation Research, we have now mutation material under way of most of the plants we work with at the Station.

Since 1952, we have had stationary cobalt 60-sources for irradiation purposes. The old one, shown in Figure 1, was enclosed in a metal tube inserted into the ground. This source was removed last year and we are now installing another one, shown in Figures 2 and 3. As we have not found any advantage in the chronic irradiation of

FIGURE 1.—*Fruit trees being irradiated at the old cobalt source. Arrow indicates the position of the source.*

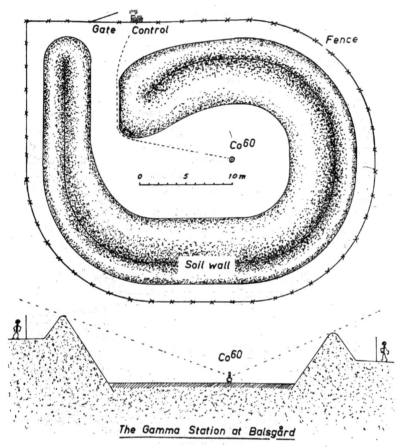

FIGURE 2.—*Plan of the new gamma irradiation field at Balsgård, Sweden.*

plants, we are going to concentrate on acute or semi-chronic exposures. Therefore, we have made the field as small as possible in order
to save space. The strength of the source is now 40 curies, which seems
suitable for our purposes. It is enclosed in a lead container sur-
rounded by a ring-shaped soil wall. The source may be lifted up out
of the container, which may also be used as a transport container, by
means of an oil piston. A small hand-driven oil pump is placed,
together with control lamps, in a box at the gate. The plexiglass tube
above the container houses the oil piston.

FIGURE 3.—*Close-up of the cobalt pot. Arrow indicates upper position of the source.*

The method mostly used for the irradiation of fruit trees is shown in Figure 1. Young, 1-year-old trees are taken from the nursery in the spring the second year after budding and placed at the cobalt source so that the source is closer to the tip than to the root. By adjusting the inclination of the trees we may get the killing dose of 10,000 r at the tips and the physiologically rather harmless dose of 2,000 r at

the roots. Somewhere between, we then have the optimal dose, which
may vary from time to time, depending on kind of fruit, chromosome
number, condition of the trees, and so on. Through repeated pruning,
first in order to remove all growth from the lower portion of the tree
leaving only the uppermost three living buds, and then again the
next year back to the basal buds of the surviving shoots, we try to
"dissolve" the tissue having got the most suitable irradiation dose.

The technique described might perhaps be improved by isolating the basal buds of the first year's growth over onto clonal rootstocks. The trees that have been exposed to irradiation or into which
a lot of irradiated scions have been grafted usually will be rather irregularly developed, and it is extremely difficult to identify these minor
changes that may be of the greatest practical interest, such as changes
in color, period of ripening, size or form of fruit, etc. When raising
new trees on clonal rootstocks by taking buds from previously irradiated material, one might expect more uniform tree material among
which it should be easier to trace slighter changes.

We have also irradiated summer buds of apples and pears which
have then been budded onto suitable rootstocks. Figure 4 shows such
a shoot coming from an irradiated bud. About the same result would
be obtained by irradiating dormant winter wood and grafting it into
a rootstock just above soil level. After pruning the next year, such a
mutated bud would result in a more or less completely changed tree
without primary effects and other disturbances in growth habit. We
have some such material with irradiated Williams pears among which
numerous minor changes have been noted. However, the deviating
trees must be tested further, budded over onto new rootstocks, together with control material, before we can be sure of the true nature of
the changes. Some drastic aberrations have been recorded, however,
like the irregularly corrugated mutants in Figure 5 and the "seedless"
mutant in Figures 6 and 7.

The irregularly furrowed fruits obviously are homologous to the
spontaneous sport Corrugated Bartlett (101) and probably also to the
irregular apple types described by Gilmer and Einset (41). A mutant
in Cortland, quite similar to the neutron-induced one of Einset
and Pratt (37), has been found in the same variety after gamma-
irradiation, Figure 8.

The seedless Williams type also deviates by its more elongated

FIGURE 4.—*X-ray-induced tissue recombination leading to chlorophyll-deficient shoot in the chimaerical sport Striped Williams.*

fruits, Figure 6, a common effect of reduced seed formation in pears (9). In the seedless fruits, shown on both sides of a normal pear in Figure 7, only the empty seed coats are to be found in the carpels. The fruit-set of this seedless type did not look markedly influenced. Another change, probably also without practical interest, is the small-flowered mutant shown in Figure 9, Cox's Pomona gamma-irradiation.

Increased russeting, i.e., a rougher fruit skin, obviously is a rather common change both spontaneously and after treatment. It has been induced in the pear varieties Seigneur d'Esperen, Graf Moltke, and Williams, as also in the apple variety Belle de Boskoop.

Of more commercial interest are the red-colored changes of

FIGURE 5.—*X-ray-induced mutations in Williams pear with irregularly furrowed fruits. Three normal pears above.*

which two are shown in Figures 10 and 11. In Cox's Orange, Figure 10, (gamma-irradiation), a whole branch carrying apples with distinctly increased color has originated. In Cox's Pomona (after pile neutrons), one branch, represented by the first two rows in Figure 11, is still sectorial in constitution, practically all fruits being sectored with deep red. The other two branches formed on the same irradiated tree, represented by the apples of the other two rows of Figure 11, are quite normal.

The chlorophyll-deficient shoot shown in Figure 4 (X-rays on summer buds) is induced in the pear clone Striped Williams (43). This Striped Williams, itself a spontaneous mutant very similar to the one of Gardner *et al.* (38), is most probably an endochimaera (Figure 12) that, besides "mutating" back to normal Williams, also gives rise to numerous such white shoots when irradiated. Thus, these changes are probably not true, original mutations, only tissue

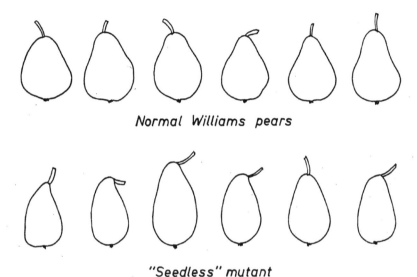

Normal Williams pears

"Seedless" mutant

FIGURE 6.—*X-ray-induced mutation in Williams pear with "seedless" elongated fruits.*

recombinations similar to those found in other plants (4, 58, 74, 87, 91, 94).

These tissue recombinations may still be of considerable interest, however, and a deliberate production of such changes is also on our irradiation program. As should be clear, most somatic mutants are

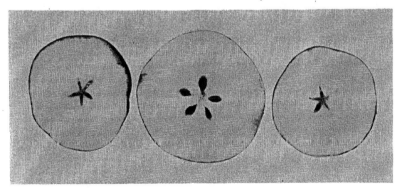

FIGURE 7.—*Cross-sections of "seedless" fruits at both sides of a normal pear.*

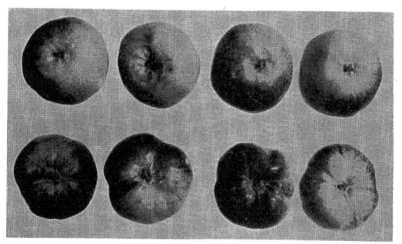

FIGURE 8.—Gamma-ray-induced mutation in Cortland apple with furrowed fruits. Four normal apples in the upper row.

FIGURE 9.—Gamma-ray-induced mutation in Cox's Pomona apple with small flowers. Mutated branch to the left, normal branch to the right.

FIGURE 10.—*Gamma-ray-induced mutation in Cox's Orange apple with increased red over-color. Normal apple to the left.*

built up as more or less stable periclinal chimaeras. Some of them make trouble by being too labile, like the reverting Starking Delicious (20), while others may be very constant. Such "ectochimaeras" (Figure 12), as for example thornless blackberry sports (26) or types with tetraploid epidermis (36), have normally not been accessible for further breeding work, because, as Darrow (26) says, "unfortunately

FIGURE 11.—*Neutron-induced sectorial mutation with increased red color in Cox's Pomona apple. The first two rows represent the fruits from an affected branch, still sectorial and unstable. The other two rows from normal branches.*

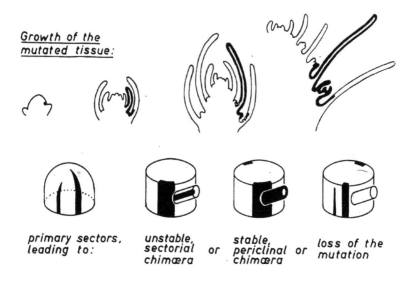

Growth of the mutated tissue:

primary sectors, leading to: unstable, sectorial chimœra   or   stable, periclinal chimœra   or   loss of the mutation

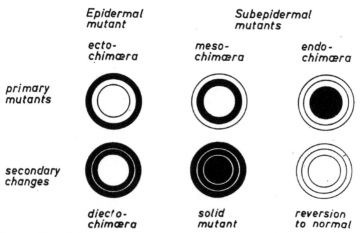

Types of periclinal mutations:

FIGURE 12.—*Schematic presentation of mode of origin and chimaerical structure of somatic mutations.*

no way has been found to use these thornless sports in breeding for thornlessnesss".

As pointed out already by Asseyeva and de Mol and also verified in later experiments, especially those of Howard (58) and Pratt (87), irradiation seems to be a potent means of producing such tissue recombinations, not only "stripping off" external mutated tissue, but also giving rise to solid mutants (Figure 12). Therefore, we have this year irradiated some ectochimaeric mutations in order to make them available for future breeding work, namely, the blackberry sport Thornless Evergreen and a colchicine-induced ecto-chimaera in *Prunus cerasifera*. The latter was irradiated as summer buds, whereas the blackberry was irradiated as leaf-bud cuttings propagated under mist.

Part of the anatomical background for such tissue recombinations has been treated by Crockett (24).

Finally, I should like to mention another case when such tissue recombinations might be the true purpose of an irradiation treatment, and that is in order to make a single layer chimaera more stable by transforming it into a diectochimaera or into a solid mutant. We have irradiated material of a spontaneous, but unfortunately not very stable nor very intensively, red sport of the James Grieve apple. If we could make it both more stable and more intensely colored, that should be a great advantage. I believe that the secondary, stable types of Delicious, like Starkrimson which has originated from the unstable Starking, very well might be examples of such tissue recombinations (1, 28). It is not clear, however, whether Starking is an ecto- or a meso-chimaera.

Reduced fertility is another kind of genetical change that may turn out to be an undesirable by-product in some cases but that might even be desired in other cases. We have been noting that the fruit-set among the irradiated trees sometimes seems to be reduced. In fact, some branches with aberrant foliage never have set any fruit. We have also found that there is a rather high frequency of reduced pollen fertility among apple trees that were irradiated 5 years earlier.

Buds from fertile and partially sterile branches have been brought over onto new rootstocks, and in time it will be possible to see whether this reduction in fertility is also reflected in reduced fruit-set and perhaps in increased fruit size. In some fruit varieties of apples, plums, and grapes, too many fruits are normally formed with reduced size and quality as a consequence. In such varieties a slight reduction in fertility could be a great advantage. This already seems to have been successfully attacked by Olmo (85). We have also had our attention directed towards the same aim and have irradiated plum varieties, like Victoria and Reine Claude d'Oullins, with the intention of inducing some degree of sterility. It may be mentioned in this connection that in many ornamental plants even complete sterility, for several reasons, might be a highly desirable change, and it should be a rather easy one to produce.

Finally, I should like to touch upon two other projects that we have also just started and therefore cannot say much about as yet. The first concerns the induction of thornless sports in blackberries. To that end we have irradiated seeds of an English variety, Merton Thornless, which is said to be apomictic (53). Irradiation of apomicts is a very interesting possibility, both because some of the induced changes might then be constantly reproduced by seeds, but also because the irradiation is known to be able to break down the apomictic mechanism itself which might be desirable in order to release the variation of apomictic species (52, 61, 62).

In addition to mutagenic treatments with various radiations, we have also performed experiments with injected radio-isotopes (33), and also more recently with mutagenic chemicals. From the point of view of mutation induction the treatments with P 32 were not very successful, but this hardly permits us to say that this kind of treatment would generally be so (111).

The treatments with chemicals involved ethyleneoxide, ethyleneimine, and ethylmethanesulphonate. Usually young trees, like those shown in Figure 1, taken just when the buds begin to swell, have been dipped in water solutions of the chemicals for 24 hours.

Various chemicals that are active mutagenes in microorganisms have turned out to be rather ineffective on higher plants as for example the mustard gas substances (48). Chemical mutagenesis in these plants has, therefore, until recently not been very promising. Ehrenberg, et al. (35) found, however, that ethyleneimine (EI) in barley was even more effective than X-rays and neutrons, 20 per cent compared with 4 to 5 per cent mutations per spike progeny. EI did produce some of the typical primary effects of radiations, e.g., yellowish patches on the leaves of young apple seedlings, but not to the same degree as X-rays. This year we made the treatments with ethylmethanesulphonate (EMS), which had been shown by Heslot, *et al.* (56) to give up to 40 to 50 per cent mutations per spike progeny. The primary effects on fruit trees were more pronounced with EMS than with the other chemicals. Even white sectors were found in Striped Williams, so this chemical seems to offer definite possibilities and should be tested further. Treatments with 0.5 to 1 per cent gave the best results; 2 per cent killed most of the buds.

Mode of Origin of Somatic Mutations

If we look back at this survey of mutation experiments we shall find certain features that are common to vegetatively propagated plants. In the first place, I think of the mode of origin and the distribution of mutated tissue. When a mutation originates at a growing point, it will be localized in a single cell. When the tissue grows, the mutation will normally be delimited to a certain one of the histogenic layers at the growing point. There are usually supposed to be three (or four) such histogenic layers, or germ layers, the dermatogen, the periblem, and the plerome, the innermost one (98). These layers are also called L I, L II, and L III, respectively (29, 30, 36).

As the mutation is only very rarely induced exactly in the central part of the growing point but more often at the side of it, it will form a more or less narrow sector at the base, or at the side, of the growing shoot (Figure 12). The shoot will become a mericlinal chimaera, and when side buds are to be formed, these may get several alternative constitutions, as shown in Figure 12. They may become completely normal again, they may retain a sectorial constitution, or they may also become complete, periclinal chimaeras. Only in the last case do we get a stable mutant more or less relieved of the competition, i.e., the intrasomatic elimination, taking place in the growing plant. In the side shoots, the mutated sectors usually will be broader, but also more rare than in the original, central shoot. Many of the induced mutants will probably be lost right in the beginning of the shoot growth and never be included in any side buds.

In some plants there is normally provision for the formation of enough basal buds in a way that effectively conserves the mutations. In the potato plant, for example, the X_2 eyes on the tubers formed by the irradiated X_1 tubers have gone through three such bud formations rather rapidly, first when the stolon is formed in the axis of a basal leaf, then when the tuber is formed on the stolon, and finally when the eyes are formed on the X_2 tubers.

However, in other plants, like fruit trees, the irradiated bud may grow vigorously, several feet long into a shoot where the mutated sectors are probably found only at the basal buds. These buds are usually sitting close together, and normally they never take up growth again due to the apical dominance. These conditions, which have already been briefly touched upon in connection with the survey of the fruit tree experiments, are no doubt the main reason for the differences in mutation yield between different mutation experiments.

Genetic Background of Somatic Mutations

With few exceptions, the genetic background of stable somatic mutations have not been the object of closer studies. This certainly is due to the fact that most of the plants concerned are very incompletely investigated genetically. An exception is formed by the studies of Blakéslee and Avery (14) on spontaneous changes in Datura. These were practically all found to be due to chromosome changes, not to gene mutations. Also in chrysanthemum spontaneous mutations have been found often to be associated with changes in chromosome number (32, 96).

The studies on radiation induced somatic aberrations in the endosperm of maize carried out by Dollinger (31) showed that most of these were losses from a genetical point of view, and that they were mostly associated with chromosome structural changes and deletion of certain chromosome segments. Some of them, however, were not cytologically detectable. This information on the cyto-genetic nature of these changes is also consistent with the repeated finding that most of them are eliminated at meiosis (97).

Although we know very little from other plants, it seems reasonable to assume that conditions would be about the same, namely, that most, though not all, of the induced somatic changes are made up of gross cytological changes with phenotypical effects, most of them, perhaps, simply being losses or duplications of gene material.

By experience we know that most of the radiation-induced mutations studied in seed-propagated plants are recessive, only quite few clear-cut dominant or semi-dominant induced changes having been found (80). Although most of the endosperm changes studied by Dollinger turned out to be losses, changes in recessive direction, he also found indications of dominant changes. In apples we know that gain in color as well as loss of color both do occur (9, 12). It is, however, not necessarily so that the former changes should be dominant. They might as well be caused by losses of suppressor factors. Cases of obvious dominant mutations are known, however. The deep-red mutant of the Williams pear, Max-Red Bartlett, for example, seems to be a really dominant subepidermal factor mutation as it gives rise to 50 per cent seedlings with the increased anthocyanin content in the shoot, typical also for the Max-Red Bartlett.

The Technique of Mutagenic Treatment

As is evident from Table 2, most of the various types of diaspores used in vegetative propagation also have been exposed to mutagenic treatment, including scions, buds, cuttings, bulbs, tubers, etc. The range of doses that have been used is, however, narrow compared with the wide range of doses tolerated by seeds (50, 82) or by continuously growing plants (81, 105, 106). The most common doses lie between 2,000 and 4,000 r units of X-rays or gamma rays. For especially sensitive, herbaceous material these doses may be halved, while for many lignified or dormant structures they may be doubled.

TABLE 2.—EXAMPLES OF RADIATION DOSES THAT HAVE BEEN USED IN ORDER TO INDUCE MUTATIONS IN VARIOUS KINDS OF VEGETATIVELY PROPAGATED PLANTS.

Apples and Pears:
  Dormant scions, 3,000 to 6,000 r, X-rays and gamma rays, Bishop, 1955; Granhall, *et. al.* 1949; Gròber, 1959; Zwintzscher, 1959; 4–6x10^{12} Nth/cm^2, Bishop, 1958
  Summer buds, 2,000 to 4,000 r, Gròber, 1959
  Growing trees, 25 to 50 r/day, several years; 50–100 r/day, 5 months, Granhall and Nybom, unpublished
  Flower buds before meiosis, 1,000 to 3,000 r, Lewis and Crane, 1954
Cherries:
  Dormant scions, 2,000 to 4,000 r, Zwintzscher, 1955
  Flower buds before meiosis, 800 r, Lewis and Crane, 1954
Plums:
  Dormant scions, 2,000 to 6,000 r, Zwintzscher, 1955
  Growing trees, 25 to 50 r/day, several years, Granhall and Nybom, unpublished
Peaches:
  Growing trees, 10 to 60 r/day, 8 to 20 months, Hough and Weaver, 1959

TABLE 2.—Continued.

Grapes:
  Dormant buds, 2,000 to 3,000 r, Breider, 1956; Olmo, 1960
Roses:
  Summer buds, 5,000 to 10,000 r, Gelin and Gustafsson, unpublished
Black currants:
  Dormant wood cuttings, 3,000 r, Bauer, 1957
Blackberries:
  Leaf bud cuttings, 4,000 to 6,000 r; dormant seeds of apomictic species, 10,000 to 20,000
    r, Koch and Nybom, unpublished
Raspberries:
  Spring suckers, 10,000 to 12,000 r, Nybom, unpublished
Potato:
  Resting tubers, 3,000 to 4,000 r, Asseyeva, et al., 1935; Hagberg and Nybom, 1954;
    Heiken, 1960; Howard, 1958
Sweet potato:
  Leaf cuttings, 3,000 to 4,000 r, Nishiyama, et al., 1959; root-tubers, 2,500 to 15,000 r,
    Matsumura and Fujii, 1957
Tulips and Hyacinths:
  Dormant bulbs, summer or early fall, 800 r, X-rays, de Mol, 1944; 2,000 to 5,000 r,
    gamma, semi-chronic, Carlsson, unpublished
Chrysanthemum:
  Stem cuttings, 1,000 to 2,000 r, Jank, 1957
Carnation:
  Stem cuttings, 5,000 r, Sagawa and Mehlquist, 1957
  Growing plants, 200 r/day, 105 days, Richter and Singleton, 1955
Camphor tree:
  Seedlings and cuttings, 2,000 to 6,000 r, Matsumura and Fujii, 1957
Coffee:
  Growing plants, 20 to 100 r/day, several months; seeds, 10,000 to 20,000 r, Moh and
    Orbegoso, 1960
Saintpaulia:
  Leaf cuttings, 2,000 to 3,000 r, Sparrow and Konzak, 1958
Various Bulbs and Corms:
  5,200 r, Spencer, 1955
Alfalfa:
  Rooted cuttings, 3,000 to 4,000 r, Murray, 1956
Antirrhinum:
  Shoots, up to 250 r/day; Cuany, et al., 1958
Nicotiana:
  Rooted cuttings, 50 to 200 r/day, several months, Sand, et al., 1960
Poa pratensis:
  Apomictically produced seeds, 20,000 to 30,000 r, Julén, 1954

In most cases slight modifications of the propagation methods
commonly used for the plants in question seem easily applied for
irradiation of the material. That large variations in the frequency

of isolated mutations might be anticipated after irradiation at different stages of development is shown by Heiken's results (55), but still we know far too little for making general recommendations. The method of decapitating rooted cuttings sometime before irradiation seem to be worth attention (59, 94). As mentioned before, we have been irradiating leaf-bud cuttings of blackberries rooted under mist. It might be added here that the modern mist-propagation methods seem to offer excellent possibilities for the management of irradiated material of vegetatively propagated plants, including the procedure of repeated tipping for the isolation of the induced changes (93).

The intrasomatic elimination is a process that should be counteracted by all means. In barley we know that this loss of mutated cells is more pronounced under some conditions than under other. It is, for example, influenced by the hydration of the seeds, by temperature, oxygen pressure, and dose, being more severe after high doses (34). It plays a much smaller role after neutron irradiation than after irradiation by X-rays or gamma rays (68, 83), which may be one of the reasons why neutrons have been considered to be a more effective mutagen than X-rays in fruit trees (10). In fact, this elimination is of great importance also for the yield of induced mutations in seed-propagated plants. In peas, for example, it will tend to reduce the part of the plant which is heterozygous for the induced mutation, and according to Blixt and Gelin (15), this part is smaller, i.e., eliminated to a larger extent after X-rays than after neutrons. The chemical mutagens tested during later years seem to follow the neutrons more than the X-rays in this respect, which is, of course, much to their credit (15).

We have reasons to believe that this intrasomatic selection is more pronounced in an object where there is great variation in radiosensitivity between the cells or between the primordia. The irradiation should, therefore, be carried out at a stage when the tissue that afterwards is to give rise to the new plant consists of as few and undifferentiated cells as possible.

By introducing chronic irradiation we originally thought we might be able to differentiate between mutation induction, showing no dose-rate dependence, and the probable physiological and dose-rate dependent damage leading to intrasomatic elimination. Thereby, we should get an accumulation of mutations above the maximal frequency obtainable by acute irradiation. For some plants, e.g., barley, we know now that this does not seem to be possible (84). Also, the results of Cuany, *et al.* (25) do not point to any conspicuous advantage of chronic irradiation over acute.

Obviously even growing plants are able to eliminate a certain fraction of the induced changes. Therefore, as pointed out by Sand, *et al.* (97), the "pertinent dose", i.e., the dose corresponding to the number of recovered mutations, is "considerably less than that administered to the whole plant as chronic irradiation during its development". Even though further studies on the possible differences in efficiency between chronic and acute exposures are required, there are no results yet to indicate any disadvantage of the more convenient method of acute or semi-chronic irradiation.

Consequences of Chimaeric Structure of Somatic Mutations

There are certain consequences of this fact that most of the induced and, probably to a still higher extent, the spontaneous somatic mutations are propagated as periclinal chimaeras (7).

One practical consequence is that some of the induced mutations will be more or less unstable, more so perhaps if the new genotype has a reduced competitive ability compared with the original one. Another consequence, also mentioned earlier in this paper, is that most of these mutations when used in crosses will turn out not to transmit the changes to the offspring unless they also comprise the subepidermal cell layer giving rise to the gametes.

However, irradiation seems to constitute a very efficient tool for bringing about tissue recombinations so that di-ectochimaeras or tissue-homogeneous plants are produced. If this succeeds, we may then expect these new types to be both more stable and perhaps more pronounced in phenotypical expression. As far as the genotypical change itself does not impair the fertility or viability of the subepidermal mutants, they should then also become accessible for further breeding work by means of crossing. The spontaneous mutations isolated hitherto present in this way a very attractive and completely unexplored raw material for further irradiation experiments.

Still another consequence of this chimaeric structure is that it permits the use of such changes that, when tissue-homogeneous, they would lead to complete sterility or complete inviability of the plant.

The genotypical background might in such cases consist of losses of large chromosome sections or other gross cytological changes, leading to profound changes of skin properties, for example. This is probably not a very rare example of variety improvement that would not easily, or not at all, be attainable by ordinary cross breeding but only by means of mutations, delimited to a special tissue of the plant. There are thornless blackberry varieties, for example, which seem to give rise to crippled dwarfs when tissue-homogeneous (26), and there are variegated Pelargonium types which, under the same conditions, either are practically lethal or not able to form any flowers (8).

One could also well imagine that many ectochimaeric mutations with retained acceptable characteristics of the deeper lying tissue might not be usable as tissue-homogeneous clones because they would impair fruit quality, flesh color, and other properties. Again, the use of somatic mutations offers quite unique possibilities perhaps somewhat hypothetical but indeed not unlikely. I might refer, for example, to the negative correlation often found between quality and disease resistance (21, 60).

Use of Mutations in Relation to Other Breeding Methods

After his review on radiation in the production of useful mutations, Smith (104) arrives at the conclusion, "The crux of the problem is—is it economically feasible to use radiations in plant improvement? There is no one answer. . . the answers will be different depending on the material, objectives, and circumstances".

I agree with this statement in all its details. The sometimes contradictory and sweeping statements about the value of the mutation method, or the irradiation method, must come from those who have not understood that the mutation method, like any other single method or tool in plant breeding, cannot possibly alone give us all the completely new and improved varieties that we need and that we are aiming at. The mutation method should be regarded as a complement not an alternative to other plant breeding methods, varietal crosses, species crosses, inbreeding, polyploidization, etc.

We may observe that in some cases the mutation method has indeed been "economically feasible". It has led to the production of varieties that are paying for themselves. That it has indeed been an "economical failure" in other cases is also evident. In addition to methodological shortcomings, the limited profit of the mutation experiments on vegetatively propagated plants reported so far may be explained by the fact that practically none of these mutation experiments were planned or carried out as real plant breeding projects. We have been lacking, and are still lacking, most of the knowledge that would permit us to do so. We still know much too little about the possibilities of the mutation method and still less about its limitations. As long as the chief aim in plant breeding is to produce basically new and, from many points of view, improved varieties, I believe the cross breeding method will still be the backbone of the breeding program. At least this appears to be the situation in fruit breeding.

But it may well turn out that after 20 years' expensive and hard work we are left with a variety that is nearly perfect in nine of ten important characteristics. In order to get the fully acceptable variety we might have needed 100,000 seedling trees instead of the 10,000 we could afford. Then, the mutation method might be the only way of coming closer to our aim.

At Alnarp, Sweden, they have had a new seedling clone of apple under test and consideration for quite a few years. It has good quality and good handling properties but lacks distinct appeal in appearance. This year we were shown a beautiful, red spontaneous mutation of this variety. Obviously the basis for releasing this new variety on the market has changed completely with the detection of this red sport.

The organizers of this conference on mutation and plant breeding have raised some questions: How fully has natural variation been used? Should more emphasis be placed on its utilization? Will mutation induction be the one method when this variation is exhausted? Again these questions can not be given one answer. If I may limit myself to fruit again, I think one can say that we have, in general, an overwhelming natural variation, a variation so rich that I think it could not possibly be exhausted within reasonable time. The difficulty is to bring all the desired characters together into one and the same variety; so generally, we would not have to irradiate material only in order to get variation.

But if we go to single fruit kinds or to single characters, the variation is not always so rich nor so easily available. In black currants

(6), the variation among available varieties with regard to size of the trusses (ease of picking) or length of the dormancy period (resistance against late frosts) is not as wide as we might wish. What do we do then, start crossing or irradiating?

In blackberries we have some relatively acceptable varieties, which are, however, all terribly spiny. There are extremely few thorn-less types available for breeding, like the diploid weak-growing and winter-sensitive *Rubus ulmifolius inermis*. It might take us 15 years to bring its recessive *inermis* gene over into acceptable polyploid varieties, if we do succeed. Which will be cheapest, crossing or irradiation? If given resources, we should of course try both ways, well knowing that not all breeding projects started so far have resulted in improved varieties.

Even after critical consideration there are applications like these and others where the mutation method indeed seems worth trying and where it no doubt in many cases will turn out to be economically feasible. With regard to the chimaeric nature of somatic mutations there are also occasions when the mutation method offers quite unique possibilities, or even turns out to be the only practicable way.

We should not forget those plants that lack regular sexual repro-duction, e.g., the apomictic types, or where this can not be utilized because of self-sterility or incompatibility or the absence of flower formation. In the imperfect microorganisms, e.g., the antibiotic pro-ducing ones, deliberate mutation breeding has turned out to be highly successful (5). Also, the possibility of using pollen irradiation in order to realize difficult species crosses certainly deserves more attention among vegetatively propagated plants (27, 78, 90).

Especially if we extend the concept to include all applications of mutagenic agents, it seems beyond doubt that the mutation method will become not only a valuable but even an indispensable tool for the plant breeder. But we shall have to learn much more before we can say when and how to apply it.

References

1. Anonymous. 1956. The Starkrimson (Bisbee Strain) Delicious apple. *Fruit Var. and Hort. Dig.*, 11: *31*.
2. Asseyeva, T. 1927. Bud mutations in the potato and their chim-erical nature. *Jour. Genet.*, 19: *1–26*.

3. ————. 1931. (Bud mutations in the potato.) *Bul. Appl. Bot. Gen. Plant Breed.,* **27:** *135–218.*

4. ———— and Blagovidova, M. 1935. (Artificial mutations in the potato.) *Bul. Appl. Bot. Gen. Plant Breed.,* **A(15):** *81–85.*

5. Backus, M. P., and Stauffer, J. F. 1955. The production and selection of a family of strains in *Penicillium chrysogenum. Mycologia,* **47:** *429–463.*

6. Bauer, R. 1957. The induction of vegetative mutations in *Ribes nigrum. Hereditas,* **43:** *323–337.*

7. Bergan, F. 1955. Einige Konsequenzen der Chimärenforschung für die Pflanzenzüchtung. *Zschr. f. Pfl. Züchtung,* **34:** *113–124.*

8. ———— and Bergan, L. 1959. Ueber experimentell ausgelöste vegetative Spaltungen and Umlagerungen an chimärischer Klonen, zugleich als Beispiele erfolgreicher Staudenauslese. *Züchter,* **29:** *361–374.*

9. Bishop, C. J. 1955. The production of bud sports in apples through the use of artificial radiation. *Proc. 14th Intern. Congr. Hort.,* *740–745.*

10. ————. 1957. Genetic changes in apples induced by thermal neutrons. *Can. Jour. Plant Sci.,* **37:** *55–58.*

11. ————. 1958. Radiation-induced morphological changes and fruit color mutations in the Cortland apple. *Proc. 10th Intern. Congr. Genet.,* **II:** *26.*

12. ————. 1959. Radiation-induced fruit color mutations in apples. *Can. Jour. Gen. Cytol.,* **1:** *118–123.*

13. ———— and Aalders, L. E. 1955. A comparison of the morphological effects of thermal neutrons and X-irradiations of apple scions. *Amer. Jour. Bot.,* **42:** *618–623.*

14. Blakeslee, A. F., and Avery, A. G. 1954. Bud sports in Datura found due to chromosomal and not to gene mutations. *Proc. Huit. Congr. Bot., Paris.* 268.

15. Blixt, S., and Gelin, O. Unpublished communication.

16. Bogdanov, P. L. 1948. Experiments with X-irradiation in poplar breeding. *Bot. Zhurnal,* **33:** *438–440.* Plant Breed. *Abst.,* **19:** *2921.*

17. Borg, G., Fröier, K., and Gustafsson, Å. 1958. Pallas barley, a variety produced by ionizing irradiation: Its significance for plant breeding and evolution. *Proc. 2nd Intern. Congr. Peaceful Uses Atomic Energy,* **15:** *2468.*

18. Breider, H. 1953. Entwicklungsgeschichtlich-genetische Studien über somatische Mutationen bei der Rebe. *Züchter,* **23:** *208–222.*

19. ————. 1956. Ueber Strahleninduzierte somatische Mutationen bei Reben. *Bayer. Landw. Jahrb.*, **33**: *515–533*.

20. Brown, D. S., Hesse, O. C., and Koch, E. C. 1959. Reversions of red sports of Delicious apple. *Proc. Amer. Soc. Hort. Sci.*, **74**: *40–46*.

21. Byrde, R. J. W. 1956. The varietal resistance of fruits to brown rot: I. Infection experiments with *Sclerotinia fructigena* Aderh. & Ruhl. on certain dessert, culinary, and cider varieties of apple. *Jour. Hort. Sci.*, **31**: *188–195*.

22. Carlsson, G. Tulpaner. *Medd. Gullåker Växtf. anst No. 16. 216 pp.*, and unpublished communication.

23. Carvalho, A., Antunes Filho, H. and Nogueria, R. K. 1954. Genética de Coffea: XX. Resultados preliminares do tratamento de sementes de café com raios-X. *Bragantia*, **13**: *17–20*.

24. Crockett, L. J. 1957. A study of the tunica corpus and anneau initial of irradiated and normal stem apices of *Nicotiana tabacum* L. *Bul. Tor. Bot. Club*, **84**: *229–236*.

25. Cuany, R. L., Sparrow, A. H., and Pond, V. 1958. Genetic response of Antirrhinum to acute and chronic plant irradiation. *Zeit. indukt. Abstamm. -u-Vererb. Lehre*, **89**: *7–13*.

26. Darrow, G. M. 1955. Nature of thornless blackberry sports. *Fruit Var. and Hort. Dig.*, **10**: *14–15*.

27. Davies, D. R., and Wall, E. T. 1960. Gamma-radiation on interspecific incompatibility within the genus Brassica. *Zeit. indukt. Abstamm. -u-Vererb. Lehre*, **91**: *45–52*.

28. Dayton, D. F. 1959. Red color distribution in apple skin. *Proc. Amer. Soc. Hort. Sci.*, **74**: *72–81*.

29. Dermen, H. 1948. Chimeral apple sports and their propagation through adventitious buds. *Jour. Hered.*, **39**: *235–242*.

30. ————. 1954. Histogenic factors in color and nectarine sports of peach. *Genetics*, **39**: *964*.

31. Dollinger, E. J. 1954. Studies on induced mutation in maize. *Genetics*, **39**: *750–766*.

32. Dowrick, G. J. 1953. The chromosomes of Chrysanthemum: II. Garden varieties. *Heredity*, **7**: *59–72*.

33. Ehrenberg, L., and Granhall, I. 1952. Effects of beta-radiating isotopes in fruit trees. *Hereditas*, **38**: *385–419*.

34. ————, Gustafsson, Å., Lundqvist, U., and Nybom, N. 1953. Irradiation effects, seed soaking, and oxygen pressure in barley. *Hereditas*, **39**: *493–504*.

35. ————, ————, ————. 1959. The mutagenic effects

of ionizing radiations and reactive ethylene derivatives in barley. *Hereditas,* 45: *351–368.*

36. Einset, J. 1952. Spontaneous polyploidy in cultivated apples. *Proc. Amer. Soc. Hort. Sci.,* 59: *291–302.*

37. ———— and Pratt, Charlotte. 1959. Spontaneous and induced apple sports with misshapen fruits. *Proc. Amer. Soc. Hort. Sci,* 73: *1–8.*

38. Gardner, V. R., Christ, J. W. and Gibson, R. E. 1933. Somatic segregation in a sectorial chimera of the Bartlett pear. *Jour Agr. Res.,* 46: *1047–1057.*

39. Gelin, O. 1955. Studies on the X-ray mutation Strål Pea. *Agr. Hort. Genet.,* 13: *183–193.*

40. ————, Gustafsson, B. Unpublished communication.

41. Gilmer, R. M., and Einset, J. 1959. Some bud mutations of apple with virus-like symptoms. *Plant Dis. Reporter,* 43: *264–269.*

42. Granhall, I. 1955. X-ray mutations in apples and pears. *Hereditas,* 39: *149–155.*

43. ————. 1954. Spontaneous and induced bud-mutations in fruit trees. *Act. Agr. Scand.,* 4: *594–600.*

44. ————, Gustafsson, Å., Nilsson, Fr., and Oldén, E. J. 1949. X-ray effects in fruit trees. *Hereditas,* 35: *267–279.*

45. Gröber, K. 1959. Mutationsversuche an Kulturpflanzen: XI. Die Erzeugung von somatischen Mutationen mittels Röntgenstrahlen beim Obst. *Die Kulturpflanze,* 7: *37–54.*

46. Gustafsson, Å. 1947. Mutations in agricultural plants. *Hereditas,* 33: *1–100.*

47. ————. 1960. The induction of mutations as a method in plant breeding. In *Mechanisms in Radiobiology, Vol. 1,* in press.

48. ———— and MacKey, J. 1948. The genetical effects of mustard gas substances and neutrons. *Hereditas,* 34: *371–386.*

49. ———— and Tedin, O. 1954. Plant breeding and mutations. *Act. Agr. Scand.,* 4: *633–639.*

50. ———— and von Wettstein, D. 1958. Mutationen und Mutationszüchtung. *Handbuch der Pfl. Züchtung,* 2, **Aufl.** 1: *612–699.*

51. Hagberg, A., and Nybom, N. 1954. Reaction of potatoes to X-irradiation and radiophosphorus. *Act. Agr. Scand.,* 4: *578–584.*

52. Hanson, A. A., and Juska, F. V. 1959. A "progressive" mutation induced in *Poa pratensis* L. by ionizing radiation. *Nature,* 184: *1600–1601.*

53. Haskell, G. 1953. Quantitative variation in subsexual Rubus. *Heredity,* 7: *409–418.*

54. Heiken, A. 1958. Aberrant types in the potato. *Acta Agr. Scand.,* **8:** *319–358.*

55. ————. 1960. Spontaneous and X-ray-induced somatic aberrations in *Solanum tuberosum* L. *Acta Acad. Reg. Sci. Upsaliensis,* **7:** *1–125.*

56. Heslot, H., Ferrary, R., Lévy, R., and Monard, Ch. 1959. Recherches sur les substances mutagénes (halogeno–2 ethyle) amines, derivés oxygénés du sulfure de bis–(chloro–2 ethyle), esters sulfoniques et sulfuriques. *Compt. Rend. Seances Acad. Sci.,* **248:** *729–732.*

57. Hough, L. F., and Weaver, G. M. 1959. Irradiation as an aid in fruit variety improvement: I. Mutations in the peach. *Jour. Hered.,* **50:** *59–62.*

58. Howard, H. W. 1958. Transformation of a monochlamydius into a dichlamydius chimaera by X-ray treatment. *Nature,* **182:** *1620.*

59. Jank, H. 1957. Experimentelle Mutationsauslösung durch Röntgenstrahlen bei *Chrysanthemum indicum. Züchter,* **27:** *223–231.*

60. Johnson, G., and Schaal, L. A. 1957. Chlorogenic acid and other orthodihydrophenols in scab-resistant Russet Burbank potatoes. *Phytopath.,* **47:** *253–255.*

61. Julén, G. 1954. Observations on X-rayed *Poa pratensis. Acta Agr. Scand.,* **4:** *585–593.*

62. ————. 1958. Ueber die Effekte der Röntgenbestrahlung bei *Poa pratensis. Züchter,* **28:** *37–40.*

63. Kaplan, R. W. 1953. Ueber die Möglichkeiten der Mutationsauslösung in der pflanzenzüchtung. *Zschr. f. Pfl. Züchtung,* **32:** *121–131.*

64. Klein, L. G. 1957. What is new in apple color sports? *New York State Agr. Exp. Sta. Farm Research,* **23:** *No. 4, 14.*

65. Knight, R. L., and Keep, E. 1957. An early sport of a red currant. *Ann. Rpt. East Malling,* **1957:** *74.*

66. Lewis, D. 1949. Structure of the incompatibility gene: II: Induced mutation rate. *Heredity,* **3:** *339–355.*

67. ———— and Crowe, L. K. 1954. The induction of self-fertility in tree fruits. *Jour. Hort. Sci.,* **29:** *220–225.*

68. MacKey, J. 1951. Neutron and X-ray experiments in barley. *Hereditas,* **37:** *421–484.*

69. Mashimo, I., and Sato, H. 1959. (X-ray-induced mutations in sweet potato.) *Jap. Jour. Breeding (Japanese),* **8:** *233–237.*

70. Matsumura, S., and Fujii, T. 1957. Induction of bud-sports by X-rays and gamma-rays. *Ann. Rpt. Nat. Inst. Genetics (Japan),* **8:** *94–95.*

71. Mehlquist, G. Unpublished communication.

72. Miller, J. C. 1954. Selection of desirable somatic mutations, a means of potato improvement. *Amer. Pot. Jour.*, **31**: *358–359*.

73. Moh, C. C., and Orbegoso, G. 1960. Efectos de radiaciones ionizantes sobre la planta de café. *Café, Turrialba*, **1**: *46. Plant Breed. Abs.*, **30**: *131*.

74. Mol, E. W. de. 1944. Dreizehn Jahre (1928–1940) Röntgenbestrahlung bei Tulpen und Hyacinthen zur Erzeugung von somatischen Mutationen. *Zschr. f. Pfl. züchtung*, **26**: *353–403*.

75. ———. 1953. X-raying of hyacinths and tulips from the beginning, before thirty years (1922), till to-day (1952). *Jap. Jour. Breed.*, **3**: *1–8*.

76. Murray, B. 1956. The X-ray sensitivity of creeping-rooted alfalfa. *Can. Jour. Agr. Sci.*, **36**: *120–126*.

77. Müller, G. 1952. Möglichkeiten der Ertragssteigerung bei der Kartoffel durch Ausnutzung der Stärkegehaltsstreuung innerhalb der Sorte. *Wiss. Z.d. Humb. Univ. Berlin*, **2**: *107–125*.

78. Nishiyama, I., and Iizaka, M. 1952. Successful hybridization by means of X-rayed pollen, in otherwise incompatible crosses. *Bul. Res. Inst. Food Sci. Kyoto*, **8**: *81–89*.

79. ———, Okamoto, M., and Teramura, T. 1959. Radiobiological effects in Plants: II. Effects of X-rays on the development of roots from leaf stalks of sweet potato. *Rpt. Kihara Inst. Biol. Res.*, **10**: *33–36*.

80. Nybom, N. 1954. Mutation types in barley. *Acta Agr. Scand.*, **4**: *430–456*.

81. ———. 1956. Some further experiments on chronic gamma-irradiation of plants. *Bot. Notiser*, **109**: *1–11*.

82. ———. 1957. Plant breeding with the aid of induced mutations. (Swedish with English summary.) *Sveriges Utsädesfören:s Tidskr.*, **1957**: *34–55*.

83. ———, Gustafsson, Å., and Ehrenberg, L. 1952. On the injurious action of ionizing radiation in plants. *Bot. Notiser*, **1952**: *343–365*.

84. ———, ———, Granhall, I., and Ehrenberg, L. 1956. The genetic effects of chronic gamma irradiation in barley. *Hereditas*, **42**: *74–84*.

85. Olmo, H. P. 1960. Plant breeding program aided by radiation treatment. *California Agr.*, **14**: *No. 7, 4*.

86. Prakken, R. 1959. Induced mutation. *Euphytica*, **8**: *270–322*.

87. Pratt, Charlotte. 1960. Changes in structure of a periclinal chromosome chimera of apple following X-irradiation. *Nature*, **186**: *255–256*.

88. Rao, B. V. 1954. A brief review of the work on the use of X-rays in sugarcane breeding in Mysore. *Proc. Conf. Sugarcane Res. Workers India Union*, 2: *11–13*.

89. Reichardt, A. 1955. Experimentelle Untersuchungen über den Effekt vom Röntgenstrahlen in der vegetativen Vermehrung einer alten Rebensorte. *Die Gartenbauwissenschaft*, 2 (20): *355–413*.

90. Reusch, J. D. H. 1956. Influence of gamma irradiation on the breeding affinities of *Lolium perenne* and *Festuca pratensis*. *Nature*, 178: *929–930*.

91. Richter, A., and Singleton, R. W. 1955. The effect of chronic gamma radiation on the production of somatic mutations in carnations. *Proc. Nat. Acad. Sci.*, 41: *295–300*.

92. Rogers, W. S., and Fromow, M. G. 1958. Royal Sovereign strawberry clone M. 415. *Ann. Rpt. East Malling*, 1958: *60–62*.

93. Rowe-Dutton, P. 1959. Mist propagation of cuttings. *Comm. Agr. Bur., England, 1959, 135 pp.*

94. Sagawa, Y., and Mehlquist, G. A. L. 1957. The mechanism responsible for some X-ray-induced changes in flower color of the carnation, *Dianthus caryophyllus*. *Amer. Jour. Bot.*, 44: *397–403*.

95. ————, ————. 1959. Some X-ray-induced mutations in the carnation, *Dianthus caryophyllus*. *Jour. Heredity*, 50: *78–80*.

96. Sampson, D. R., Walker, G. W. R., Hunter, A. W. S., and Bragdø, M. 1958. Investigations on the sporting process in greenhouse chrysanthemums. *Can. Jour. Plant Sci.*, 38: *346–356*.

97. Sand, S. A., Sparrow, A. H., and Smith, H. H. 1960. Chronic gamma irradiation effects on the mutable V and R loci in a clone of Nicotiana. *Genetica*, 45: *289–308*.

98. Satina, S., Blakeslee, A. F., and Avery, A. G. 1940. Demonstration of the three germ layers in the shoot apex of Datura by means of induced polyploidy in periclinal chimaeras. *Amer. Jour. Bot.*, 27: *895–905*.

99. Schander, H. 1954. Untersuchungen über die Gestalt der Frucht bei Kernobst. *Die Gartenbauwissenschaft*, 1: *313–324*.

100. Schamel, A. D. 1946. Bud variation and bud selection. *The Citrus Industry (edited by H. J. Webber and L. D. Batchelor), Univ. Calif. Press*, 1946: *915–952*.

101. ————, Pomeroy, C. S., and Harmon, F. N. 1931. Bud variation in Bartlett pear trees. *Jour. Heredity*, 22: *81–89*.

102. ————, ————. 1936. Bud mutations in horticultural crops. *Jour. Heredity*, 27: *486–494*.

103. Shapiro, S. Unpublished communication.

104. Smith, H. H. 1958. Radiation in the production of useful mutations. *Bot. Rev.*, **24**: *1–24.*

105. Sparrow, A. H., and Christensen, E. 1953. Tolerance of certain higher plants to chronic exposure to gamma radiation from cobalt-60. *Science,* **118**: *697–698.*

106. ———— and Gunckel, J. E. 1955. The effects on plants of chronic exposure to gamma radiation from radiocobalt. *Pròc. Intern. Conf. Peaceful Uses Atomic Energy,* **12**: *52–59.*

107. ———— and Konzak, C. F. 1958. The use of ionizing radiation in plant breeding: Accomplishments and prospects. In *Camellia Culture. New York MacMillan Co., 425–452.*

108. ————, Sparrow, R. C., and Schainer, L. A. 1960. The use of X-rays to induce somatic mutations in Saintpaulia. *African Violet Mag., 13: 32–37.*

109. Spencer, J. L. 1955. The effect of X-radiation on the flowering of certain cultivated bulbs and corms. *Amer. Jour. Bot.,* **42**: *917–920.*

110. Williams, W., and Brown, A. G. 1960. Breeding new varieties of fruit trees. *Endeavour,* **19**: *147–155.*

111. ———— and Dowrick, G. J. 1958. The uptake and distribution of radioactive phosphorus (^{32}.P) in relation to the mutation rate in plants. *Jour. Hort. Sci.,* **33**: *80–95.*

112. Zwintzscher, M. 1955. Die Auslösung von Mutationen als Methode der Obstzüchtung: I. Die Isolierung von Mutanten in Anlehnung an primäre Veränderungen. *Züchter,* **25**: *290–302.*

113. ————. 1959. Die Auslösung von Mutationen als Methode der Obstzüchtung. *Proc. II Kongr. EUCARPIA, 202–211.*

Comments

CALDECOTT: You have indicated that you obtain as many mutants from acute irradiation as from chronic irradiation. If this is true, does the gamma field still serve a useful purpose in a plant breeding program—note a plant breeding program, not a basic research program?

NYBOM: I think we know still too little to distinguish between a mutation breeding program and a basic research program. Certainly more studies are required to elucidate the different effects of chronic and acute irradiations, irradiations at different stages, etc. Also, for the irradiation of bulky objects, like fruit trees and others, a gamma source seems to be a most suitable tool, so I do not think such a source can be said to be out-of-date.

MacKey: In the discussion as to preference of chronic or acute radiation, I think it is very dangerous to generalize. Species with different speeds in cell division and different patterns of diplontic selection are bound to respond very differently to the two kinds of treatment.

Hough: In reference to the use of chronic irradiation in gamma field versus acute irradiation, we have obtained useful mutations in peaches from both chronic irradiation of whole trees and acute irradiation of buds. We think that acute irradiation of buds and acute and semi-chronic irradiation of dormant or potted trees will be most useful in the future.

Gamma irradiation appears to be quite valuable in irradiating some Prunus species in which microsporogenesis occurs in midwinter and in which the buds can be maintained only on the tree for the remainder of the dormant season. Thus, for irradiation during microsporogenesis in these Prunus species, whole-tree gamma irradiation is required.

Singleton: Concerning chronic versus acute radiation it has also been our experience that a Co^{60} source in a gamma field is useful in providing gamma radiation to a wide variety of plants brought near the source for short periods of time similar to the way Nybom uses his gamma field. The gamma field at the Blandy Experimental Farm has been used since 1957 to give semi-acute doses of radiation in a 24-hour period. This machine is an economical radiation source. The machine is on loan to us by the Atomic Energy Commission. Total cost of the gamma field, 1 acre in extent, exclusive of the radiation machine, was $5,554.

Strauss: It seems that the terms micromutation and macromutation have been used also by you in an operational field sense only. Should not these terms be defined more carefully in terms of actual gene changes?

Nybom: I have used the terms in the same sense as Doctor Gaul, just for the sake of convenience only. In higher plants we still know very little about the actual genetical changes underlying the phenotypical effects observed.

Dermen: In the research of the African Violet are the mutations in the control or irradiated plants from adventitious buds?

Nybom: According to the paper of Doctor Sparrow, quoted in the list of references, the new plantlets are derived from single cells of the leaf petiole. They are, therefore, not chimaeric in structure but represent homogeneous changes.

DERMEN: What is the chimeral make-up of yellow variegation in the pear?

NYBOM: I believe the pear clone in question has normal epidermis and second layer and a chlorophyll deficiency in the third cell layer. It remains to be investigated further, however.

Discussion of Session III[1]

W. M. MYERS

University of Minnesota, St. Paul, Minnesota

IN THE EARLY papers of this symposium, and particularly in Doctor Rhoades' discussion, reference has been made to different kinds of events which collectively may be referred to as mutation. These may include true changes in the molecular structure of the gene as well as minor chromosome changes such as deletion, duplication, or position effects resulting from inversions. They may also include major chromosomal derangements, especially in the polyploids and in vegetatively propagated organisms, including losses or duplications of large chromosome segments or entire chromosomes. Finally, recombination by normal crossing-over may result in effects which can be distinguished from true mutation only by refined genetic and cytological analysis, if at all. These may include sudden "large" changes due to crossing-over between members of a compound locus or, on the other hand, merely increases in genetic variability in characters controlled by polygenes.

In the experiments on the role of mutations in plant breeding, it is evident that the very old and very broad definition of mutation as a sudden heritable change has been applied. The necessary cytogenetic tests have not been used in most studies to classify in a more specific fashion than this the kinds of mutations observed. Indeed, it is evident that with presently available knowledge sufficiently critical tests are not possible, in most instances, to distinguish between minute duplications or deficiencies on the one hand and true changes in the physical-chemical structure of the gene on the other. Furthermore, for the purely practical assessment of the value of mutations in plant breeding, it is of little or no consequence whether the heritable changes observed were the result of changes in the DNA molecule itself or of some chromosomal aberration. To be sure, there are good reasons why the plant breeder wants to know the basic nature of the genetic change. But it is not inappropriate that in many of the plant breeding studies, no effort is expended in determining it.

[1] Paper No. 1062, Miscellaneous Journal Series, Minnesota Agricultural Experiment Station.

The title of this section of the program is "The Evaluation of Mutations in Plant Breeding". During the discussion so far, there has been consideration of the role of so-called "mutation breeding" and some concern regarding the relative efficacy of "mutation breeding" versus conventional cross breeding. Perhaps a consideration of the essential phases involved in plant breeding will aid in answering these questions. The three essential phases are:

1. Finding or developing populations in which there is adequate and appropriate genetic variation,
2. Effective selection, including critical evaluation of the selections,
3. Appropriate use of the selections, either directly as the varieties of commerce or indirectly in building such varieties.

In connection with the discussions in this symposium the third phase is not a matter of concern, but the first two are and therefore warrant some further consideration.

Consideration of the first phase is based on the premise that all genetic variability is dependent, basically, on differences in function at corresponding loci of homologous chromosomes; in other words, in the classical genetic terminology, that two or more alleles exist of one or more genes. The premise is also held that these differences at various gene loci have resulted from mutation. Therefore, all genetic variation must have had its origin in mutation. If these premises are valid, and this discussant can see no obvious alternative, the question posed by the title of this program can be given a definitive answer—mutation is a necessary prerequisite to successful plant breeding for, without mutation, there could be no genetic variation and without genetic variation plant breeding would be impossible.

The words adequate and appropriate were used to describe the required genetic variation. By appropriate genetic variation, it is meant that for success in the plant breeding endeavor there must be in the population with which the breeder is working the genes which will, in proper combination, condition the expression of the characteristics sought in the improved variety. By selection, the breeder can do no more than extract from the population lines or new populations which have combinations of the alleles that existed in the population. Genetic advance, that is the success of the breeding effort, is a function of the selection differential times the heritability. Both of these factors are determined in part by the extent of genetic variability in the population. Therefore, within limits, the greater the appropriate genetic variability, the better the average chance and magnitude of success.

Before considering the tools available to the plant breeder for developing populations within which he practices selection, it will be desirable to consider the kinds of genetic variation, from the standpoint of source, available to him. For purposes of consideration, these may be divided into the following four classes:

1. Spontaneous mutations that were in the wild ancestor of the species or that have occurred and been preserved since its domestication,
2. Spontaneous mutations that have occurred and been preserved in related species,
3. Spontaneous mutations that arise in the breeding cultures,
4. Induced mutations produced by any of the mutagenic agents.

From the considerations given above, it becomes clear that the only basic difference between so-called "mutation breeding" and "conventional breeding" is in the source of genetic variability used by the breeder. It would seem preferable, therefore, not to speak of "mutation breeding" as a distinct method, but simply to recognize induction of mutations as a tool for obtaining genetic variability. For the population with which he works, the plant breeder may accept those available, either from natural sources or from mutagen-treated materials. On the other hand, he may build populations by cross-breeding, using parents that differ either because of natural mutations, induced mutations, or both. Finally, he may produce new populations with more desirable frequencies and combinations of genes by such tools as backcrossing or recurrent selection. Here again, he will be using mutations as the sources of genetic variability and these mutations may be spontaneous in origin, induced by mutagenes, or both. Induction of mutations as a breeding tool must be considered for efficiency or value, therefore, not in comparison with "conventional" breeding methods, but rather in terms of induced versus spontaneous mutations. It is appropriate, therefore, to consider in more detail the kinds of variation outlined above.

No one who has worked with the crop plants can help but be impressed with the enormous natural variation that exists in most species. Doctor Quinby has pointed out the variation in sorghum. The

3,000 forms studied by Snowden were classified into 31 species on morphological grounds; yet all of these may be freely interbred and, on a genetic basis, can be considered as one species. Doctor Quinby has reviewed for this symposium the accomplishments in remodeling this species from a too tall, too late, tropical forage grass into short, early maturing, grain-producing varieties adapted throughout most of the Great Plains of the United States. In this biological engineering program, the basic materials were mutations, particularly at three loci controlling floral initiation, four loci controlling plant height, and several loci which condition grain color, grain type, and disease and insect reaction. Finally, a gene which reacts in the appropriate cytoplasm to produce male sterility has provided the mechanics for hybrid seed production, thus enabling the plant breeder to reap the benefits of heterosis on a practical scale. In connection with this story of the use of "macro" mutations in remodeling the sorghum crop, it is important to be aware also of the innumerable minor genes of which suitable combinations have provided for varietal superiority and heterosis within the larger groups determined by the major genes.

Variability of the magnitude found in sorghum is also encountered in other crops. Corn, wheat, and alfalfa are notable examples. All students of plant breeding are familiar with the writings of Vavilov based on most extensive plant explorations and collections by him and other Russian scientists. A recent, posthumous publication (8)[2] dealing with world resources of cereals, grain, leguminous crops, and flax appeared in 1957. Vavilov, as noted by Hayes, Immer, and Smith (5), for example, defined eight principal regions of origin of crop plants. It is in these regions that the greatest genetic diversity in a crop is found.

Many others, including Harlan (4), have also written about the enormous natural variation found in most crop plants. It has been estimated that domestication of most of the present-day crops began 50 to 70 centuries ago and the species from which domestication started may well have arisen many centuries or millennia before that. One could assume in the thousands of generations and vast populations of a crop that have existed that every possible genetic change has occurred, not once but probably many times. To be sure, it is probable that most of these spontaneous changes have been lost by

---

[2]See References, page 305.

natural elimination or by chance. Even so, it is not surprising that so much variability exists. Even with the greatly accelerated incidence of mutations induced by mutagenic agents, the plant breeder may, with his limited time and facilities, be unable by controlled mutagenesis to exceed or even to equal the work of nature.

Faced with this fact and realizing that mutations provide the "life blood" of plant breeding, it seems appropriate to sound again the alarm regarding the danger of losing our natural sources of germ plasm. In a sense, the plant breeder is his own worst enemy. Improved varieties developed by him replace the natural stands, farmer's seed lots, and land races that are the reservoirs of the accumulated spontaneous mutations. It is fortunate for the future of plant breeding that the eight regions of origin, as defined by Vavilov, are in the less-developed areas of the world. But this situation is changing rapidly. Some of the great gene centers may have already been destroyed by the advancing science of plant breeding (4). So-called "World Collections", valuable as they are, represent only a fraction of the world's germ plasm. The "Germ Plasm Bank" at Fort Collins, Colorado, is only a beginning of the facilities required if the best of the world's germ plasm is to be preserved. Plant breeders throughout the world must increase their efforts to assemble and save the storehouse of variability which has accumulated over so many centuries.

This discussant believes that in many crop plants, natural variation can be found for almost any characteristic that might be sought in the crop. Obviously, there are limitations. One would probably not expect to find a genotype in wheat capable of symbiosis with Rhizobium bacteria. One of the interesting stories of extensive search for and the eventual finding of mutations required is that of the work done to convert the *Lupinus* spp. from wild plants to cultivated. This may be found in the publications of Von Sengbusch and his German associates which are reviewed by Hackbarth and Troll (3). As one example, Von Sengbusch and his staff examined more than one and one-half million plants searching for low alkaloid content. A total of six such plants were found.

In search for genetic variability in one crop, the plant breeder can use as a guide the range of variability found in related species and genera. This is in accordance with the Law of Homologous Series, proposed by Vavilov (5). It is interesting that Doctor Quinby has pointed out the parallelism between mutant types found in sorghum and corn, i.e., endosperm color, waxy endosperm, sugary endosperm, etc.

Added to the genes of the cultivated species as possible sources of variability are the vast array of genes in related species. Classical examples of transfer from a related wild species of genes not known to be available in the cultivated species are found in wheat, potatoes, tobacco, and other crops. Modern advances in embryo culture, polyploidy, use of bridging species, alien chromosome and gene substitution, and other techniques are expanding the range of species from which desired genes can be obtained (6).

The third class of variation, as outlined above, consists of the spontaneous mutations that occur in the breeding cultures. These probably occur more frequently than was once thought. An early recognition of the frequency of spontaneous mutations was by East. From studies of maternal diploids of *Nicotiana rustica,* obtained in attempts at interspecific hybridization, he found that each progeny row was astonishingly alike, more so than "any ordinary inbred population that I have ever examined". When these lines were continued by self-fertilization, they were, within three or four generations, as variable as ordinary inbred populations (5). Despite their frequency, however, spontaneous mutations do not occur often enough in the breeding cultures to provide the amount and kinds of variation required in plant breeding programs.

The papers of Doctors Nybom and Gaul have dealt in considerable detail with the potentialities of using induced mutations. Doctor Nybom has emphasized particularly the use of induced mutations in improving varieties of fruits, bulb flowers, and other perennial ornamentals that are vegetatively propagated. In these crops, in the absence of a sexual cycle, the only genetic variation that is found in a variety must arise from mutation. When a sexual cycle is used to provide for variation, the extreme heterozygosity of the parent variety or varieties results in populations within which the favorable gene recombinations can rarely be found in the size of sample with which the breeder can work. Furthermore, the long life cycle of many of these species limits the number of generations which can be grown. It is not surprising, therefore, that in these crops so many of the new varieties are mutant types from older varieties. Obviously, in such a crop, agents that will increase the frequency of mutation can be a powerful tool. This is particularly true with the ornamentals in which deviations from existing varieties have immediate appeal and in which vigor and reproductive capacity are not characters of prime importance.

Two points made by Doctor Nybom, in addition to the documentation of varietal improvements resulting from induced mutations, seem to deserve special mention. One of these is his suggestion that mutation induction may be expected in the vegetatively reproduced crops to pave the way for production of new commercially important varieties only if generally "good" varieties are used in mutation induction. The other point is that what appear to be mutations can be produced in periclinal chimeras by tissue alterations.

Doctor Gaul has cited cases of favorable mutations induced in seed-propagated crops and of varieties released which owed their improvements apparently to induced mutations. From these reports one could conclude that induced mutations are a valuable addition to the natural variability that exists in the crop species. In fact, one is tempted to say that these reports are "too good to be true". This is not a valid reason for failure to accept the documented results of research. It does seem appropriate, however, to point out some precautions that must be used both in conducting mutation experiments and in evaluating reported successes.

The first precaution is the need for adequate controls. The varieties of self-pollinated crops are not, contrary to a commonly held opinion, pure lines. Rather they are mixtures of pure lines and, although they may appear to be uniform, they exhibit variability in physiological character and often, under new environments, in morphological characters as well. It is not sufficient proof of mutation-induced variability merely to have selections from a treated variety which differ from the variety in physiological characteristics. The treated population must be, in fact, a pure line or an untreated portion of the population must be subjected to selection pressure comparable to that used in the treated material.

The nature of varieties of self-pollinated crops, namely, that they are mixtures of pure lines, is probably in some cases responsible for their broad adaptation in years and locations. When single lines are selected from such a variety following mutagen treatment, there is real danger that the derived lines, being more nearly "pure lines", will have a narrower range of adaptation than the original variety.

A second precaution required is the prevention of contamination due to cross-pollination. Some natural cross-pollination occurs in almost every self-pollinated crop and the amount of such contamination is sharply increased by the sterility in the first generation following treatment with the mutagene.

Caldecott, et al. (2) have suggested that many of the variant types found in populations following radiation may actually have been of this origin. Subsequent studies by Ausemus (unpublished) have, in fact, shown that the rust-resistant segregates found in wheat populations following radiation (7) were probably the result of cross-pollination with other rust-resistant varieties, even though the X_1 generation had been spatially isolated in the breeding nursery.

A third precaution has to do with the validity of plot trials at one or two locations in 1 or 2 years. Such preliminary data provide a very uncertain basis for concluding that true increases have been obtained in yield or other characters having low heritability.

An example of efficacy of "mutation breeding" that might be viewed with some reservation because of the precautions stated is the case reported by Doctor Gaul of Jutta winter barley obtained as an induced mutant from Kleinwanzlebener. The new variety is said to have improved winterhardiness, straw-stiffness, and yielding ability. This seems to be a large order of favorable effects to have occurred in a single line from treated material.

Doctor Gaul has suggested that particular emphasis be given to "micro" mutations involving the genes of polygenic systems. In this connection it should be noted that so-called micro mutations might result either from a change in a minor gene (polygene) or from a small change in a gene having ordinarily major effect. As Doctor Gaul has pointed out, most of the characters of major importance in plant breeding are quantitative in nature and dependent upon polygenes. Whether or not there is more likelihood of favorable mutations and less of strongly deleterious mutations in these numbers of polygenic systems and whether or not, because of the number of polygenes involved, the chances of mutations significantly affecting quantitative characters are greater, are questions which must be settled by further experimentation. Nevertheless, as Doctor Gaul has suggested, more attention in mutation experiments needs to be given to changes in the polygenic systems.

In connection with the studies of "micro" mutations, Doctor Gaul has pointed out the need for better selection procedures that will permit accurate mass screening of populations for heritable variations in the quantitive characters. Returning for a moment to our three phases of plant breeding, it is evident that this suggestion applies to the second phase, i.e., selection. Efficient selection methods are, of course, just as important in populations where the variability is due to accumulated spontaneous mutations as in those where it is due to induced mutations.

The suggestion that efficiency of selection for yield and other less readily evaluated characters might be increased by selection in mutagene-treated materials for more readily observable changes, should be viewed with some reservation. When the more readily observable character is a component of the less readily observable one, the method seems to have potentialities. Selection for increased seed size with the objective of getting increased yield, as illustrated by Doctor Gaul, is a good example. On the other hand, there seems in many cases to be no *a priori* reason to assume that the pleiotropic effect of a macromutation on a quantitative trait such as yield will always be favorable. It might just as logically be expected to be unfavorable.

As part of a discussion of the papers presented in this program, it seems appropriate to list some of the questions which are pertinent to the problem of the efficacy of induction of mutations to provide genetic variability for use in plant breeding. Some of these questions, as yet unanswered, are:

1. Are the induced mutations the same as spontaneous ones? Related to this are the questions of whether mutations can be induced at loci where such mutations are not available in natural populations or whether new kinds of mutant alleles can be induced at loci at which alleles from mutation are already known. It is evident that a definitive answer to this question is not yet available. Also related is the question of whether true changes in molecular gene structure (true gene mutations) actually can be induced in higher plants by mutagenic agents. There seems to be fairly general acceptance of the fact that they can be. Yet of the induced

mutations in corn which have been subjected to sufficiently critical genetic analysis none has been of this type.

2. What is the frequency of total mutations? What proportion of the total mutants are favorable? Doctor Gaul has reported some of the information now available in answer to these questions, but it is evident that much more information is required before a definite answer can be given for any crop species.

3. Are more efficient methods of treatment possible so that the total percentage of mutations can be increased? Doctor Gaul has dealt with this problem, as have others in this symposium. It is obvious that this is still a fertile field of research.

4. Can we look forward to directed mutations? There is already some evidence that, by appropriate treatments with different mutagenes, the relative frequencies of "mutations" versus gross chromosomal changes and of various kinds of "mutations" can be altered. Even if some day true directed mutations become possible, we might then ask how useful this will be to the plant breeder.

5. Are the results of mutation, recombinations, etc., in viruses and bacteria completely applicable to the higher plants? There have been suggestions at this symposium and elsewhere (9) that differences in chromosomal organization and gene action might exist which would limit the validity of extrapolation of results from the lower to higher organisms.

Another consideration, not directly related to mutations but certainly involved in the extent of genetic variability, is that of linkage and recombination. Numerous experiments have shown something of the magnitude of the genetic variability, particularly for characters determined by polygenes, that is "locked up" in chromosome blocks by linkage and that is slowly released in succeeding generations by recombination. Anderson (1) has written about the hinderance to recombination imposed by linkage and refers to the cohesiveness of the germplasm. It has already been suggested in this discussion that sudden (or gradual) changes that appear to result from mutation may actually be caused by recombination. One might indeed question whether the results of the selection experiment in Drosophila, referred to by Doctor Gaul, might have been due to increased genetic recombination rather than to mutations as was suggested.

It is known that treatment with gamma- or X-rays or with high temperature results in small increases in frequency of crossing-over at least in some regions of the chromosome, notably in regions proximal to the centromere (10). Most chemical mutagenes that have been tested cause some crossing-over in the male Drosophila as also does radiation, according to Doctor Auerbach in personal conversation. The possibility, as yet virtually unexplored, exists that certain chemicals or other treatments (not necessarily mutagenic) have the capacity greatly to increase crossing-over. The natural variability that could be "unlocked" by radiation, chemical, or other treatment might equal or exceed any new variability that could be induced by mutagenic agents.

In conclusion the following points might be made:

1. There is a growing body of information pointing to the possibilities of inducing favorable mutations in crop plants.

2. Most plant breeders, working with crops in which an enormous amount of genetic variability exists, should concentrate at present on more effective use of that natural variability.

3. Plant breeders should not, with available information and techniques, use mutagenic agents in their breeding programs with only the vague objective of increasing variability.

4. Plant breeders who have specific reasons for increasing genetic variability may find mutagenic agents are useful tools. An example might be the case where resistance to a specific race of a disease organism is not known and is needed. Another example would be with crops which seem to have limited natural genetic variation such as peanuts. (See Gregory, this symposium, for example.)

5. Research on the questions listed above and others relating to the efficacy of induced mutation versus use of naturally occurring genetic variation should be carried out intensively by plant breeders, geneticists, and cytogeneticists who have the facilities and opportunities to do a competent job of such research.

References

1. Anderson, Edgar. 1949. Introgressive Hybridization. *New York: John Wiley & Sons, Inc.*

2. Caldecott, Richard S., Stevens, Harland, and Roberts, Bill J. 1959. Stem rust resistant variants in irradiated populations—mutations or field hybrids. *Agron. Jour.*, **51**: *401–403.*

3. Hackbarth, Jaochim, and Troll, Hans-Jürgen. 1959. Lupinen als Körnerleguminosen und Futterpflanzen. *Handbuch der Pflanzenzüchtung*, 4: *1–51. Berlin: Verlag Paul Parey.*

4. Harlan, Jack R. 1956. Distribution and utilization of natural variability in cultivated plants. In *Genetics in Plant Breeding. Brookhaven Symp. in Biology, No. 9, 191–206.*

5. Hayes, H. K., Immer, F. R., and Smith, D. C. 1955. *Methods of Plant Breeding. New York: McGraw-Hill Book Co., Inc. Ed. 2.*

6. Myers, W. M. 1960. Some limitations of radiation genetics and plant breeding. *Indian Jour. Gen. & Plant Breed.*, 20: *89–92.*

7. ————, Ausemus, E. R., Koo, F. K. S., and Hsu, K. J. 1956. Resistance to rust induced by ionizing radiations in wheat and oats. *Proc. of the Intern. Conf. on the Peaceful Uses of Atomic Energy, Geneva. New York: United Nations, Vol. 12: 60–62.*

8. Vavilov, N. I. 1957. World resources of cereals, grain, leguminous crops, and flax and their utilization in plant breeding. In *Agroecological Survey of the Principal Field Crops. Izdatel stvo Akademii Nauk. SSSR Moskva-Leningrad. (Plant Breed. Abs., 28: 3576. 1958.)*

9. Westergaard, M. 1960. A discussion of mutagenic specificity: 1. Specificity on the "geographical" level. *Chem. Mutagenese Erwin-Baur-Gedächtnisvorlesungen. I. Abhandlungen der Deutschen Akademie der Wissenschaften zu Berlin, 116–121.*

10. Whittinghill, Maurice. 1955. Crossover variability and induced crossing-over. *Symp. on Gen. Recombination, Jour. Cellular and Comparative Physiol., 45:Suppl. 2, 189–217.*

Comments

GAUL: If I understood it correctly, you have the feeling that the example of Professor Vettel which I gave might not be representative. Vettel indicated that only 0.3 per cent of his 5,045 F_2 populations resulted in new varieties. You mentioned that professor Hayes needed fewer F_2 populations for the development of new varieties. Such a comparison is, however, very limited because it depends on the breeding procedure. I would like to repeat that Vettel is one of the most successful German breeders. In Europe many breeders prefer to have many F_2 populations and analyze only a few very carefully. If I understand, you at Minnesota did the contrary. You had very few hybrid populations which you analyzed very carefully. Thus, I think that a comparison of Vettel's results and yours is very difficult, if not impossible.

MYERS: I quite agree with Doctor Gaul that a comparison of Doctor

Vettel's results and Doctor Hayes' results is of limited validity, since one has the choice of growing many F_2 populations with few plants in each or few F_2 populations with many plants in each. Perhaps the most important single factor in success is the total number of plants that can be grown. Nevertheless, the plant breeder who selects carefully the parents he uses, as did Doctor Hayes, will make fewer crosses and have fewer F_2 populations. If, however, he was successful in selection of the parents, the average value of his F_2 populations would exceed that of the F_2 populations of the breeder who did not select the parents so intensively.

The real point I was trying to make, however, is that the low percentage of Vettel's crosses from which new varieties were obtained is not, in itself, a valid reason for condemning the pedigree method. If, to take an extreme example, one grew only one F_2 plant of each cross, there would be little likelihood that any superior varieties would result from 5,045 F_2 populations.

MacKey: The spontaneous mutations occurring in the breeding cultures mentioned by Doctor Myers as one cause of variability should definitely not be overlooked. In wheat, using a marker gene on chromosome IX, I found 0.7 per cent mutations (speltoids) in a homozygous stable variety, close to 20 per cent after optimal X-irradiation, close to 40 per cent after optimal neutron irradiation, but more than 50 per cent in heterozygous F_1 material, all calculated on a per plant progeny basis.

Davies: In relation to Doctor Myers' comment on artificially stimulating crossing over, we have already some results in this field. If we expose certain early meiotic stages to low doses of gamma radiation, a definite increase in the number of true chiasmata is observed in the bivalents at metaphase.

Langham: There seems to be a tendency of new breeders to ignore partially the natural genetic variability in a species (such as sesame) and go directly to radiation equipment to look for induced mutations. I believe emphasis should be placed in (1) obtaining a world collection of genetic variability and (2) looking for new methods of screening this material for useful genes. In this connection, the High-Low method of breeding, as used in our sesame program, has given positive results and offers a useful tool in screening available breeding material. The High-Low method is as follows:

A. It assumes that genes with strong potentials:
   — exist in the population under study
   — are not showing their maximum expression due to specific buffers
   — are maintained in a buffered condition in High × High or Low × Low crosses (such crosses give *some* transgressive segregation)
   — segregate free of their specific buffers as a result of High × Low crosses (such crosses give *strong* transgressive segregation)
   — receive an enhancing effect from specific buffers of the other extreme

B. It consists of:
   — crossing the best parent with the worst parent for the character under study (High × Low)
   — growing large F_2 populations from such crosses
   — selecting at both ends of the curve for plants showing extreme transgressions (H' and L'); H' is H minus its buffers, and L' is L minus its buffers
   — intercrossing favorable segregates (H' × H', or L' × L') to accumulate genes affecting same character or complementary characters
   — crossing H' with L and selecting in F_2 for H'' (H'' = H' plus enhancers from L)
   — crossing L' with H and selecting in F_2 for L'' (L'' = L' plus enhancers from H)
   — crossing H'' × H'' and L'' × L'' for further advance in respective directions

C. It works as demonstrated by results of sesame breeding programs involving the following characters:
   — length of pods
   — gland number
   — disease resistance (Alternaria and Cercospora)
   — insect resistance (aphids)
   — number of branches
   — dehiscence

Session IV

Utilization of Induced Mutations

F. L. PATTERSON, *Chairman*
Purdue University,
Lafayette, Indiana

Screening Methods in Microbiology

THOMAS C. NELSON

Eli Lilly and Company, Indianapolis, Indiana

S CREENING methods as used in both basic and applied microbiology are designed to uncover organisms possessing specific physiological characteristics. The infrequent occurrence of these organisms, existing as only a small fraction of the population, requires that an effective screen be devised. Efficient screening techniques have been developed for various types of organisms of interest in basic research. Problems arise in adapting such techniques for industrial use, since the aims of the two types of investigation are different. In basic research the process and not the result is often of sole interest, the yield of variant types being used as a measure of effectiveness of variables, as in studies on mutagenesis. When it is the variant organism that is of interest, as in studies on the pathways of biosynthesis using nutritionally deficient mutants, the property sought is a lack rather than a gain of physiological function. Industrial screening searches for a gain of function, either qualitative, in the production of a new antibiotic, or quantitative, in yield improvement. Screens to detect such changes, especially quantitative variations, are difficult to design. It is however possible that some screening methods for readily obtained variants can be adapted to industrial problems. This article will discuss modifications of some screening methods of basic research and the applicability of the resulting readily obtained variants to antibiotic production. Other industrial fermentations and aspects of production are reviewed yearly (26).[1]

Variants with Increased Yield

The bulk of developmental microbiology in the antibiotics industry consists in screening for increased yield, "potency" in industrial jargon. More effort is expended in this phase of development than in experimentation with media composition, conditions of fermentation, and purification. The reasons for this are, first, increase in yield through culture selection has been continuous and usually larger in

---

[1]See References, page 327.

magnitude than improvements by other modifications and, secondly, chemical purification can only remove what is already present and now approaches complete recovery. With culture development responsible for yield improvement from the 1 to 10 gamma/ml to the 1,000 gamma/ml range, further increase is expected.

Practically no experimental methods and results are published from industrial laboratories on this topic, since these are trade secrets (59). A complete and detailed account of strain selection for increased penicillin yield has been published (8). The absolute yields are, however, below present industrial production levels. References to published papers concerning strain selection of commercially important antibiotics have been published (2); other papers appear in symposia (14, 41) and review volumes (62).

No published work clearly separates the effects of culture selection from changes in media, methods of growing inocula, and conditions of fermentation on yield of commercially important antibiotics during industrial production. An interesting account of a strain selection program is given by Kashii, *et al.* (67) using successive mutagenic treatments, isolation, and application of statistics.

When a decision is made to produce pilot plant quantities of a new antibiotic for pharmacological and clinical tests and determination of structure, a program of strain selection and media development is usually begun on a laboratory scale. This program then expands to include conditions of inoculum development and fermentor tank operation in the pilot plant. Strain selection programs in the laboratory follow a sequence of steps, regardless of the specific organism and antibiotic. These steps are routine and have been in use throughout the period of industrial antibiotic production. These will be described with some speculations on different approaches suggested by recent genetic findings.

A word of caution concerning measuring quantitative differences in antibiotic yield needs introduction here. The only proof of the worth of an antibiotic-producing culture is the yield obtained under actual production conditions. Since strains with an improved yield from a production level culture are rare, a less expensive system than tank fermentations returning results rapidly on a large scale is necessary. The "shake flask" system is commonly used. The antibiotic yield of various isolates is determined in a small volume of medium in "fermentor flasks", inoculated from a previously grown "vegetative flask", stirred and aerated on rotary shakers at constant temperature. The misnomer "fermentation" is industrial jargon for a highly aerobic process. The activity of the antibiotic is measured by bioassay which has a readily determinable error, usually less than 5 per cent. Problems of bio-assay are discussed in a recent article (68).

An apparent increase in yield may be due to the production of related antibiotics, hence all new strains must be tested for the presence of other components. The major problem in shake flask testing has been the large variation between separate runs under apparently identical conditions. Attention to many details is necessary to reduce this variation below 30 per cent when determined by an analysis of variance between and within runs. While different antibiotic fermentations will each be found to have their own peculiarities, the major factor causing variation between runs has been the different states of the vegetative growth. The age and extent of vegetative growth can be determined and controlled on pilot plant and production scales but not in shake flasks, where smaller volumes of media and larger numbers of tests are required. Several inoculations of vegetative growth at different times may be used to compensate for different ages of growth.

Any claims of increased antibiotic yield should be accompanied by a measure of reliability based on repetitive determinations in separate experiments, preferentially using pilot plant scale equipment. Use of sequential analysis as an aid in making decisions has been described (91). An increase in yield may occur on "scaling-up" the fermentation from the smaller laboratory equipment through the pilot plant. Since the same culture and media are used in both shake flasks and fermentor tanks, the fortuitous increase in yield is an indication that the medium can support more antibiotic production than is found in shake flasks. The limiting factor in laboratory scale fermentations may be the physical characteristics of the equipment, often giving insufficient aeration and agitation. Strains selected for higher yield in shake flasks may possess a different response to these physical limitations and not scale-up to larger equipment.

Purification of Culture

A new antibiotic-producing culture is usually "unstable", producing colonies with different morphology and variegated sectors within individual colonies. These different morphological types may be "purified" by any of several methods for obtaining single cultures. A spectrum of stable but different morphological types will be obtained along with the unstable prototype. The yield of different types will often be different and a pattern of production associated with morphology may be found. The variability may be due to hetero-karyosis (57). An isolate with stable morphology and high yield is established to supply inocula. The original culture should be retained since some derivatives are found to "degenerate", that is, decrease in yield, to be susceptible to actinophages, or fail to survive mainte-nance techniques such as lyophilization.

Isolation of Colonies with Different Morphology

Usually, some variation in morphology can be found in cultures that have been repeatedly "purified". New morphological types can be uncovered by plating on different media. An increase in morpho-logical variation follows mutagenic treatment and may be used as a measure of effectiveness. Increased variation may not be due to muta-tion, however. Dissociation of homokaryotic sectors from hetero-karyotic growth, induction and lysis by temperate phage or bacterio-cins, changes in cytoplasmic determinants of morphology, if such exist, and changes of balance of nuclear types may produce variants that are not mutants. Thus, the use of morphological variation as a measure of induced mutation is not valid unless proved by genetic analysis. Such variation may be a partial explanation of pleomorph-ism found among the actinomycetes as well as the higher fungi (13).

Experience with morphological variation in antibiotic-producing cultures can be used to predict what types do not produce desirable yields, but not what types will produce higher yields than the cultures established as the norm or "controls". Two types common to both the actinomycetes and higher fungi and producing no or very little yield are aconidial and colorless forms.

Random Isolation

Isolation of large numbers of colonies derived from individual spores surviving various mutagenic treatments is the next phase of a strain selection program. It is essential that these isolates be pure clones and not mixtures of mutant and nonmutant genomes. Muta-genic treatments applied to suspensions of nongerminating conidia followed by dilution and plating to obtain well separated colonies will usually give pure colonies since the conidia probably contain a single haploid genome. Mutagenic treatments applied to germinating conidia and to mycelial growth will give genetically mixed colonies derived from several genomes. Testing such isolates is fruitless. An intermediate growth step with sporulation and replating is necessary to eliminate this error when the mutagenic treatment is applied to growing cultures. It is convenient to use this step to measure the yield of several classes of easily recovered variants (resistance to antibiotics and phage) as a function of the treatment method. These "mutagenicity indices" may be used to maximalize production and recovery of mutants. While a generalization of effective conditions to mutations affecting yield may not be valid, it is possible to eliminate inefficient methods which would prevent the recovery of any isolates with improved yield.

Small yield improvements may sometimes be obtained by random selection of colonies plated out from production fermentor tanks. Such isolates may be returned to production fermentor tanks without more than one shake flask test showing yields at or above production levels. This program can often be coupled with continuous changes in operation of fermentor tanks and minor changes in media constitution in a form of evolutionary operations. Unfortunately, a statistical interpretation of results on either a laboratory or a production scale is not often possible since projected demands for material for processing and the cost of operating nonproductive equipment as "controls" may prevent carrying out valid experiments.

Random isolation methods are distinguished from other techniques by the fact that the culture isolated is not known to be a variant. So many tests of different isolates may have to be made as to make random isolation an inefficient program. It is advisable to set a limit to the number of isolates to be run with various treatments to avoid an unrewarding search.

Isolation of Variants with Known Physiological Differences

The possibility of determining the potential yield of isolates without an initial quantitative test in shake flasks has occupied most industrial workers, but no adequate method has yet been reported. A technique commonly used is the "streak plate", on which several isolates of an antibiotic-producing culture are grown using a suitable solidified medium, and their antibiotic production measured by streaking a sensitive indicator organism on the uninoculated portion of the plate. Various refinements of this method have been devised, the most elaborate is a "top-layering" method in which the well developed colonies of the producing organism are covered with an inoculum of a sensitive organism. The size of the zone of inhibition of the sensitive organism around a colony of the antibiotic producer, or the ratio of zone to colony size, is used as a measure of production. A positive growth response may be substituted for inhibition by using an antibiotic-requiring organism, such as streptomycin-dependent *Escherichia coli* (36) or macrolide-dependent Micrococcus (64).

While logical in theory, these rapid test methods are difficult in practice. The antibiotic-producing organism must be plated in convenient number, develop at a constant rate to the same extent, position of colonies on the test plates must not influence results, a medium adequate for growth of both the antibiotic-producing and sensitive organisms chosen, and the producing organism must be isolated free of the test organism. This method has been disappointing with the exception of hyphal tip isolations of Penicillia (44). Nonproducing isolates can be recovered. There is usually little correlation between increased inhibition zone diameter and the yield in submerged fermentation (14). Again, as with correlations of yield and morphology, one can go backwards but not forwards. Possible explanations for an absence of correlation are that the test organism is inhibited by factors other than the antibiotic, and that antibiotic yields obtained by colonial growth on the surface of media are not responsive to the same factors controlling yield in submerged fermentation.

A popular approach to isolates with increased yield has been the selection of antibiotic-resistant variants (113). An isolate producing a specific level of antibiotic is assumed to be resistant to this level, regardless of whether the antibiotic is being produced. Isolates of greater resistance are assumed to be immune by virtue of increased yield. As an initial screening method for new cultures or for "degenerated" cultures, this technique is effective but fails for variants resistant to greatly increased amounts of antibiotic. These highly resistant isolates produce no antibiotic and are often asporogenous.

A method based on a different rationale consists in selecting variants resistant to related antibiotics. Mutants resistant to amino acid analogs have been found to excrete small quantities of the homologous amino acid due to a lack of repression of enzyme synthesis (1, 31, 32). Assuming that similar enzyme repression mechanisms control antibiotic synthesis, it should be possible to select variants lacking suppression, and therefore yielding more antibiotic. The problem of producing analogs of antibiotics is solved for those antibiotics existing in families, such as the inositol amines streptomycin, neomycin, kanamycin, paromomycin, and the hygromycins, and the macrolide family. Streptomycin- and hygromycin-producing actinomycetes were found to be sensitive to other inositol amine antibiotics and resistant variants easily obtained on gradient plates, but no increase in production occurred. Erythromycin-producing actinomycetes proved to be resistant to other macrolides so that no test was possible.

A general repressor effect of excess glucose on synthesis of enzymes in bacteria may explain the delay in antibiotic synthesis that occurs until growth is nearly complete (34). Occasional increases in antibiotic yield have been obtained by feeding carbohydrate slowly, maintaining the measurable external level low. However, the effect of sugar feeds may be due to better pH stabilization rather than to absence of glucose repression. Mutants of bacteria have been found that are not suppressed by glucose (46, 78, 82). Similar variants have not been described in actinomycetes. No increases in yield were found by substituting carbon sources that do not produce glucose on hydrolysis, however.

A recently described group of mutants lacks permease activity for inorganic ions (76) or amino acids (77, 92). It is difficult to devise a method increasing antibiotic synthesis by interfering with the utilization of amino acids, the principal nitrogenous components of fermentor media. An inability to concentrate phosphate ion might shift metabolism to antibiotic synthesis earlier with a subsequent yield improvement (39, 61). The ability of certain Penicillia strains to accumulate sulfate favors penicillin synthesis (105). "Boostable enzymes" have been described, but restricted growth is necessary for their occurrence (5, 55, 103).

Variants resistant to actinophages as well as the phage are readily obtained. Various results have been obtained with phage resistant strains, high yields or no yield (85, 90). Complex patterns of resistance to virulent phage, production of temperate phage, and growth inhibition by bacteriocins occur among actinomycetes (17, 23, 98, 112). The occurrence of lysogeny raises the possibility of transferring characteristics between strains by transduction or conversion with temperate phage (7, 101). Survivors of a phage exposed population have a varied cobalamin yield (65). The usefulness of transduction is unexplored in strain selection. The success of such a program would depend upon the efficiency of the screening technique used to detect change. As an example, consider a series of strains resistant to high levels of an antibiotic that is produced by most of the strains at low levels, and at a high level by at least one strain. Production by the low level strains may be inhibited by genetic factors replaceable with different alleles from the production level strain. With a different genetic constellation, these strains may produce antibiotic at yields above the level of the former production strain.

Various correlations between physiological properties during fermentation and yield have been noticed but are usually specific to the strain and conditions used and cannot be generalized or adapted for use as a screening method. Correlations between catalase activity and streptomycin production (72) and fluorescence and tetracycline production (88) have been described.

Nutritionally exacting variants, auxotrophs, of antibiotic-producing microorganisms can be obtained by modifications of screening methods applied to Neurospora and Escherichia. Auxotrophs described in the literature have been obtained for use as selective and differential genetic markers in recombination studies, and groups of mutants blocked in the production of known metabolites have been used to determine pathways of biosynthesis. Two differences exist between these studies and possible applications to antibiotic production. Biosynthetic pathways blocked in auxotrophs lead to metabolites that are completely utilized during growth for further synthesis of more complex substances, there being little or no overproduction and excretion of the metabolite into the medium by nonmutants, as occurs with antibiotics, and none of these metabolites serve as necessary intermediates in the biosynthesis of antibiotics of major commercial importance with the exception of penicillin. Thus, the selection of auxotrophs of a given biochemical class cannot be predicted to affect yield.

Excluding a direct link between the specific block in the auxo-troph and the antibiotic synthesized, is there any reason to suspect a change in yield? Continued improvements in yield have been con-sidered to result from mutations preventing diversion of energy and substrate into side reactions (41). Thus, the more nutritional defi-ciencies introduced into a strain, the higher the expected yield. providing the fermentation medium is supplied with a sufficiency of the required nutrients. A corallary of this hypothesis is the prediction that strains with high yields are auxotrophs. No significantly improved yields have been found among auxotrophs. However, the levels of loss of biochemical function which affect antibiotic synthesis may be more subtle than those existing in all-or-none auxotrophs.

Another line of reasoning predicting yield increases in auxo-trophs follows from the suggestion that a medium with a surplus of energy sources but limiting in nutrients essential to growth, but not antibiotic synthesis, results in antibiotic production (61). Thus, anti-biotic production can be viewed as a consequence of unbalanced growth (33). While phosphate is usually the growth limiting factor, a lack of phosphate may depress the rate of production of high energy intermediates and limit antibiotic production as well as the level of growth. Furthermore, continuing ribose nucleic acid synthesis may be a prerequisite for enzyme synthesis and maintenance. Limiting phosphate would adversely affect both energy generation and enzyme synthesis. A specific method for blocking deoxyribose nucleic acid synthesis is required. Physical (ultraviolet irradiation) or chemical (addition of pyrimidine analogs or transmethylation inhibitors) methods are too cumbersome for industrial use. A thymine-requiring auxotroph would have a built-in method for growth level adjust-ment by control of the thymine concentration of the medium.

Mutants blocked in alternate pathways of dissimilation and energy generation, such as the "poky" strains of Neurospora (58), would be expected to have changed rates of synthesis. Strains resistant to acridine dyes, which may clear the cell of plasmid-like portions of the cytochrome system (47), proved to have lower growth rates and antibiotic synthesis.

Mutagenic Techniques

There are no clues to suggest that one mutagen is more effective than another in producing variants with improved yield. The action of mutagens has been shown to be "geographically" specific but not functionally specific (6). When a mutagen is found more effective in producing one rather than another physiological property, and when the effect is not due to differential sensitivity of one locus primarily affecting synthesis or to plasmid clearing, a selective effect of the system for recovering variants should be suspected.

Although no specific mutagenic treatment can be singled out as most effective, certain pitfalls of inefficient use must be avoided. Some mutagens require special conditions for maximal effectiveness. Thus, ultraviolet irradiation is most effective when applied to growing cells containing large pools of nucleic acid intermediates, followed by incubation in a medium deficient in these compounds but enriched with amino acids. Conditions may be allele as well as mutagen specific (40, 99, 102, 114, 115). Certainly a mutagen will be ineffective if it or the products of its action cannot penetrate the cell. Pretreatment of bacteria is necessary to obtain maximum effectiveness of manganous ion due to its low permeability (38).

The discovery of "hot spots" within a locus more sensitive to one than another mutagen, as well as the locus specificity of mutagens (probably due to such regions), favors a cyclical application of different mutagens (9, 37, 52). Continued application of the same mutagen may be fruitless if the most probable mutations inducible by the mutagen have already been recovered. Besides the use of a variety of mutagens and careful design of the techniques of application optimal to induction, detection, and recovery of mutants, the order of application is important. Thus, X-ray and ultraviolet irradiation may be initially used to obtain radio-kinetic data on the genetic constitution (30, 89), and to obtain auxotrophs for mutagenicity indices and incorporation of mutagenic purine and pyrimidine base analogs.

Recombination Between Antibiotic-Producing Strains

An approach to higher yields lies in the use of existing variation among different antibiotic-producing strains. A combination of desirable characteristics and elimination of unwanted properties might occur during joint cultivation of two different strains (80), but such a "mixed fermentation" would be difficult to stabilize. A single strain possessing only the desirable properties should be obtained through some form of genetic recombination (73) between differing isolates, the form of recombination depending upon the plasticity of the organism and the technical competence of the worker. The possibility of improved yield through recombination has been recognized in a patent (84) but no examples of increased yield have been reported. An attempt to improve yields of organic acids by the use of heterokaryons did not succeed (29). Parasexual recombination in penicillin-producing strains demonstrated the genetic control of penicillin synthesis but did not result in yield improvement (28). A diploid recombinant between variants of the Wisconsin family of Penicillia separated by several mutational steps was reported to give a 50 per cent yield improvement (93, 94).

The discovery of recombination between actinomycetes leading to heterokaryons and stable recombinants (22, 96) opened the way to possible industrial application of such techniques to antibiotic-producing microorganisms. A symposium on the genetics of actinomycetes contains further papers (104). The formation of antibiotic-producing recombinants from nonproducing mutants has been described (21). In summary, these and other papers (4, 15, 16, 18, 19, 20, 25, 63, 89, 97) report the possibility of forming heterokaryons within but not between "species" of streptomycetes and the occasional strain-specific formation of stable recombinants, but no recovery of industrially useful derivatives.

This work on genetic interaction among actinomycetes provides the basis for the possible production of isolates with increased antibiotic yield obtained as recombinants between strains of diverse origin. Antibiotic production is probably dependent upon a complex of genetically separable factors, affecting an estimated 30 different biosynthetic steps through their enzymes and control systems. Recombinants obtained between genetically different strains may give higher yields than either parental strain by reassortment of factors limiting and enhancing these individual biosynthetic steps. An essential character of the strains selected for this analysis, other than the necessity for the production of the same antibiotic, is divergent geographic origin, ecology, and physiology. Recombinants derived from mutants selected from the same line, of similar "pedigree", would not be expected to have an enhanced yield and no increases with measures of significance have been reported for such recombinants (4, 93). Changes in ploidy within the same strain could affect yield without the necessity of "outcrossing" (95).

A strain improvement program following this principle has been carried out by the author and associates. The highest yielding strain derived through a linear sequence of mutagenic treatments from the original erythromycin-producing isolate of *Streptomyces erythreus* (81) was selected as one parent. Other erythromycin-producing strains of differing geographic origin, morphology, physiology, and levels of synthesis have occasionally been found by the new antibiotics screening program. A number of single and double auxotrophs were selected in these strains. In addition, it was necessary to prove the different strains were not mutually inhibitory due to cross-sensitivity and production of temperate phage, bacteriocins, or other antibiotics. Various genetic markers other than nutritional deficiency were used as aids to the recovery of infrequent recombinant classes or for proof of recombination. Interaction between strains, probably due to heterokaryosis, but no formation of stable recombinants has been found.

Screening for New Antibiotics

Lack of knowledge of the role of antibiotics in ecology, paths of biosynthesis, and mode of action prevent a rational as opposed to an accidental approach to uncovering new antibiotics. Résumés of screening methods for soil microorganisms are given by Waksman (110) in several of his books with extensive literature references. The approach of present day screening programs is not very different from that described eight years ago (87). An outline of a generalized screening method will acquaint the reader with the procedure:

(a) Soil samples are obtained from diverse sources and individual organisms selected by plating suspensions on solid media. A preliminary growth step, in selective liquid media, usually to suppress molds and bacteria and favor actinomycetes, is often used.

(b) Individual organisms are grown in a variety of media favoring antibiotic synthesis.

(c) The spent culture media, sterilized, concentrated, or extracted, are tested for antibiotic activity against a variety of bacteria and molds, and possibly viruses and tumors (43, 108, 116).

(d) Organisms producing activity in this initial screen are regrown under a larger variety of conditions and the antibiotic categorized by paper chromatography to determine whether it is known or new. A disappointing feature of this method is not the rarity but the prevalence of organisms showing activity, most of them too weak to identify further or consisting of known antibiotic complexes (109). It is difficult to specify the exact number of known antibiotics due to incomplete chemical characterization, overlapping families, and use of different names for the same substance (100).

Either an extensive or an intensive approach may be used to increase the effectiveness of the screen. A more extensive program would increase the combination of isolates, media, conditions, and tests, usually by an increase at the input end, and the number of new cultures tested. This number is already large, 20,000 per year in one company's program, and the recovery of new clinically useful antibiotics is low—3 in 10 years (74). More extensive testing of new isolates may not be effective if the same organisms are merely being encountered more often. This suspicion is supported by the increasingly frequent recovery of the same antibiotic types.

An increase in the variety of media used for fermentation may yield more active isolates. The effectiveness of a given medium may not be due to specific precursors but to the lack of inhibiting substances. However, the effect of a medium is usually quantitative and the variety of media is probably not as great as the variety of organisms. Detection of activity depends upon inhibition or killing of sensitive bacteria, but more subtle effects on cell metabolism may be used to screen for useful metabolites that no longer fulfill the strict definition of antibiotic. No changes in media, conditions of fermentation, or methods of detection can be expected to replace extensive testing of many different microorganisms.

The intensive approach depends upon concentrating effort on different groups of microorganisms. Elective culture methods are often proposed to isolate one group of microorganisms from a mixed collection in their natural habitat. Difficulties encountered here are, first, elective culture methods are designed to recover a single type of organism rather than all members of its physiological group, and, second, such methods depend upon the selective use and not production of a metabolite. Thus, it is possible to isolate one or a few actinomycetes but not all actinomycetes by repetitive passage through appropriate media favoring actinomycete growth, without a preferential selection of antibiotic-producing types, however.

Elective culture methods, operating by reduction of populational variability through selective growth, diminish the number of different types of organisms originally present in a soil sample. Various methods have been proposed to retain and separate physiological and taxonomic groups *in toto*. Besides the physical difficulty in separating all organisms from their micro-environments in a heterogeneous soil suspension, the microorganisms will be in different physiological states and growth phases from active metabolism and multiplication to dormancy. Separation of the various groups assumes a greater difference between groups than between members within the same group. Thus, various physical separation techniques may be effective in resolving artificial mixtures of different organisms during "reconstruction experiments" but fail when applied to soil samples.

Besides methods favoring the growth of one group of microorganisms, there is the reverse technique of inhibiting all but one group. Addition of antibiotics to plating media is common (42). This method may be adequate for the recovery of new strains producing known antibiotics but could not be expected to uncover organisms producing new antibiotics. A variation could be used to isolate strains producing unknown antibiotics in such low yield as to make their initial identification impossible. Spent broths showing activity in step (c) but too dilute to classify in step (d) could be concentrated and added to plating media in step (a) to recover, by a system of positive feedback, strains producing higher yields or similar antibiotics of greater specific activity.

A separate intensive approach concentrates upon different groups of microorganisms isolated without an initial regard to their ability to produce antibiotics. The most intensively studied group has been the aerobic sporulating actinomycetes since success breeds success. Related groups, either less differentiated, such as the nocardias, corynebacters, and propionic acid bacteria, or more complex forms, such as the actinoplanes and streptosporangia, would be logical departures. Other groups might be chosen on the basis of similar habitat, growth forms, physiology, production of organic compounds, or possible unbalanced growth.

It is impossible to predict what group of organisms or set of conditions will yield new antibiotics. Even macroorganisms may produce useful substances (70, 106). This is due to a lack of connection between antibiotic production as an industrial process and production of similar substances in a natural habitat. Even though no convincing demonstration of production with chemical isolation of a major commercially important antibiotic under natural conditions has been reported, the suspicion remains that antibiotic production is but a magnified and variant expression of a normal biochemical function (27). Experiments to determine whether antibiotic production occurs in the soil or to determine the effects of antibiotics in soil have been performed, but none have suggested methods for the ready recovery of new antibiotic-producing organisms.

The premise that microorganisms produce antibiotics in the soil as a mechanism conferring adaptive value in competition with other soil organisms does not have to be accepted before speculating on the function of antibiotic production. Antibiotic activity may not be the prime property of these substances conferring an adaptive advantage upon the producing organism. The principal groups of antibiotic-producing microorganisms, the actinomycetes, possesses a little but not a lot of morphological differentiation during the growth cycle, both during colony formation on solid media and during submerged growth in liquid media. Antibiotic production may represent a sloughing off of structures in passing through stages of differentiation (12), such as discarded or abnormal cell wall fragments or other components of the vegetative cell or spore forming cells. Production of bacitracin by Bacillus species, a group of microorganisms that may lie on the evolutionary path of actinomycetes, occurs at sporulation (10, 11). Antibiotic activity may be a secondary and accidentally derived property of these substances. If this view is correct, then substances chemically similar to antibiotics but inactive should occur. None have been found, but the present biological screen would not be expected to uncover them. Chemical tests for antibiotics do not depend upon biological activity but are too insensitive and nonspecific. Although antibiotics can be classed with major biochemical groups, none have known roles in biochemistry even when related structurally to normal metabolites.

An approach to new antibiotics, alternative to random screening of newly isolated cultures, is induction of antibiotic synthesis in nonproducing organisms. Such proposals take one of two forms, *viz.*, (a) the potential to produce an antibiotic may be considered already present but dormant and the correct conditions, usually a substrate or "precursor", or a "challenge" with a competing organism or deleterious environment must be chosen to "induce" production; or (b) a culture may be considered to lack the immediate potential to produce but may be "forced" to evolve antibiotic production in order to survive by subjecting it to an artificial selection system with competing microorganisms.

The second proposal can be disposed of by some calculations on the time necessary for the evolution of an estimated 30 enzymes in the biosynthesis scheme of an antibiotic. The first proposal, restated in operational terminology and stripped of unpalatable and uncritical nonmechanistic allusions, is more attractive on our limited time scale. The difficulty arises in devising a method for "inducing" and recovering such microorganisms from a complex soil population or in determining which, of a large number of pure cultures, is capable of "induction".

Attempts to isolate new antibiotic-producing microorganisms from soil seeded with pathogenic bacteria have failed (111). Various mutagenic methods that have been proposed are probably selective rather than inductive in mechanism (69). Examples of gain of function are cited as supporting evidence, but critical evaluation of these weigh against such proposals (45, 49, 50, 51, 56, 79).

Any effective change in screening methods in the near future will probably come by way of increased efficiency of the screen coupled with a wider range of sources of organisms. Specific changes in screening methods can be tested by "reconstruction" experiments. Soil samples can be salted with a known number of spores of an antibiotic-producing actinomycete, productive at the same low levels typical of new isolates, and the efficiency of recovery of the added organism determined. This form of proofing of a new method would be most effective when the introduced strain is genetically marked to allow its identification and recovery at various stages of the screen (54).

Success in uncovering new antibiotics and strains with improved yields has been dependent upon empirical methods. When continued application of routine and standardized methods does not produce results, the usual solution has been to increase the scope of the operation. Any approach but this empirical one has been closed by lack of knowledge of the biosynthetic paths and mode of action of antibiotics at the molecular level. A beginning to the solution of these problems is now being made (35, 48, 53, 60, 71, 83, 86, 117). It is possible that the empirical approach has been over extended, and an amount of effort comparable to that expended on repetitive screening but directed towards these two problems would provide methods for designing rather than discovering new antibiotics.

References

1. Adelberg, E. A. 1958. Selection of bacterial mutants which excrete antagonists of antimetabolites. *Jour. Bact.*, **76**: *326–328*.

2. Alikhanyan, S. I. 1959. The radioselection of antibiotic-producing strains. *Antibiotics (Russian)*, 4: *770–775*.

3. ———— and Mindlin, S. Z. 1957. Recombinations in *Streptomyces rimosus. Nature*, 180: *1208–1209*.

4. ————, Mindlin, S. Z., Goldat, S. U., and Vladimizov, A. V. 1959. Genetics of organisms producing tetracyclines, *Ann. N. Y. Acad. Sci.*, 81: *914–949*.

5. Ames, B. N., Garry, B., and Herzenberg, L. A. 1960. The genetic control of the enzymes of histidine biosynthesis in *Salmonella typhimurium. Jour. Gen. Microbiol.*, 22: *369–378*.

6. Auerbach, C., and Westergaard, M. 1960. A discussion of mutagenic specificity. *Abh. dtsch. Akad. Wiss. Berl.*, *30–44*.

7. Azarowicz, E. N. 1959. Transduction of citrate utilization by nocardiophage. *Bact. Proc., Abs. 59th Gen. Meet., Soc. Amer. Bact.*, 60.

8. Backus, M. P., and Stauffer, J. F. 1955. The production and selection of a family of strains in *Penicillium chrysogenum. Mycologia*, 47: *429–463*.

9. Benzer, S., and Freese, E. 1958. Induction of specific mutations with 5–bromouracil. *Proc. Nat. Acad. Sci.*, 44: *112–119*.

10. Bernlohr, R. W., and Novelli, G. D. 1960. Bacitracin biosynthesis and spore formation in *Bacillus licheniformis. Bact. Proc., Abs. 60th Gen. Meet., Soc. Amer. Bact.*, 149.

11. ————, ————. 1960. Uptake of bacitracin by sporangia and its incorporation into the spores of *Bacillus licheniformis. Biochim. Biophys. Acta*, 41: *541–543*.

12. Bisset, K. A. 1959. The morphology and natural relationships of saprophytic actinomycetes. In *Hockenhull, D. J. D. (editor), Progress in Industrial Microbiology*, 1: *29–43*.

13. Bistis, G. N. 1959. Pleomorphism in the dermatophytes. *Mycologia,* 51: *440–452.*

14. Borenstajn, D., and Wolf, J. 1956. Réchèrches sur la variabilité de *Streptomyces rimosus. Ann. Inst. Pasteur,* 91 *(Suppl.): 62–71.*

15. Bradley, S. G. 1957. Distribution of lysogenic Streptomyces. *Science,* 126: *558–559.*

16. ———. 1958. Genetic analysis of segregants from heterokaryons of *Streptomyces coelicolor. Jour. Bact.,* 76: *464–470.*

17. ———. 1959. Cross-resistance patterns of mutants resistant to Streptomyces phage. *Mycologia,* 51: *125–131.*

18. ———. 1960. Reciprocal crosses in *Streptomyces coelicolor. Genetics,* 45: *613–619.*

19. ——— and Anderson, D. L. 1958. Compatibility system controlling heterokaryon formation in *Streptomyces coelicolor. Proc. Soc. Exp. Biol. and Med.,* 99: *476–478.*

20. ———, Anderson, D. L., and Jones, L. A. 1959. Genetic interaction within heterokaryons of streptomycetes. *Ann. N. Y. Acad. Sci.,* 81: *811–823.*

21. ——— and Johnson, T. C. 1960. Physiology and genetics of antibiotic production by *Streptomyces violaceoruber, Bact. Proc., Abs. 60th Ann. Meet., Soc. Amer. Bact., 71–72.*

22. ——— and Lederberg, J. 1956. Heterokaryosis in Streptomyces. *Jour. Bact.,* 72: *219–225.*

23. Braendle, D. H., Gardiner, B., and Szybalski, W. 1959. Heterokaryotic compatibility in Streptomyces. *Jour. Gen. Microbiol.,* 20: *442–450.*

24. ——— and Szybalski, W. 1957. Genetic interaction among Streptomycetes: heterokaryosis and synkaryosis. *Proc. Nat. Acad. Sci.,* 43: *947–955.*

25. ———, ———. 1959. Heterokaryotic compatibility, metabolic cooperation, and genic recombination in Streptomyces. *Ann. N. Y. Acad. Sci.,* 81: *824–853.*

26. Bresch, S. C., and Tanner, F. W. 1959. Chemical engineering, reviews: Fermentation. *Ind. and Eng. Chem.,* 51: *1086–1098.*

27. Brian, P. W. 1957. The ecological significance of antibiotic production. In *Williams, R. E. O., and Spicer, C. C. (editors), Microbial Ecology, 7th Symp. Soc. Gen. Microbiol., Cambridge: The University Press.*

28. Caglioti, M. T., and Sermonti, G. 1956. A study of the genetics of penicillin-producing capacity in *Penicillium chrysogenum. Jour. Gen. Microbiol.,* 14: *38–46.*

29. Ciegler, A., and Raper, K. B. 1957. Applications of heterocaryons of Aspergillus to commercial type fermentations. *Appl. Microbiol.,* **5:** *106–110.*

30. Clark, J. B. 1959. Radiation as a cytogenetic tool. *Ann. N. Y. Acad. Sci.,* **81:** *907–912.*

31. Cohen, G. N., and Adelberg, E. A. 1958. Kinetics of incorporation of fluorophenylalanine by a mutant of *Escherichia coli* resistant to this analog. *Jour. Bact.,* **76:** *328–330.*

32. ———— and Jacob, F. 1959. On the repression of synthesis of enzymes mediating tryptophane formation in *Escherichia coli. C. R. Acad. Sci. (Paris),* **248:** *3490–3492.*

33. Cohen, S. S., and Barner, H. D. 1954. Studies on unbalanced growth in *Escherichia coli. Proc. Nat. Acad. Sci.,* **40:** *885–893.*

34. Cohn, M., and Horibata, K. 1959. Inhibition by glucose of the induced synthesis of beta-galactoside-enzyme system of *Escherichia coli:* I. Analysis of maintenance. *Jour. Bact.,* **78:** *601–623.*

35. Corcoran, J. W., Kaneda, T., and Butte, J. C. 1960. Actinomycete antibiotics: I. The biological incorporation of propionate into the macrocyclic lactone of erythromycin. *Jour. Biol. Chem.,* **235:** *PC 29.*

36. Demerec, M., *et al.* 1949. The gene. *Yearbook Carnegie Inst.,* **48:** *164–166.*

37. ————. 1950–51. *Yearbook Carnegie Inst.,* **50:** *167–195.*

38. ———— and Hanson, J. 1951. Mutagenic action of manganous chloride. *Cold Spr. Harb. Symp. Quant. Biol.,* **16:** *215–228.*

39. Doskocil, J., *et al.* 1959. Development of *Streptomyces aureofaciens* in submerged culture. *Jour. Biochm. Microbil. Technol. Eng.,* **1:** *261–271.*

40. Doudney, C. O., and Haas, F. L. 1959. Mutation induction and macro-molecular synthesis in bacteria. *Proc. Nat. Acad. Sci.,* **45:** *709–722.*

41. Dulaney, Eugene L. 1954. Induced mutation and strain selection in some industrially important microorganisms. *Ann. N. Y. Acad. Sci.,* **60:** *155–163.*

42. ————, Larsen, A. H., and Stapley, E. O. 1955. A note on the isolation of microorganisms from natural sources. *Mycologia,* **47:** *420–422.*

43. Ehrlich, J. 1960. Problems in detection and isolation of antitumor antibiotics. In *Miller, B. M. (editor), Developments in Industrial Microbiology,* **1:** *81–85.*

44. Elander, R. P. 1960. Studies on antibiotic-producing species and

varieties of Emericellopsis and Cephalosporium. *Doctoral Thesis, Univ. Wisconsin.*

45. Englesberg, E. 1957. Physiological basis for rhamnose utilization by a mutant of *Pasteurella pestis. Arch. Biochem. Biophys.,* **71:** *179–193.*

46. ———. 1959. Glucose inhibition and the diauxie phenomenon. *Proc. Nat. Acad. Sci.,* **45:** *1494–1507.*

47. Ephrussi, B., Hottinguer, H., and Chimenes, A. M. 1949. Action de l'acriflavine sur les levures: 1. La mutation "petite colonie". *Ann. Inst. Pasteur,* **76:** *351–367.*

48. Erdos, T., and Ullmann, A. 1959. Effect of streptomycin on the incorporation of amino acids labelled with carbon–14 into ribonucleic acid and protein in a cell-free system of a Mycobacterium. *Nature,* **183:** *618–619.*

49. Flaks, J. G., and Cohen, S. S. 1957. The enzymatic synthesis of 5–hydroxymethyldeoxycytidylic acid. *Biochim. Biophys. Acta,* **25:** *667–668.*

50. ———, ———. 1959. Virus-induced acquisition of metabolic function: I. *Jour. Biol. Chem.,* **234:** *1501–1506.*

51. ———, Lichtenstein, J., and Cohen, S. S. 1959. Virus-induced acquisition of metabolic function: II. *Jour. Biol. Chem.,* **234:** *1507–1511.*

52. Freese, E. 1959. The difference between spontaneous and base-analogue induced mutations of phage T4. *Proc. Nat. Acad. Sci.,* **45:** *622–633.*

53. Gale, E. F., Shepherd, C. J., and Folkes, J. P. 1958. Incorporation of amino acids by disrupted staphylococcal cells. *Nature,* **182:** *592–595.*

54. Galzy, P., and Dupuy, P. 1957. Perspectives d'utilisation d'une levure mutante a la canavanine pour le controle du levurage. *Ann Tech. Agr.,* **2:** *235–242.*

55. Gorini, L. 1960. Antagonism between substrate and repressor in controlling the formation of a biosynthetic enzyme. *Proc. Nat. Acad. Sci.,* **46:** *682–690.*

56. Gross, S. R. 1957. Enzymatic autoinduction in a mutant of Neurospora. *Genetics,* **42:** *374.*

57. Haas, F. L., Puglisi, T. A., Moses, A. J., and Lein, J. 1956. Heterokaryosis as a cause of culture rundown in Penicillium. *Appl. Microbiol.,* **4:** *187–195.*

58. Haskins, F. A., Tissieres, A., Mitchell, H. K., and Mitchell, M. B.

1953. Cytochromes and the succinic acid oxidase of poky strains of Neurospora. *Jour. Biol. Chem.,* **200:** *819–826.*

59. Hastings, J. J. H. 1958. Present trends and future developments. In *Steel, R. (editor), Biochem. Eng., New York: The Macmillan Company.*

60. Hlavka, J. J., and Buyske, D. A. 1960. Radioactive 7–iodo–6–deoxy-tetracycline in tumour tissue. *Nature,* **186:** *1064–1065.*

61. Hockenhull, D. J. D. 1958. The biosynthesis of streptomycin. *Abs., VII Intern. Cong. Microbiol., Stockholm, 385–386.*

62. ————. 1959. *Progress in Industrial Microbiology. New York: Interscience Pub., Inc. Vol. 1.*

63. Hopwood, D. A. 1958. Linkage and the mechanism of recombination in *Streptomyces coelicolor. Ann. N. Y. Acad. Sci.,* **81:** *887–898.*

64. Hsie, Jen-Yah, Kotz, R., and Epstein, S. 1957. Carbomycin requiring mutants of *Micrococcus pyogenes* var. *aureus. Jour. Bact.,* **74:** *159–162.*

65. Il'ina, T. S., and Alikhanyan, S. I. 1959. The use of actinophages in the selective cultivation of actinomycetes. *Antibiotics (Russian),* **4:** *530–533.*

66. Kanazir, D. 1958. Apparent mutagenicity of thymine deficiency. *Biochim. Biophys. Acta,* **30:** *20–23.*

67. Kashii, K., *et al.* 1958. Statistical study in the mutation of *Streptomyces racemochromogenus,* producing racemomycin O, by the irradiation of ultraviolet light and the contact of nitrogen mustard. *Jour. Antibiotics, Ser. B;* **11:** *277–283.*

68. Kavanagh, F. 1960. A commentary on microbiological assaying. In *Umbreit, W. W. (editor), Advances in Appl. Microbiol.,* **2:** *65–93.*

69. Kelner, A. 1949. Studies on the genetics of antibiotic formation: The induction of antibiotic-forming mutants in actinomycetes. *Jour. Bact.,* **57:** *73–92.*

70. Kirsanov, G. P. 1959. An antibiotic preparation from the web of the common synanthropic spiders. *Antibiotics (Russian),* **4:** *123–125.*

71. Koffler, H. 1959. Chemistry, site of action, and biosynthesis of the circulins. *Science,* **130:** *1419–1420.*

72. Kovacs, E., and Matkovacs, B. 1957. Veränderung der katalseaktivität in *Streptomyces griseus*-Külturen. *Acta Univ. Szeged,* **7:** *343–347.*

73. Lederberg, J. 1955. Recombination mechanisms in bacteria. *Jour. Cell. Comp. Physiol.,* **45** *(Suppl.): 75–107.*

74. Lilly Review. 1959. *Vol.* **19:** *15–18.*

75. Litman, R. M., and Pardee, A. B. 1956. Production of bacteriophage mutants by a disturbance of deoxyribonucleic acid metabolism. *Nature*, **178**: *529–531*.

76. Lubin, M., and Kessel, D. 1960. Preliminary mapping of the locus potassium transport in *Escherichia coli*. *Biochem. Biophys. Res. Comm.*, **2**: *249–255*.

77. ————, ————, Budreau, A., and Gross, J. D. 1960. The isolation of bacterial mutants defective in amino acid transport. *Boichem. Biophys. Acta*, **42**: *535–538*.

78. Magasanik, A. K., and Bojarska, A. 1960. Enzyme induction and repression by glucose in *Aerobacter aerogenes*. *Biochem. Biophys. Res. Comm.*, **2**: *77–81*.

79. Marmur, J., and Hotchkiss, R. D. 1955. Mannitol metabolism, a transferable property of Pneumococcus. *Jour. Biol. Chem.*, **214**: *383–396*.

80. McCormick, J. R. D., Hirsch, U., Sjolander, N. O., and Doerschuk, A. P. 1960. Cosynthesis of tetracyclines by pairs of *Streptomyces aureofaciens* mutants. *Jour. Amer. Chem. Soc.*, **82**: *5006–5007*.

81. McGuire, J. M., *et al.* 1952. "Ilotycin", a new antibiotic. *Antibiotics and Chemotherapy*, **2**: *281–283*.

82. Neidhardt, F. C. 1960. A mutant of *Aerobacter aerogenes* lacking glucose repression. *Bact. Proc.*, *Abs. 60th Gen. Meet., Soc. Amer. Bact., 173.*

83. Park, J. T., and Strominger, J. L. 1957. Mode of action of penicillin. *Science*, **125**: *99–101*.

84. Pontecorvo, G. P. A., and Roper, J. A. 1958. Synthesis of strains of microorganisms. *U. S. Patent #2,820,742.*

85. Reilly, H. C., Harris, D. A., and Waksman, S. A. 1947. An actinophage for *Streptomyces griseus*. *Jour. Bact.*, **54**: *451–466*.

86. Roth, H., Ames, H., and Davis, B. D. 1959. Purine nucleotide excretion by *Escherichia coli* in the presence of streptomycin. *Biochim. Biophys. Acta*, **37**: *398–405*.

87. Routien, J. B., and Finlay, A. C. 1952. Problems in the search for microorganisms producing antibiotics. *Bact. Rev.*, **16**: *51–67*.

88. Rudaya, S. M. 1958. Luminescence of *Actinomyces rimosus* and its variants. *Microbiology (Moscow)*, **27**: *588–592*.

89. Saito, H., and Ikeda, Y. 1959. Cytogenetic studies on *Streptomyces griseoflavus*. *Ann. N. Y. Acad. Sci.*, **81**: *862–878*.

90. Saudek, E. C., and Collingsworth, D. R. 1947. A bacteriophage in the streptomycin fermentation, *Jour Bact.*, **54**: *41–42*.

91. Schultz, J. S., Reihard, D., and Lind, E. 1960. Statistical methods

in fermentation development. *Ind. Eng. Chem.*, **52:** *827–830.*

92. Schwartz, J. H., Maas, W. K., and Simon, E. J. 1959. An impaired concentrating mechanism for amino acids in mutants of *Escherichia coli* resistant to L–canavanine and D–serine. *Biochim. Biophys. Acta*, **32:** *582–583.*

93. Sermonti, G. 1956. Complementary genes which affect penicillin yields. *Jour. Gen. Microbiol.*, **15:** *599–608.*

·94. ————. 1957. Produzione di penicillina da diploidi eterozigoti. *Ric. Sci.*, **27** *(Suppl.): 3–10.*

95. ————. 1959. Genetics of penicillin production. *Ann. N. Y. Acad. Sci.*, **81:** *950–966.*

96. ———— and Spada-Sermonti, I. 1956. Gene recombination in *Streptomyces coelicolor. Jour. Gen. Microbiol.*, **15:** *609–616.*

97. ————, ————. 1958. Genetics of *Streptomyces coelicolor. Ann. N. Y. Acad. Sci.*, **81:** *854–861.*

98. Shirling, E. B. 1956. Studies on relationships between actinophages and variation in Streptomyces: 1. Morphology of phage production and lysis in a streptomycete carrying a temperate phage. *Virology*, **2:** *272–283.*

99. Sinsheimer, R. L. 1957. The photochemistry of cytidylic acid. *Radiation Res.*, **6:** *121–125.*

100. Spector, W. S. 1957. Handbook of Toxicology: Vol. 2, Antibiotics. *Philadelphia: W. B. Saunders and Company.*

101. Spizizen, John. 1958. Transformation of biochemically deficient strains of *Bacillus subtilis* by deoxyribonucleate. *Proc. Nat. Acad. Sci.*, **44:** *1072–1078.*

102. Strauss, B., and Okubo, S. 1960. Protein synthesis and the induction of mutations in *Escherichia coli* by alkylating agents. *Jour. Bact.*, **79:** *464–473.*

103. Szilard, Leo. 1960. The control of the formation of specific proteins in bacteria and in animal cells. *Proc. Nat. Acad. Sci.*, **46:** *277–292.*

104. Szybalski, W. 1959. Genetics of Streptomyces and other antibiotic-producing microorganisms. *Ann. N. Y. Acad. Sci.*, **81:** *805–1016.*

105. Tardrew, P. L., and Johnson, M. J. 1958. Sulphate utilization by penicillin-producing mutants of *Penicillium chrysogenum. Jour. Bact.*, **76:** *400–405.*

106. Thompson, J. A. 1960. Inhibition of nodule bacteria by an antibiotic from legume seed coats. *Nature*, **187:** *619–620.*

107. Treffers, H. P., Spinelli, V., and Belser, N. O. 1954. A factor (or mutator gene) influencing mutation rates in *Escherichia coli. Proc. Nat. Acad. Sci.*, **40:** *1064–1071.*

108. Umezawa, H. 1956. Antitumor substances of antibiotics. *Gen. Micro-biol.*, **2:** *160–193.*

109. Vanek, Z., Dolezilova, L., and Rehacek, Z. 1958. Formation of a mixture of antibiotic substances, including antibiotics of a polyene character, by strains of actinomycetes freshly isolated from soil samples. *Jour. Gen. Microbiol.,* **18:** *649–657.*

110. Waksman, S. A. 1959. The Actinomycetes. *Baltimore: Williams and Wilkins Co. Vol. 1.*

111. ———. 1959. Biochemistry of antibiotics. *Sym. V, Proc. 4th Intern. Cong. Biochem., London: Pergamon Press.*

112. Welsch, M. 1959. Lysogenicity in streptomycetes. *Ann. N. Y. Acad. Sci.,* **81:** *974–993.*

113. Wilson, E., Koffler, H., Coty, V. H., and Tetrault, R. A. 1951. The effect of streptomycin on the respiration and growth of various strains of *Streptomyces griseus. Bact. Proc., Abs. 51st Gen. Meet., Soc. Amer. Bact., 15.*

114. Witkin, E. 1956. Time, temperature, and protein synthesis: A study of ultraviolet-induced mutation in bacteria. *Cold Spr. Harb. Symp. Quant. Biol.,* **21:** *123–140.*

115. ——— and Theil, E. C. 1960. The effect of post-treatment with chloramphenicol on various ultraviolet-induced mutations in *Escherichia coli. Proc. Nat. Acad. Sci.,* **46:** *226–231.*

116. Yamazaki, S., *et al.* 1956. Cylinder plate method of testing the anti-cell effect. *Jour. Antibiotics (Japanese), Ser. B,* **9:** *135–140.*

117. Yarmolinsky, M. B., and de la Haba, G. L. 1959. Inhibition by puromycin of amino acid incorporation into protein. *Proc. Nat. Acad. Sci.,* **45:** *1721–1729.*

Comments

AUERBACH: Did you carry out cytological observations? When I was in Moscow this year, I saw very beautiful preparations made by Prokofjewa-Belgowskaja, which showed a close correlation between the stages of growth and antibiotic production in Streptomyces and Actinomyces with the appearance of the structure of nuclear bodies. On the basis of these correlations she has been able to advise on the use of culture methods for antibiotic production.

NELSON: Inspection of stained or unstained mycelia is often used to determine when a pilot plant or production scale vegetative growth tank is ready for use as an inoculum. It has not been used as a method of strain selection since we don't know what changes to look for nor how to avoid confusion of hereditary changes with physiologically induced morphological changes as the culture ages.

LANGHAM: You defined *potency* as *yield*. If you consider potency for potency's sake, it is known that the potency of insecticides, such as pyrethrum, can be increased about 100 per cent by the use of synergids. This was discovered during the last war when sesame oil was used in aerosol bombs. Chemical tests showed that both sesamin and sesamolin were responsible for the observed synergism, and further that the methylenedioxyphenyl group of these compounds was the effective one. Later, other compounds with this same group proved effective and are now used in commercial insecticides. (These are piperonyl butoxide, sulfoxide, etc.) Have these chemicals been tested for a possible synergistic effect on antibiotics?

NELSON: Those specific compounds have not been tested and I suspect that they would be too toxic for intravenous administration. However, a large part of clinical research concerns synergisms and antagonisms between different antibiotics and potentiating effects of derivatives upon antibiotic effectiveness.

Methods of Utilizing Induced Mutation in Crop Improvement

JAMES MacKEY

Swedish Seed Association, Svalöf, Sweden

T HIRTY YEARS of applied mutation work have gradually sapped the first pessimistic idea that induced mutations represent only the deleterious types of spontaneous mutation. More and more, genet- icists have become convinced that the fundamental evolutionary force, the creation of variability, can now be controlled by man. From our present knowledge, it seems theoretically not too incon- siderate to accept ideas like the one held by Gaul (26, p. 276),[1] when he says, "Today there is little doubt that all the genes involved in the world collections of our cultivated plants can be reproduced by induced mutations". But theory is one thing and practical plant breeding another. For that reason, it is still a debate, but it is now less concerned with whether the imitation is possible or not than with whether it is economical or not.

Proper economical evaluations need more evidence and experi- ence than does a statement of purely academic significance. However, since plant breeding is nothing more than controlled and directed evolution, I think it is easier for us to make the correct evaluation if we always keep the theoretical aspects in mind. If we compare with nature, I have no feeling that we use the two evolutionary forces, recombination and selection, in a different way in our plant breeding.

Can we today make the same statement in connection with the third force, mutation? I do not think so, unless we accept such an extreme idea that macromutations are of overwhelming importance and are independent of background genotype. Most of our evidence that mutation breeding is a reality is based on distinct off-types either from a morphological or from a physiological standpoint, and these off-types have mostly been evaluated as raw mutants. It is logical first to build up proof in this way, since a macromutation in a self- fertilized crop plant can be made to constitute absolute evidence of

---

[1]See References, page 354.

the progressive value of induced mutation if proper precautions are taken. If we now believe in a more balanced, more subtly differentiated, and a more dynamic evolution, such proofs will, however, be insufficient to rate mutation experiments as an ingredient in plant breeding. The meagre data available on induced mutation in quantitative traits and on the combining ability of induced mutations make it very difficult completely to understand the evolutionary implication of artificial mutation. The very promising results obtained in the sparse experiments along these two lines (2, 3, 17, 33, 34, 37, 76, 81, 91, 92, 104, 106), however, seem to justify more optimism.

Our deficient knowledge of how to utilize mutation in crop improvement must be borne in mind when we try to evaluate mutation experiments as a technique in plant breeding. It seems very likely that mutation work in the future will be more completely interwoven with the so-called conventional methods, with which it now too often is set in opposition. We have also to recall that most of the mutation experiments done so far have been biased by theoretical considerations and not planned as strict breeding projects.

In addition, our knowledge of the relation between genotype and the possibility of inducing progressive mutations is most meagre. Nearly all our efforts have been concentrated on methods to increase mutation rate and to find mutagens with a specific and effective mode of action. For a plant breeder, it seems just as important to explore the specific response of genotype to mutation, since the careful choice of parental material is perhaps the most decisive part of the program. Not even the most clever and observant handling can compensate for unsuitable material.

Table 1 lists some of the more important factors influencing the genotypic response to progressive mutation. Available information is too scant on many points to allow me to present more than self-evident items of knowledge in this paper, and some of the problems will be taken up by other participants in this symposium. For these reasons, I will discuss only some factors related to the choice of parental material which may be important for progressive mutation.

From this special point of view, inherent mutagenic resistance with respect to survival is only indirectly important as long as it influences the total yield of mutations. Drastic interspecific differences in radio-resistance have been demonstrated in connection with both

TABLE 1.—FACTORS INFLUENCING THE GENOTYPIC RESPONSE TO PROGRESSIVE MUTATION
IN CULTIVATED PLANTS.

Nutrimental and industrial or ornamental use
Age from standpoint of evolution, cultivation, regional adaptation, and scientific breeding
Extent and availability of natural variation, including possibilities for interspecific gene
    transfers
Anatomy and differentiation of meristematic tissue
Annual and biannual or perennial habit of growth
Autogamy, allogamy, or clonal propagation
Genic duplication and tolerance to chromosomal rearrangements
Inherent mutagenic resistance and environmental possibilities to its modifications
Degree of heterozygosity
Phenotypic buffering in relation to characteristics
Presence of desirable precursor genes
Combining ability with genotypic background
Goal of the specific breeding program and availability of effective selection techniques

acute and chronic radiation with chromosome size and level of ploidy
as the most decisive characteristics (39, 52, 72, 73, 80, 100). It is interesting, though by no means conclusive, that practical results in mutation breeding were early reported for rape and mustard, two extremes
in high seed radio-resistance among agricultural crops (2, 3).

Clear differences in primary radio-resistance are also found
between closely related genotypes. The degree, and thus the chance
to obtain maximum yield of mutation, may be controlled by inherent
differences in simple metabolic activities (51). It may depend on individual chromosomes (114) or even genes (30, 61, 62, 99). It has been
proved to vary witth degree of genotypic stability (8), can easily be
recombined, and is dependent on type of radiation (34, 35). From
Drosophila experiments, we can expect varietal differences in the
response to chromosome breaks (16, 18). Gregory (34, 35) has even
shown that it is possible to induce mutants with improved resistance
to primary radiation injury.

An example on induced decrease in radio-resistance is given in
Table 2. Such a phenomenon has implication also in connection with
recurrent radiation experiments in the same or successive generations.
In the latter case, Shestakov, et al. (cited by Hoffmann, 48) found an
increased sensitivity to radio-phosphorus in the second generation of·
treatment, and the same trend was found by Hoffmann (48) in acute
X-irradiation of successive generations of wheat.

With respect to the degree of heterozygosity, it appears that not

TABLE 2.—ARTIFICIALLY INDUCED DECREASE IN RADIO-RESISTANCE.

	Control	10,000 r	15,000 r	20,000 r
No. of seeds sown.....................	150	300	300	300
No. of seedlings, %				
Skandia III winter wheat............	86.0	86.3	84.0	80.7
X-ray-induced deficiency–speltoid of				
Skandia III......................	92.0	79.7	78.3	52.7
Difference against difference in control	–	–12.6	–11.7	–34.0
mdiff...............................	–	±4.72^{xx}	±4.80^x	±5.14^{xxx}

Seed moisture: 14.7 per cent   Seed size: 2.50 – 2.75 mm

only radio-resistance but also width of mutation spectrum may be influenced. Unfortunately, no definite information is yet available on this aspect on genotypic response to mutability. At present, there are thus only some rather vague statements on the advantage of using heterozygous parental material in mutation work (36, 40, 69), and the problem can therefore only be pointed out here. Heterozygosity implies, however, more unlike genes that can be changed, rearranged in blocks, or split up. The fact that heterozygosity in itself implies an enhanced pressure and chance of mutation (69) may also increase the efficiency of the applied mutagen. Such an idea is in a way supported by the fact that rare and new mutations become gradually more frequent in experiments including radiation of successive generations (48, 103).

Phenotypic buffering, i.e., the ability to resist and absorb mutation without severe deviations from type, is another genotypic factor controlling rate and spectrum of progressive mutation. Duplication, epistasis, polymery, and other cumulative or complementary gene effects are added to allelism to form a very complex system of interference and interaction within the germ plasm. The extent, type, and direction of this interdependence of genes, as well as their position and stability, will vary with genera, species, and biotype and also with character to form different patterns of buffering.

The example I present is somewhat special, since it deals with macromutations in a ploidy series, but I am sure of its wider application. In Table 3, the relative prevalence of chlorophyll mutations per plant progeny in relation to all other phenotypically distinct mutations in M₂ of 2x *monococcum*, 4x *dicoccum*, and 6x *vulgare*

TABLE 3.—THE RELATIVE PREVALENCE OF CHLOROPHYLL AND OTHER PHENOTYPICALLY
DISTINCT MUTATIONS IN M_2 OF X–RAY- AND NEUTRON-TREATED 2x, 4x, AND 6x WHEAT.

Species	N	The relative prevalence in M_2 of	
		Chlorophyll mutations, %	Other phenotypically distinct mutations, %
2x *monococcum* wheat..............	282	*66.3*	*33.7*
4x *dicoccum* wheat....	389	43.7	*56.3*
6x *vulgare* wheat................	983	0.7	*99.3*

Test of heterogeneity	χ^2	d.f.	P
2x – 4x wheat	32.7	1	<0.001
4x – 6x wheat............ ..	687.7	1	<0.001
2x – 6x wheat......	721.2	1	<0.001

wheat is calculated from a set of X-ray and neutron experiments
(72, 73).

A sharp decrease in the relative rate of chlorophyll mutations
with increasing level of ploidy is associated with the fact that the
total mutation rate increased twofold in the X-ray experiments and
threefold in the neutron experiments from diploid to hexaploid
wheat. What is interesting from our present approach is that the data
demonstrate that the ability to produce chlorophyll, and presumably
also other characteristics essential for the ability to complete life
and to reproduce, are better buffered in the polyploid wheats. The
induced macromutations in these wheats center more in characteristics which are not decisive for life or death but rather interfere with
ecological adaptation, etc.

As evident from Table 4, the same trend can be observed even
within the category of mutations influencing the genetic control of
the chlorophyll apparatus. Mutations harmful to the carrier already
at the young seedling stage, like *albina, xantha*-types and *viridoalbina,* decrease in relative frequency with increasing level of ploidy in
contrast to types like *tigrina* and *viridis* where the production of
chlorophyll is less completely blocked (38).

This observation that the majority of mutations in one species
can be centered on more harmful changes than in another is not only
observed in ploidy series. If we still keep to the cholorophyll muta-

TABLE 4.—TYPES OF CHLOROPHYLL MUTATIONS AND THEIR RELATIVE PREVALENCE IN M_2 OF X-RAY- AND NEUTRON-TREATED 2X, 4X, AND 6X WHEAT.

Species	N	The relative prevalence in M_2 of					
		albina	xantha xan.alb. alb.xan.	virido- albina	tigrina	viridis	All other
2x monococcum wheat...	295	42.4	6.4	9.8	2.7	34.6	4.1
4x dicoccum wheat.....	191	27.8	4.2	6.3	18.3	40.8	2.6
6x vulgare wheat......	15	6.7	—	—	26.7	53.3	13.3

	Test of heterogeneity		
	χ^2	d.f.	P
2x − 4x wheat.............	43.4	5	<0.001
4x − 6x wheat.	3.9	1*	<0.05
2x − 6x wheat............ ..	12.5	1*	<0.001

*Albina, xantha-types and viridoalbina is one group; tigrina and viridis is another. "All other" omitted.

tions, which generally though not always (10, 43, 64, 74, 111) are detrimental, and refer only to diploids, we will find great differences in the relative commonness especially of *albinas* between cereals, on the one hand (72), and large-seeded leguminous plants, such as peas (89), lupines (110), soybeans (117), and beans (Sjödin, unpublished), on the other. It is also symptomatic that Steuckhardt (103) found the *albina* mutations gradually increasing in relative frequency when irradiation of millet was repeated over successive generations and the buffering in this way gradually broken down.

The varying degree of progress in the use of mutation experiments in crop improvement may partly be due to differences like those discussed here. The relatively high frequency of lethal or subvital mutations and a very strict, morphological architecture of, e.g., the barley plant, may partly account for the comparatively meagre results in contrast to those with peas, beans, and peanuts. Similar differences but of different magnitude can also be found on an intervarietal level within species. To take only one example, Hagberg (46) found different *erectoides* loci more or less stable with a different pattern for different varieties. In the variety Gull 4 out of 8 induced *erectoides* mutations involved locus *b* which never has been found to mutate in the varieties Maja and Bonus, where 7 and 87 *erectoides* mutations, respectively, have been analyzed. As a general rule for all characteristics, it seems more difficult to induce and observe an improvement in a direction to which the parental variety has already been intensively bred (70).

It is thus evident that the possibility to be able to induce, observe, and properly evaluate a mutation is highly dependent on parental genotype, and for that reason the breeder should not restrict his basic materal too drastically (15) and only to his very best strain. The choice of material especially suitable for the detection of the desired mutations can sometimes greatly improve the experiment. Thus, Melchers (75) demonstrated the use of haploid plants in Antirrhinum for detection of recessive mutations. Monosomics or other types with known deficiencies could be used in a similar manner. Mertens and Burdick (76) used the technique of irradiating autodiploids (2n ex-haploids) and then backcrossing and comparing them with an untreated control in order to detect mutations in quantitative traits that could exhibit a dominance type of heterosis in tomato to F_1's. Genotypes with appropriate marker genes or translocations can greatly improve the detection of desirable chromosomal rearrangements in an irradiated material (19, 46, 93).

The choice of the most suitable genotype in accordance with the intention of the mutation experiment is, however, only one side of the problem of how to render mutation breeding as efficient as possible. The chance to induce the desired mutations also depends much on the metabolic stage of the treated cells, the mutagen, and the dose applied as well as on the handling of the material before, during, and after treatment. These problems are, however, taken up by other speakers on this symposium, and we will here continue to center our interest in pure breeding techniques.

In the treated M_1 generation, the breeder has certain possibilities to influence the final yield of induced mutations. As in a hybrid bulk population, competition between genotypes will occur, but this competition will start already on the cellular level if plants or seeds are treated. Such an intrasomatic or diplontic selection is desirable as long as only deleterious mutations are screened away as in the very first cell divisions after treatment. Later on, the risk increases that cells carrying valuable mutations may also be suppressed and lost, and it is, therefore, important to overcome this phenomenon of intraplant selection.

In principle, this can be achieved in two different ways. One is to hinder or eliminate competition between primordia or shoots already differentiated before the mutagenic treatment. In radiation experiments with seed of cereals, Freisleben and Lein (24, 25) suggested and Gaul (28, 29) more definitely proved that the highest relative yield of mutation will be recorded if the M_1 plants are permitted to develop a maximum of one tiller. In this way only one of the three to five differentiated ear primordia of the treated embryo can develop. The simplest method to suppress tillering is dense and late sowing, preferably with an ordinary drill and with proper calculation with respect to lowered seed vitality. High temperature and light deficiency, as in ordinary greenhouse conditions, are also efficient. In vegetatively propagated plants, Bauer (5) showed that the detection of mutations greatly improved if bud competition was reduced. This can be done by cutting back the treated shoot successively or dividing it into small cuttings which are allowed to root and develop separately.

The other method of overcoming the disadvantages of diplontic selection is to adjust the dose and its time of application. If only one initial is allowed to develop, a stronger dose will increase the frequency of mutated over normal meristematic cells. The relative rate of mutation will thus increase much longer with dose than if competition between affected and unaffected initials is allowed to counterselect (29). A treatment in a late ontogenetic phase of plant development will decrease both inter- and intrameristematic competition, since the floral shoots gradually lose their interdependence and the time for the diplontic selection to work will be shortened. The most radical solution along this line will be to treat meiotic stages or even gametes and then, preferably, pollen (13, 21, 29). In connection with ionizing radiation and treatment with radiomimetic chemicals, such extreme efficiency may, however, at least in certain species, be counteracted by the exceptional response of very condensed chromosomes resulting nearly entirely in chromosome aberrations (98, 101, 102). The delay in treatment should, under such circumstances, be restricted preferably to the somatic stages of the plant ontogeny.

As Mericle (unpublished) has shown, treatment in early embryonal stages offers similar advantages without the undesirable effects of treating gametes. A higher mutation frequency and a higher ratio of mutant to normal offsprings can thus, in this way, be combined with a much lower demand on dose.

A chronic or recurrent treatment will act similar to a delayed application, but offers, in addition, chance for a continuous selection against very harmful mutations and a successive accumulation of more vital changes. Pursuant to this reasoning, Mikaelsen (77) found chronic radiation of barley definitely more effective than the comparative, acute dose already given to the seed. By applying a recurrent treatment, gross deleterious effects may be eliminated before new valuable mutations are added step by step. In vegetatively propagated plants, where a shift in generation over a sexual phase may break down a complex but well-balanced, heterozygous genotype, such a recurrent treatment may have great possibilities even to inactivate a dominant gene in homozygous position. As will be discussed by Caldecott (11) in this symposium, recurrent treatment of successive generations of seed-propagated plants has special merit, since a mutation can here be removed from competition with the parental genotype before new mutations are induced.

In certain situations, mutation rate can be enhanced by proper selection of M_1 individuals. Selection can directly be made for dominant mutations or recessive changes in heterozygous position. Induced mottling or streaking of primary leaves or other signs of injury are generally correlated with mutagenic effect (55, 68, 117) and can preferably be used as a basis for selection when treatment is not uniform among individuals. Thus, Blixt; et al. (6) were able greatly to increase mutation rate by selecting mottled pea seedlings treated with ethyleneimine. A similar improvement is reported by Gregory (33) after selection for X_1 seedling injury in peanuts. Since there does not appear to exist any correlation between chromosome and point mutations if competition between shoot initials is eliminated (27, 29), a shift to either category can be obtained by selecting semifertile or fertile M_1 plants, respectively.

Even with the best choice of parental material, the highest efficiency in mutagenic application, and the most appropriate handling of the treated generation, the vast majority of scorable mutations will be harmful or practically uninteresting. The economical success of mutation breeding will, therefore, depend greatly on the efficiency of selection in the segregating generations. It is also symptomatic that the most rapid and elegant completions in mutation breeding are generally directly associated with a specific screening technique by which a most rare mutant can easily be detected in a large population.

Screening for disease resistance by means of artificial infection with known races (58, 59) or extracted toxin (116) allows enormous amounts of individuals to be handled. Rapid chemical selection methods like those elaborated for detection of sweet lupines (97) or coumarin-free sweet clover (88, 90) will also greatly improve efficiency in mutation breeding. The same is true for micro-methods for quality analyses of individual plants like the new Svalöf method for oil determination which has radically changed the possibilities in breeding oil crops for higher yield of fat (82, 112, 113).

Properties like frost, heat, drought, and sprouting resistance, where artificial selection methods are elaborated (57, 65), have preference in mutation breeding. An interesting work along these lines is reported by Zacharias (117). By simply germinating X_2 material of soybeans at low temperatures, he was able to select mutants with the ability to develop and start growth earlier, which is of greatest importance for extending soybean cultivation to cool climates. In crops with distinct response to day length or vernalization, mass screening may also be used at low cost. By spring sowing of X-material of Skandia III winter wheat, off-types were easily selected which were able to shoot and ripen before the end of the vegetation period (69). This technique is now used by Konzak (personal communication) in large-scale breeding projects for better spring wheats.

In all situations where aim and detection technique are precise, selection can very well have set in already in M_2, the first segregating generation. If the inventory is laborious or otherwise expensive, selection in M_2 will even be preferable, since it will imply the highest chance to find the desired mutation among the smallest number of plants (103). If the screening is easily done and the treatment given to seeds or seedlings, it may, however, be advisable to select first in M_3, where every mutation is no longer represented by single plants but rather by a whole group of plants. By testing only a part of each M_2 progeny, screening can also work efficiently even if the tested material has to be destroyed during the analysis. All M_2 lines proved to include the desired off-type can in this way be picked out and propagated. In most situations, where the desired mutations are difficult to detect due to imperfect selection methods or too vague a phenotypic penetration, or the aim of the mutation experiment is indeterminate, the advantage of screening groups instead of single plants will often be decisive. For that reason, more and more mutation experts recommend selection first in M_3 (24, 25, 26, 29, 49, 50, 103), a trend definitely stimulated by the promising experiments on induced mutations in quantitative traits.

The rapid increase in size of population from M_1 to M_3 can partly be eliminated by proper sampling. Since mutation in a uniform parental material is a chance event but seeds traced back to one and the same germ line more or less interrelated, the sampling should be based on as many independent inflorescence units and as few seeds from each such unit as possible (24, 25, 56). The more seeds taken per independent unit, the lower relative frequency of mutant individuals will appear in the next generation, but the higher will be the chance to detect all mutations induced in the given material. The last statement, however, has a limit set by the segregation ratio between normal and mutant types. In barley, for example, the average chlorophyll mutant frequency per segregating X_1 head progeny has been found to vary between 13 to 18 per cent, depending on tillering (28, 68). The corresponding figure for X_2 is about 20 per cent (78). Within the 99 per cent probability limit, therefore, not more than about 30 vital seeds per unit should be tested. The simple system of sampling by means of dense sowing under unfavourable conditions, resulting in few seeds per plant, can be used also in M_2, but this has to be weighed against the risk of interplant competition (107). Negative or positive mass selection in M_2 is another method to keep down size of population. Natural selection under more or less extreme conditions as to overwintering, disease infection, day length, vegetation period, etc., can also be of great help. Highly sterile plants may be discarded if only point mutations are desired. Selection on the basis of morphological properties may generally be effective for physiological features as well as due to the high chance for character association.

Unless the goal of the breeding project is not very determinate, one of the greatest difficulties with practical mutation experiments is to decide which mutants should be selected and incorporated in future work. Induction of mutation as an independent method of plant breeding is only restricted to cases where the plant is vegetatively propagated and where a mutation is able to change a highly valuable autogamous genotype in a positive direction. In all other cases, the new mutant will be utilized through recombination, and combining ability will be about as difficult to see "from outside" as it is between entries in a world collection. This limitation of mutation breeding as an independent method is often overlooked, or perhaps it is better to say that there has so far been a tendency to restrict mutation breeding to such specific uses where it is sufficient by itself. It is true that one of the great promises in mutation breeding is the possibility to add a single characteristic to a delicate system of genic balance in which recombination may cause a breakdown. It is, however, just as true that a mutation will seldom be induced in its own optimal genic environment. The extensive work at Svalöf on stiff-strawed *erectoides* mutants in barley may be a good example in this discussion.

We now know that it was a fairly gross simplification, when Gustafsson and MacKey in 1948 (43) stated that "strength of straw can be produced at will" in barley radiation experiments. The promised stiffness of straw was no mistake, but from hitherto 166 *erectoides* mutations analysed extremely few have given a direct practical result. Only one, the Pallas barley, has been released (7). The overwhelming majority was induced in inferior germ plasm or had pleiotropic by-effects which made them either lower in yield, more sensitive to drought, more specialized in soil and nutritional demands, more liable to stay in boot, etc. It is also important to observe that, the *erectoides* factor in Pallas shows a different pleiotropic pattern, both in relation to straw and head, if transmitted from the original Bonus genotype. Thus, it interferes generally not as well with the brittle straw of Carlsberg II but even better with the elastic straw of Rika (Hagberg, personal communication).

The *erectoides* mutations, as well as other categories of induced mutations, have shown us that it is often much more efficient to transfer a successful mutation than to try to induce it anew in another genic background. It is further just as likely to make progress by trying to transfer interesting mutations into new genotypes with the hope for improved combining ability than to start searching in new M_2 material. The recent introduction of growth chambers to speed up the backcross technique has added further arguments along this line, since this method is especially suitable for the transfer of the distinct types of mutation here discussed. We need, however, a more detailed understanding about the combining ability of induced macromutations, since there are many apparent differences between the genetic behavior of natural and induced factors. Thus, we have produced a great many mutations resistant to diseases. Almost all of these mutations are recessive, while dominant mode of inheritance is more common among natural genes of the same kind. If Fisher (23) is right, this difference may depend on genic environment and co-adaptation.

The combining ability of a mutation has been proved to vary greatly with environment and background genotype (4, 10, 41, 42, 105). Only one example, cited by Stubbe (106), will be given here. In the snapdragon, a mutation called *eramosa* was characterized by a nearly complete loss in the ability to branch and by a very erect growth, desirable properties for an ornamental plant. Unfortunately, however, less pleasant features were also associated with the *eramosa* mutation, *viz.*, considerable inhibition and deformation of the flowers. By recombining the pathological mutant with other genotypes, it was possible to neutralize parts of the pleiotropic complex and to find a genic background where the desirable but not the undesirable characteristics of the *eramosa* mutation could be established (Vogel, unpublished). Induction of new neutralizing mutations can be considered another variation on the same thema (34, 37, 44).

In principle, there is no difference in the behavior of macro- and micromutations. They gradually pass into each other, and their effects may be cooperant or opposed. Quantitative variation can seldom definitely be proved to depend on multigenic differences of small magnitude only. The smaller phenotypic contribution of individual micromutations necessitates, however, recombination or recurrent induction, since only their additive effects can be phenotypically scored. Recombination, which allows accumulation of valuable and omission of negative modifiers, occurs to great advantage automatically in allogamous plants. In autogamous plants, it must, however, be artificially stimulated if the evolutionary so important micromutations are to be fully utilized. For this reason and also for the chance to evaluate the induced macromutations in different genic environments, mutation experiments with self-fertilizers should preferably be combined with heterozygosity and/or outcrossing. Instead of carefully purified parental stocks and rigorous isolation in order to meet high demands on scientific exactness (12), the purely practical mutation breeder should stimulate outcrossing and recombination above what decreased fertility can give. It would thus be interesting to try artificial induction of mutations in connection with composite crosses or multi-cross bulks where male sterility was added to augment genic exchange.

In evolution, the mutation processes are not only restricted to interior changes of the genes. Their deletion, duplication, and position, alone or in blocks, interfere with evolutionary fitness. The loss of genic material is definitely the most common type of chromosome mutation, and the rare evolutionary advantage of such a process is largely responsible for the discredit of this whole group of gross mutations. A loss may have a positive effect as a sequence of duplication (70, 72, 94). Duplications are, however, definitely more interesting, since they may imply a cumulative effect and also a buffering which allows vital genes to mutate in directions that might have been impossible otherwise. In genetically well-studied objects like maize and barley, systematic production of duplications starts to become a reality (1, 31, 45, 46).

In barley, the breeder is interested in duplicating a segment of chromosome 6 carrying the gene "orange lemma". This gene is strongly associated with high α–amylase activity and a duplication may thus offer a chance to improve malting properties. Another idea is to duplicate parts of the 40 centimorgans long segment of chromosome 5, where 13 of the 14 known loci for mildew resistance are incorporated (22). Such a procedure may augment the degree of resistance of some of the weaker genes, and it may offer a possibility to overcome difficulties with very close linkage. It would also allow more than one allele to be present simultaneously in the homozygous condition necessary for a stable barley variety. The last-mentioned possibility to fix a heterozygous condition would give a chance to utilize superdominance also in autogamous plants. Such a relation is found by Wiebe (cited by Hagberg, 46) to exist between the alleles V for 2-row and v for 6-row barley. By a proper duplication in chromosome 2, a genotype of the constitution VV, vv would allow the advantageous interaction between V and v without the risk of segregation and separation.

The above-mentioned examples are taken from the work of my Svalöf colleague Hagberg (46). Since duplication followed by differentiation must be considered as one of the most constructive phases in evolution, works of this kind may soon drastically change our ideas of induced mutation in plant breeding. The principle of producing well-defined duplications will have to go via the induction of translocations involving the same two chromosomes but with different positions of the breakage points. The combination of two such translocations will imply a duplication of the segment between the breaks. For the proper use of this technique in plant breeding, a large set of well-defined translocation lines and a detailed gene map are essential.

Induced chromosomal rearrangements may be valuable in many situations. They may be useful in splitting up gene blocks and character associations which had an evolutionary advantage in the wild plant but are undesirable from the demand of man. In the opposite way, valuable gene blocks may be inverted to prevent crossing-over or moved inside localized chiasmata. A fascinating way of using radiation experiments for incorporation of small segments from non-pairing chromosomes is well demonstrated by Sears' (93) already classical transfer of leaf-rust resistance from *Aegilops umbellulata* to common wheat. A similar approach is successfully accomplished by Elliott (19, 20) in transferring stem rust resistance from *Agropyron elongatum* to wheat, and the attempts to transfer bunt resistance from the same donor species are progressing (60, 63). Also, in Europe, this technique is now applied (unpublished by McKelvie, and Wienhues-Ohlendorf).

The great merits of induced chromosome mutation are definitely in a well-defined chromosome engineering work. Repeatedly, however, suggestions have been put forward to use recurrent radiation experiments for the diploidization of artificial polyploids, where differentiation of identical or too closely related chromosomes was thought to improve disomic pairing (26, 48, 71). No definite report of success along these lines is yet available, if we except the observation by Vettel (115) in Triticale that morphological X-mutants with long, dense heads had an improved seed setting. The Svalöf work along this line, including 4x barley, flax, and rye, does not indicate any successive improvement with recurrent radiation, but there is some hope that suddenly types with improved fertility may appear. In the light of the recent finding of a genetic control of the disomic behavior of hexaploid wheat chromosomes (86, 87, 95, 96), it seems just as possible that one mutational step would be enough to improve fertility as a successive accumulation of many small changes of the chromosome structure.

All through the above presentation, the importance of an inter-fusion between mutation experiments and conventional breeding methods is stressed. Mutation is the first step in a creative process followed by efforts to bring it in progressive interaction with the background genotype. This dependence on other breeding methods, however, can also be reversed. Mutations may open possibilities to apply new approaches in breeding a specific crop. Thus, Lewis (66, 67) and Pandey (83) were able to increase the mutation rate of self-incompatibility alleles in *Oenothera organensis* and Trifolium species, respectively, by the use of X-irradiation. Similar phenomenon can also be produced as primary and thus nonheritable effects of radiation (9). It may also be of interest to mention that gamete irradiation is a possibility to overcome certain interspecific barriers (14, 84, 85, 108, 109). Primary irradiation is also known to induce intersexuality in hemp (79).

The ability of radiation to disrupt a biochemical sequence and to induce a heritable shift in the same metabolic pathway is also demonstrated by the interesting transfer from apomixis to sexuality in *Poa pratensis* (32, 53, 54) and Potentilla species (Asker, unpublished). Apomixis is evidently dependent on a rather complex gene balance where many different genetic events may interfere and induce a higher or lower degree of sexuality. Due to the high tolerance to chromosome disturbances in the multiploid Poa, even losses of whole chromosomes were ultimately found to result in sexuality. The artificial induction of such a shift enables the breeder to utilize very strictly apomictic clones as parental components in crosses or merely to explore the variability inherent to an extreme heterozygote. In both cases he may select further among more or less sexual segregates or isolate fixed, apomictic types, where the balance is restored (Figure 1).

The above review is an attempt to understand mutation experiments and their possibilities in plant breeding. Just as the induced mutation can be evaluated only in relation to its background genotype, mutation breeding can be evaluated only as an integral part

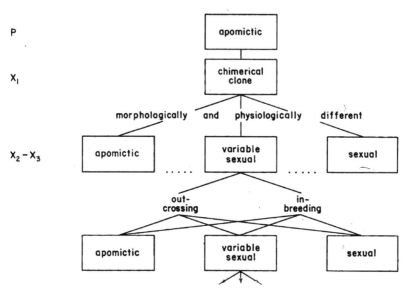

FIGURE 1.—*X-ray-induced shift to sexuality in* Poa pratensis. *After Grazi, et al. (32).*

in our efforts to improve our cultivated plants. An evolutionary system based on mere addition of mutations under different selective pressure would be a rigid and uneconomical mechanism of adjustment to environment. Mutation breeding must be considered as *one* method of approach, sometimes practically impossible, sometimes inferior, sometimes insufficient but definitely promoting, and sometimes superior or the only solution. Its ability to furnish basic variability should not be considered as a substitute, but rather as a complement to our world collections and other sources of natural variation (47).

Our skepticism and hesitation today can partly be referred to the novelty of the method and to the feeling that nature should first be exhausted of its resources, but partly also to our incomplete genetic knowledge in relation to our specific breeding objects. The work on maize and barley shows doubtless that a more profound knowledge and a more systematic accumulation of gene and chromosome markers will greatly improve our chance to use mutation experiments in plant breeding. In the collection of all these data and marker types, mutation induction will be one of our best helps. This indirect contribution of mutation genetics to crop improvement is a large and fascinating subject worth another story. It is evident, however, that we have to possess ourselves in patience before we will be able fully to understand the merits and drawbacks of induced mutation as a method of augmenting the food production of the world.

Summary

Besides the metabolic stage at time of treatment, the mutagen and dose applied, and the method of handling before, during, and after treatment, the genotypic constitution of the material subjected to mutation experiments plays an important role. The total yield of mutations depends on inherent resistance to the mutagen applied, probably to degree of heterozygosity, to extent and direction of phenotypic buffering, and to the presence of necessary precursor genes. From all these aspects, inter- as well as intraspecific differences occur, which greatly influence the yield of desirable mutations from case to case. Material especially suitable for the detection of desired mutations may improve the chance for success.

Maximum yield of interesting mutations is also dependent on handling of M_1. A diplontic selection over the first cell divisions after seed or plant treatment should be prohibited by special arrangements. Different kinds of positive or negative mass selection may also increase the mutation rate in desired directions.

Due to the very low frequency of valuable mutations, the success in mutation breeding depends largely on efficient screening. Selection in M_2 is preferred for very easily detectable mutations. In other cases, selection in M_3 may be better, since every mutation will here be represented by a whole group of off-type individuals. Proper sampling in M_2 can keep down population size.

Valuable mutations seem often so rare that it may be easier to transfer them to other genotypes than to try to induce them anew. Greater emphasis should be laid on micromutations. The combining ability of induced mutations varies drastically with background genotype.

Among chromosome mutations, translocations may be very useful for a systematic production of duplications, for splitting up or building up gene blocks, and for interspecific gene transfers. Their role in diploidizing artificial polyploids and inducing disomic pairing and better fertility is not yet well understood.

Induced mutations may open possibilities to apply new breeding methods to specific crops. Induced self-compatibility in self-sterile plants and induced sexuality in apomictic plants are given as examples.

It is essential not to set mutation breeding in opposition to more conventional breeding methods but rather to evaluate this novelty as an integral and interdependent part. In this way, and with more complete knowledge in relation to the specific breeding objects, the value of induced mutation in crop improvement will certainly increase.

References

1. Anderson, E. G. 1956. The application of chromosomal techniques to maize improvement. *Brookhaven Symp. Biol.,* 9: 23–35.
2. Andersson, G. 1953. Svalöf's Regina vårraps II. *Sveriges Utsädesf. tidskr.,* 63: 201–204.
3. ———— and Olsson, G. 1954. Svalöf's Primex White Mustard—a market variety selected in X-ray-treated material. *Acta Agr. Scand.,* 4: 574–577.
4. Bandlow, G. 1959. Mutationsversuche an Kulturpflanzen: X. Über Pleiotropie und eine zweifache Mutante bei Wintergerste. *Züchter,* 29: 123–132.
5. Bauer, R. 1957. The induction of vegetative mutations in *Ribes nigrum. Hereditas,* 43: 323–337.
6. Blixt, S., Ehrenberg, L., and Gelin, O. 1960. Quantitative studies of induced mutations in peas: III. Mutagenic effect of ethyleneimine. *Agr. Hort. Genet.,* 18: 109–123.
7. Borg, G., Fröier, K., and Gustafsson, Å. 1958. Pallas barley, a variety produced by ionizing radiation: Its significance for plant breeding and evolution. *2nd Int. Conf. Peaceful Uses Atomic Energy, UN, Geneva, A/Conf. 15/P. 2468.*
8. Breider, H., and Reichardt, A. 1955. Neue Wege in der Pflanzenzüchtung mittels Röntgenstrahlen. *Strahlentherapie* 97: 151–158.
9. Brewbaker, J. L., Shapiro, N., and Majumder, S. 1958. The incompatibility inhibition in flowering plants. *Proc. 10th Int. Congr. Genet.,* 2: 33.
10. Brücher, H. 1943. Experimentelle Untersuchungen über den Selektionswert Künstlich erzeugter Mutanten von *Antirrhinum majus. Zeit. f. Botanik,* 39: 1–47.

11. Caldecott, R. S., and North, D. T. 1961. Factors modifying the radio-sensitivity of seeds and the theoretical significance of the acute irradiation of successive generations. *This Symposium, 365–404.*

12. ————, Stevens, H., and Roberts, B. J. 1959. Stem rust resistant variants in irradiated populations—mutations or field hybrids? *Agron. Jour., 51: 401–403.*

13. Conger, A. D. 1957. Some cytogenetic aspects of the effects of plant irradiations. *Proc. Oak Ridge Reg. Symp., 9: 59–62.*

14. Davies, D. R. 1960. Induced mutations in plant crops. *Atom No., 45: 13–19, 31–32.*

15. Davies, R., and Wall, E. T. 1958. Artificial mutagenesis in plant breeding. *Nature, 182: 955–956.*

16. Demerec, M. 1938. Hereditary effect of X-ray radiation. *Radiology, 30: 212–220.*

17. Down, E. E., and Andersen, A. L. 1956. Agronomic use of an X-ray-induced mutant. *Science, 124: 223–224.*

18. Dubovskij, N. V. 1935. On the question of the comparative mutability of stocks of *Drosophila melanogaster* of different origins. *Compt. Rend. Acad. Sci. URSS, 4: 95–97.*

19. Elliott, F. C. 1957. X-ray-induced translocation of Agropyron stem rust resistance to common wheat. *Jour. Hered., 48: 77–81.*

20. ————. 1959. Further information on an X-ray-induced translocation of Agropyron stem rust resistance to common wheat. *Wheat Inform. Serv., 9–10: 26–27.*

21. Fabergé, A. C. 1957. A method for treating wheat pollen with ultraviolet radiation for genetic experiments. *Genetics, 42: 618–622.*

22. Favret, E. A. 1960. Spontaneous and induced mutations of barley for the reaction to mildew. *Hereditas, 46: 20–28.*

23. Fisher, R. A. 1930. The genetical theory of natural selection. *Oxford: The Clarendon Press.*

24. Freisleben, R., and Lein, A. 1943. Vorarbeiten zur züchterischen Auswertung röntgeninduzierter Mutationen. *Zeit. Pfl. zücht., 25: 235–283.*

25. ————, ————. 1944. Möglichkeiten und praktische Durchführung der Mutationszüchtung. *Kühn-Archiv, 60: 211–225.*

26. Gaul, H. 1958. Present aspects of induced mutations in plant breeding. *Euphytica, 7: 275–289.*

27. ————. 1958. Über die gegenseitige Unabhängigkeit der Chromosomen und Punktmutationen. *Zeit. Pfl. zücht., 40: 151–188.*

28. ————. 1959. Über die Chimärenbildung in Gerstenpflanzen nach

Röntgenbestrahlung von Samen. *Flora,* 147: *207–241.*

29. ———— and Mittelstenscheid, L. 1960. Hinweise zur Herstellung von Mutationen durch ionisierende Strahlen in der Pflanzenzüchtung. *Zeit. Pfl. Zücht.,* 43: *404–422.*

30. Gelin, O., Ehrenberg, L., and Blixt, S. 1958. Genetically conditioned influences on radiation sensitivity in peas. *Agr. Hort. Genet.,* 16: *78–102.*

31. Gopinath, D. M., and Burnham, C. R. 1956. A cytogenetic study in maize of a deficiency duplication produced by crossing interchanges involving the same chromosome. *Genetics,* 41: *382–395.*

32. Grazi, F., Umaerus, M., and Åkerberg, E. 1961. Observation on the reproductional behavior and embryology of *Poa pratensis. Hereditas,* 47. In press.

33. Gregory, W. C. 1955. X-ray breeding of peanuts *(Arachis hypogaea* L.). *Agron. Jour.* 47: *396–399.*

34. ————. 1956. Induction of useful mutations in the peanut. *Brookhaven Symp. Biol.,* 9: *177–190.*

35. ————. 1956. Radiosensitivity studies in peanuts *(Arachis hypogaea* L.). *Proc. Int. Genet. Symp. Japan, Caryologia Suppl. Vol.,* *243–247.*

36. ————. 1956. The comparative effects of radiation and hybridization in plant breeding. *Proc. Int. Conf. Peaceful Uses Atomic Energy, UN, Geneva,* 12: *48–51.*

37. ————. 1957. Progress in establishing the effectiveness of radiation in breeding peanuts. *Proc. Oak Ridge Reg. Symp.,* 9: *36–48.*

38. Gustafsson, Å. 1940. The mutation system of the chlorophyll apparatus. *Lunds Univ. Arsskr. N.F. Avd. 2,* Bd 36: *Nr. 11.*

39. ————. 1944. The X-ray resistance of dormant seeds in some agricultural plants. *Hereditas,* 30: *165–178.*

40. ————. 1947. Mutations in agricultural plants. *Hereditas,* 33: *1–100.*

41. ————. 1951. Mutations, environment, and evolution. *Cold Spr. Harb. Symp. Quant. Biol.,* 16: *263–281.*

42. ————. 1954. Mutations, viability, and population structure. *Acta Agr. Scand.,* 4: *601–632.*

43. ———— and MacKey, J. 1948. Mutation work at Svalöf. *Svalöf 1886–1946 (editors: Å. Åkerman, et al.), C. Bloms Boktr., Lund, 338–357.*

44. ———— and Wettstein, D. von. 1958. Mutationen und Mutationszüchtung. *Handb. Pfl. züchtung (editors: H. Kappert and W. Rudorf), Berlin, Verlag P. Parey, 2nd Ed.,* 1: *612–699.*

45. Hagberg, A. 1958. Cytogenetik einiger Gerstenmutanten. *Züchter*, **28**: *32–36*.

46. Hagberg, H. 1959. Barley mutations used as a model for the application of cytogenetics and other sciences in plant breeding. *Proc. 2nd Congr. Eucarpia, Cologne, 235–248*.

47. Harlan, J. R. 1956. Distribution and utilization of natural variability in cultivated plants. *Brookhaven Symp. Biol.*, **9**: *191–206*.

48. Hoffmann, W. 1959. Neuere Möglichkeiten der Mutationszüchtung. *Zeit. Pfl. Zücht.*, **41**: *371–394*.

49. Hänsel, H., and Zakovsky, J. 1956. Röntgeninduzierte Mutanten der Vollkorngerste *(Hordeum distichum nutans)*: I. Bestrahlung und Auslese auf Mehltauresistenz. *Bodenkultur*, **9**: *50–64*.

50. ───── and Zakovsky, J. 1956. Mildew-resistant barley mutants induced by X-rays. *Euphytica*, **5**: *347–352*.

51. Johnson, E. L. 1933. The influence of X-radiation on *Atriplex hortensis* L. *New Phytol.*, **32**: *297–307*.

52. ─────. 1936. Susceptibility of seventy species of flowering plants to X-radiation. *Plant Physiol.*, **11**: *319–342*.

53. Julén, G. 1954. Observations on X-rayed *Poa pratensis*. *Acta Agr. Scand.*, **4**: *585–593*.

54. ─────. 1958. Über die Effekte der Röntgenbestrahlung bei *Poa pratensis*. *Züchter*, **28**: *37–40*.

55. Kaplan, R. W. 1954. Beeinflussung des durch Röntgenstrahlen induzierten mutativen Fleckenmosaiks auf den Blättern der Sojabohne durch Zusatzbehandlung. *Strahlentherapie*, **94**: *106–118*.

56. Kappert, H. 1953. Die vererbungswissenschaftlichen Grundlagen der Züchtung. *Berlin: Verlag P. Parey. 2nd Ed.*

57. ───── and Rudorf, W. 1958–61. Handbuch der Pflanzenzüchtung. *Berlin: Verlag P. Parey. 2nd Ed. Vols. 1–6.*

58. Konzak, C. F. 1956. Induction of mutations for disease resistance in cereals. *Brookhaven Symp. Biol.*, **9**: *157–171*.

59. ─────. 1959. Induced mutations in host plants for the study of host-parasite interactions. *Wash. Agr. Exp. Sta. Sci. Paper No. 1764.*

60. ───── and Heiner, R. E. 1959. Progress in the transfer of resistance to bunt *(Tilletia caries, and T. foetida)* from Agropyron to wheat. *Wheat Inform. Serv., 9–10: 31.*

61. Lamprecht, H. 1956. Röntgen-Empfindlichkeit und genotypische Konstitution bei Pisum. *Agr. Hort. Genet.*, **14**: *161–176*.

62. ─────. 1958. Röntgen-Empfindlichkeit und genotypische Konstitution von Phaseolus. *Agr. Hort. Genet.*, **16**: *196–208*.

63. Larter, E. N., and Elliott, F. C. 1956. An evaluation of different ionizing radiations for possible use in the genetic transfer of bunt resistance from Agropyron to wheat. *Can. Jour. Bot.,* **34:** *817–823.*

64. Levan, A. 1944. Experimentally induced chlorophyll mutants in flax. *Hereditas,* **30:** *225–230.*

65. Levitt, J. 1956. The hardiness of plants. *Advances in Agronomy, Vol. 6. New York: Academic Press Inc.*

66. Lewis, D. 1946. Useful X-ray mutations in plants. *Nature, 158–519.*

67. ————. 1951. Structure of the incompatibility gene: III. Types of spontaneous and induced mutation. *Heredity,* **5:** *399–414.*

68. MacKey, J. 1951. Neutron and X-ray experiments in barley. *Hereditas,* **37:** *421–464.*

69. ————. 1954. Neutron and X-ray experiments in wheat and a revision of the speltoid problem. *Hereditas,* **40:** *65–180.*

70. ————. 1954. Mutation breeding in polyploid cereals. *Acta Agr. Scand.,* **4:** *549–557.*

71. ————. 1956. Mutation breeding in Europe. *Brookhaven Symp. Biol.,* **9:** *141–152.*

72. ————. 1959. Mutagenic response in Triticum at different levels of ploidy. *Proc. 1st Int. Wheat Genet. Symp., Winnipeg, 88–111.*

73. ————. 1960. Radiogenetics in Triticum. *Genet. Agraria,* **12:** *201–230.*

74. Matsumura, S., and Fujii, T. 1957. Mutants in tobacco plants induced by X-rays and their application. *Ann. Rpt. Nat. Inst. Genet. Japan,* **7:** *87–88.*

75. Melchers, G. 1960. Haploide Blütenpflanzen als Material der Mutationszüchtung. Beispiele: Blattfarbmutanten und mutatio wettsteinii von *Antirrhinum majus. Züchter,* **30:** *129–134.*

76.' Mertens, T. R., and Burdick, A. B. 1957. On the X-ray production of "desirable" mutations in quantitative traits. *Amer. Jour. Bot.,* **44:** *391–394.*

77. Mikaelsen, K. 1956. Studies on the genetic effects of chronic gamma radiation in plants. *Proc. Conf. Peaceful Uses Atomic Energy, UN, Geneva,* **12:** *34–39.*

78. Moh, C. C., and Smith, L. 1951. An analysis of seedling mutants (spontaneous, atomic bomb-radiation-, and X-ray-induced) in barley and durum wheat. *Genetics,* **36,** *629–640.*

79. Moutschen, J., and Dahmen, M. 1955. Intersexualité induite par les rayons gamma chez *Cannabis sativa* L. *Bul. Soc. Roy. Belg.,* **87:** *21–28.*

80. Nybom, N. 1956. Some further experiments on chronic gamma-irradiation of plants. *Bot. Notiser,* **109:** *1–11.*

81. Oka, H. I., Hayashi, J., and Shiojiri, I. 1958. Induced mutation of polygenes for quantitative characters in rice. *Jour. Hered.,* 49: *11–14.*

82. Olson, G. 1960. Some relations between number of seeds per pod, seed size, and oil content and the effects of selection for these characters in Brassica and Sinapis. *Hereditas,* **46:** *29–70.*

83. Pandey, K. K. 1956. Mutations of self-incompatibility alleles in *Trifolium pratense* and *T. repens. Genetics,* **41:** *327–343.*

84. Reusch, J. D. H. 1956. Influence of gamma radiation on the breeding affinities of *Lolium perenne* and *Festuca pratensis. Nature,* **178:** *929–930.*

85. ————. 1960. The effects of gamma radiation on crosses between *Lolium perenne* and *Festuca pratensis. Heredity,* **14:** *51–60.*

86: Riley, R. 1958. Chromosome pairing and haploids in wheat. *Proc. 10th Int. Congr. Genet., Montreal,* 2: *234–235.*

87. ———— and Chapman, V. 1958. Genetic control of the cytologically diploid behavior of hexaploid wheat. *Nature,* **182:** *713–715.*

88. Roberts, W. L., and Link, K. P. 1937. Determination of coumarin and melilotic acid: A rapid method for determination in Melilotus seed and green tissue. *Ind. Eng. Chem., Anal. Ed.,* 9: *438–441.*

89. Rosen, G. von. 1942. Röntgeninduzierte Mutationen bei *Pisum sativum. Hereditas,* **28:** *313–338.*

90. Scheibe, A., and Hülsmann, G. 1958. Mutationsauslösung beim Steinklee *(Melilotus albus). Zeit. Pfl. Zücht.,* **39:** *299–324.*

91. Scossirolli, R. E. 1953. Effectiveness of artificial selection under irradiation of plateaued populations of *Drosophila melanogaster. I.U.B.S. Symp. Genetics of Population Structure, Pavia, 42–66.*

92. ————. 1959. Selezione artificiale per un carattere quantitivo in popolazioni di *Drosophila melanogaster* irradiate con raggi X. *Com. Naz. Ric. Nucl., Div. Biol., No. 4.*

93. Sears, E. R. 1956. The transfer of leaf-rust resistance from *Aegilops umbellulata* to wheat. *Brookhaven Symp. Biol.,* 9: *1–22.*

94. ————. 1959. The systematics, cytology, and genetics of wheat. *Handb. Pfl. Zücht. (editors: H. Kappert & W. Rudorf), Berlin: Verlag P. Parey.* 2nd Ed., 2: *164–187.*

95. ————. 1959. The aneuploids of common wheat. *Proc. 1st Int. Wheat Genet. Symp., Winnipeg, 221–228.*

96. ———— and Okamoto, M. 1958. Intergenomic chromosome relation-

ships in hexaploid wheat. *Proc. 10th Int. Congr. Genet., Montreal,* **2:** *258–259.*

97. Sengbusch, R. von. 1942. Süsslupinen und Öllupinen: Die Entstehungsgeschichte einiger neuen Kulturpflanzen. *Landw. Jahrb.,* **91:** *719–880.*

98. Singleton, W. R. 1957. Gamma radiation of growing plants for mutation induction. *Proc. Oak Ridge Regional Symp.,* **9:** *46–57.*

99. Smith, L. 1942. Hereditary susceptibility to X-ray injury in *Triticum monococcum. Amer. Jour. Bot.,* **29:** *189–191.*

100. Sparrow, A. H., and Gunckel, J. E. 1956. The effects on plants of chronic exposure to gamma radiation from radiocobalt. *Proc. Int. Conf. Peaceful Uses Atomic Energy, UN, Geneva,* **12:** *52–59.*

101. Stadler, L. J. 1932. On the genetic nature of induced mutations in plants. *Proc. 6th Int. Congr. Genet., Ithaca, N.Y.,* **1:** *274–294.*

102. ———— and Roman, H. 1948. The effect of X-rays upon mutation of the gene *A* in maize. *Genetics,* **33:** *273–303.*

103. Steuckhardt, R. 1960. Untersuchungen über die Wirkung von Röntgenstrahlen auf Rispenhirse *(Panicum miliaceum L.)* nach einmaliger und mehrfacher Bestrahlung. *Zeit. Pfl. Zücht.,* **43:** *85–105; 297–322.*

104. Stubbe, H. 1934. Einige Kleinmutationen von *Antirrhinum majus* L. *Züchter,* **6:** *299–303.*

105. ————. 1950. Über den Selektionswert von Mutanten. *Sitzungsber. Deutsch, Akad. Wiss. Berlin, Kl. f. Landw. wiss.,* **1:** *1–42.*

106. ————. 1958. Advances and problems of research in mutations in the applied field. *Proc. 10th Int. Congr. Genet., Montreal,* **1:** *247–260.*

107. Suneson, C. A., and Stevens, H. 1953. Studies with bulked hybrid populations of barley. *U.S. Dept. Agr. Tech. Bul. No. 1067.*

108. Swaminathan, M. S., and Murty, B. R. 1959. Effect of X-radiation on pollen tube growth and seed setting in crosses between *Nicotiana tabacum* and *N. rustica. Zeit. Vererbungsl.,* **90:** *393–399.*

109. Takana, M. 1957. The application of X-rayed pollen in crosses of Nicotiana: I. The effect on reciprocal crosses between tobacco and some species of the section Alatae. *Jap. Jour. Breed.,* **7:** *39–44.*

110. Tedin, O., and Hagberg, A. 1952. Studies on X-ray-induced mutation in *Lupinus luteus* L. *Hereditas,* **38:** *267–296.*

111. Tollenaar, D. 1938. Untersuchungen über Mutation bei Tabak: II. Einige künstlich erzeugte Chromosom-Mutanten. *Genetica,* **20:** *285–294.*

112. Troëng, S. 1955. Oil determination of oilseed: Gravimetric routine method. *Jour. Amer. Oil Chem. Soc.*, **32:** *124–126.*

113. ———. 1955. Oljehaltsbestämning i svenskt oljeväxtfrö. *Kungl. Lantbr. akad. tidskr.,* **94:** *125–140.*

114. Tsunewaki, K., and Heyne, E. G. 1959. Radiological study of wheat monosomics. *Genetics,* **44:** *933–954.*

115. Vettel, F. K. 1959. Mutationsversuche an Weizen-Roggenbastarden (Triticale): I. Mutationsauslösung bei Triticale Rimpau. *Züchter,* **29:** *293–317.*

116. Wheeler, H. E., and Luke, H. H. 1955. Mass screening for disease-resistant oats. *Science,* **122:** *1229.*

117. Zacharias, M. 1956. Mutationsversuche an Kulturpflanzen: VI. Röntgenbestrahlungen der Sojabohne *(Glycine soja* (L.) Sieb. et Zucc.). *Züchter,* **26:** *321–338.*

Comments

BREWBAKER: May I comment first that there are two misconceptions that appear in the discussions of mutations involving incompatibility alleles. First, that a curious sort of revertible mutation can occur in which the gene mutates to an inactive form to permit fertilization and then reverts in the X_1 embryo to its original form. The evidence supporting this double mutation event is extremely tenuous, and gains no support in our Petunia work. Second, there is no evidence whatsoever of "point mutations" resulting in the formation of a new S allele (our studies alone exceeding 30 million pollen grains) and reports of mutation by inactivation of subunits in the S gene (as in the cherries) can be suggested to result from chromosomal aberrations, such as translocation or centric fragments.

Now to my question—having screened a great number of such self-fertile mutants in Petunia, in which S locus duplication is an elegant theoretical way to produce "mutants", we have no evidence at all for such duplications, although irradiations have been applied premeiotically. Unless one has a particularly effective screen for duplications, how optimistic do you feel we can be regarding such duplications which might be, as you suggest, of considerable value in mutation breeding?

MacKEY: I share your skepticism on the correct interpretation of the mechanism behind the induced compatibility in some cases. This will, however, not interfere with the fact that the plant breeder can make use of this shift in his breeding procedure, even if a chromosome mutation may be a little more complicated to work with than a point mutation.

As to your second problem, the possibility is certainly extremely remote of being able to induce a very specific type of duplication directly by prophase irradiation. Thus, the success will depend more or less entirely on the efficiency of the available screening technique. For that reason, the method must have a limited, practical value. However, this does not necessarily mean that it must always be preferable to work via translocation lines. I would definitely prefer more data to be compiled before any clear answer can be given to your question.

CALDECOTT: The need for attention being directed to the production of duplications is most important. Today we are relying almost entirely on translocation between the same arms of homologous chromosomes that were broken at dissimilar points.

What we need is a good method of obtaining cells in which the chromosomes are uniformly in the bipartite condition before irradiation. This would enchance sister strand reunions and in instances where the breaks were at dissimilar points should prove very effective.

Dr. B. H. Beard is studying the problem using the 2–row vs. 6–row gene in barley. His procedure is to irradiate seeds in which the genes are in the heterozygous state and screen in the X_3 for "nonsegregating permanent heterozygotes".

AUERBACH: The two-step process of the production of duplications, which Doctor Caldecott outlined for X-rays, occurs normally at a high frequency after treatment with chemical mutagens whose breakage often is delayed until after the treated chromosome has divided into chromatids. Ford found many small duplications after treatment of Vicia with nitrogen mustard, and Slizynska found many after formaldehyde treatment on Drosophila.

LEWIS: In cotton many introductions cannot be evaluated in the latitude of the United States because of a short-day photoperiod requirement. Do you think irradiation to break this reaction might be feasible?

MACKEY: The answer to this question greatly depends on the genetic complexity of the photoperiodic response. A complex genetic background greatly impairs the chance to break the short-day reaction by induced mutation, while a simple genetic control would greatly favor a program along this line. A shift in photoperiodic response can easily be detected in a large material and a simple mutation is likely sometimes to hit just right. The chance that many mutational events should happen simultaneously according to a fixed pattern is, however, practically nil.

Judging from other crops, the genetic control of photoperiodism is likely to be rather simple. Thus, it would seem worth trying this experiment in cotton.

PATTERSON: The pleiotropic effects of the *erectoides* mutations on straw strength are of special value for use in cross breeding for straw strength in barley. Selection for the *erectoid* type results in selection for straw strength, a character usually difficult to obtain.

MACKEY: I think that the Swedish barley breeding along conventional lines, with results like Rika, Herta, and Ingrid, proves that straw strength can be achieved also in other ways than through inducing *erectoides* mutations. The important thing is that the two ways are different with possibilities for transgression when combined.

KONZAK: Since terminology in this new field has an important bearing on the understanding of methods and treatments, I wish to recommend that we utilize a standard designation system for describing the generation following mutagen treatment as the M_1, M_2, etc., designation system. In addition, I would suggest that we consider an idea expressed to me by Doctor Wellensiek recently—this is to use a superscript to denote the mutagen and the treatment used, as $M_1^{EMS\ 4hr.02}$ for the first generation following ethyl methane sulfonate treatment or to designate the first generation from a 15 Kr gamma radiation treatment as $M_1^{15\ Kr\ G}$.

In a case where repeated treatments are studied, the designation might be modified thus:

M^x_{1-2} first generation,   x = kind of treatment and dose, if desired
          second cycle   M = mutagen treatment
M^x_{2-2} second generation,   1 = 1st generation
          second cycle   2 = 2nd generation treatment

MACKEY: I think it is highly desirable to agree upon a standard designation system, but the abbreviations should not be so exclusively used that people outside our group will not be able to understand our publications.

KONZAK: It should be recorded here for the benefit of those interested that we put into use two additional microanalyses techniques in our mutation and plant breeding programs with wheat.

One of these involves a rapid microtest for protein analyses. The equipment used is an Udy protein analyzer which, using a dye binding method, can give protein analyses with accuracy comparable to or better than, Kjeldhals in about $3\frac{1}{2}$ to 5 minutes time per sample. We now consider it more accurate than the Kjeldahls because it measures only amino nitrogen and does not measure nitrate. This is important, especially for the analysis of plots fertilized with nitrogen. This machine was developed at the Regional Wheat Quality Laboratory at Pullman, Washington, and is now being manufactured and further developed by Dr. Doyle C. Udy, its inventor at Pullman.

Using this equipment, we have sorted out induced variation in wheat and will be able to evaluate material on a large scale.

Another machine for rapid analyses also was developed in the same laboratory. This machine, the Micromill, permits isolation of better milling selections using visual identification of selections according to bran cleanup. Two operators can study about 1,000 samples per day. In breeding, we use 1.5 to 5 grams of seed, but we hope that we can identify lines carrying promising material by recognizing one or few clean nearly endosperm-free bran flakes among a bulk sample from an M_2 progeny.

OLMO: In retrospect, would it have been more economical to produce a Pallas-type barley by conventional breeding methods than by the radiation breeding technique?

MACKEY: A strict answer to your question cannot be given, since the two kinds of approaches have not been tried. I have the feeling that *erectoides* types became of interest to the barley breeders first when induced in a highbred genic environment. As with many other things hidden in our world collections, they did not attract the breeders when combined with otherwise undesirable genes and this the more so when their mode of inheritance was less well explored. Now that the information is available, transfers by backcross are likely to be more efficient than trying to induce just the right type of *erectoides* characters anew in another variety.

Factors Modifying the Radio-Sensitivity of Seeds and the Theoretical Significance of the Acute Irradiation of Successive Generations[1]

RICHARD S. CALDECOTT AND D. T. NORTH[2]

University of Minnesota, St. Paul, Minn.

THE PURPOSE of this report is twofold, namely, to present data which demonstrate the use of dormant seeds in biophysical studies and to discuss the possible genetic and applied significance of the irradiation of successive seed generations of the small-grained cereals.

Part I. Modification of the Radiosensitivity of Seeds

For years it has been obvious to investigators working with seeds that these biological structures apparently represent an unique system in that they can be subjected to extremes of environment without impairment of function when restored to the conditions required for normal growth and development. Using seeds, it has been possible to study the biological consequences of treatments that were otherwise only possible *in vitro*. This was perhaps best shown in studies where, by decreasing the water content well below levels at which physiological activity was possible, seeds tolerated temperature extremes ranging from those of liquid nitrogen to 112° C.

It also was shown that under conditions where physiological activity was not detectable, the expression of damage to X-rayed seeds could progressively increase for weeks after initial photon absorption (1, 2, 5, 12).[3] Furthermore, the rate and degree of this injury enhance-

[1]This work was conducted under Contract No. AT (11–1)–332 between the University of Minnesota and the U. S. Atomic Energy Commission. The report is a compilation of two reports previously presented at symposia sponsored by the IAEA and FAO in Karlsruhe, Germany, August, 1960, and the AAAS in Chicago, Ill., in December, 1959. Contribution from the U. S. Department of Agriculture, Field Corps Research Branch, ARS.

[2]The writers are pleased to acknowledge the help of Dr. Alessandro Bozzini on the studies relating to seedling height and genetic injury, Miss Victoria L. Bergbush on all phases of the temperature investigations, and Miss Louise Heine and Mr. Fa-ten Kao for assistance with the cytological investigations.

[3]See References, page 398.

ment was shown to be dependent on temperature. Thus, the post-irradiation increase in injury to seeds was negligible over a period of 96 hours at the temperature of solid carbon dioxide.

Studies complementary to those on the relation of time and temperature to the manifestation of post-irradiation injury have shown that the availability of oxygen to the seed following X-irradiation also has a profound effect on the expression of injury. Thus, seeds that are stored or hydrated in the presence of oxygen immediately after X-irradiation are more severely injured than are seeds that are stored or hydrated anaerobically (5, 10). Significantly, sensitivity to oxygen following irradiation diminishes as a function of time and can be completely eliminated by a brief post-irradiation temperature treatment administered at 75° C.

In a recent series of comprehensive and refined experiments, using bacterial spores which will tolerate desiccation to levels comparable to those commonly used in the seed work, Powers, et al. (21) have corroborated and expanded upon much of the work that has been done with seeds. This parallel response, at least in the more basic considerations, of a multicellular and a unicellular organism is a fortunate circumstance in that it gives credence to the wealth of data accumulated in biophysical studies with seeds that has often been considered exceptional and has, accordingly, too often been ignored. Furthermore, analyses of these two organisms supplement one another. The bacterial spore techniques are amenable to rapid analysis of large populations to determine the relative lethality of specific treatments; whereas the seed material is ideal for the most refined cytogenetic and genetic studies, which have resulted in pinpointing the lethal consequences of the treatments as originating in the chromosome.

The work with seeds was given further significance in an original experiment by Zimmer, et al. (31) in which it was shown that the paremagnetic resonance spectra obtained in X-rayed barley seeds was similar to that obtained from free radicals. Equally relevant was the demonstration that irradiation in air gave rise to more magnetic centers than irradiation in the presence of nitrogen. Later studies have confirmed and extended this work (11, 13).

It is of interest that much higher dosages were necessary to obtain clear electron spin resonance signals (ESR) than were necessary to demonstrate the storage phenomenon and other after-effects of irradiation with seeds. If the ESR methods enable the detection of the same kinds of events that initiate detectable biological injury, it seems apparent that the biological system has a greater resolving power than the physical system by several orders of magnitude. In this connection, because the production of some reactive entities may be dose-dependent, there is a distinct need for the physicist and the biologist to maintain the closest liaison in all of their work on pre- and post-irradiation effects. Only in this way can it be certain that the biological significance of a particular event observed with electron spin resonance methods will be placed in proper perspective.

Material and Methods

In the entire series of experiments reported, dormant seeds of Himalaya barley were used as the test material. In all instances they were of uniform size from the same harvest year and had an embryo water content of about 4 per cent before being used experimentally.

The X-rays used in the studies were unfiltered and were generated with a constant potential machine operated at 100 KV and 7 ma. For irradiation, the seeds were placed on a turntable 9.7 inches from the target, with the embryo oriented toward the beam. Under these conditions the seeds received a dose of approximately 3,000 r per minute.

In different experiments the seeds were subjected to a variety of pre- and post-irradiation treatments before germination. To eliminate confusion, these particular conditions are indicated in the experimental results. However, in all experiments, hydration immediately before the initiation of germination was accomplished by steeping the seeds for 45 minutes in boiled distilled water through which either oxygen or nitrogen was continuously flushed. This was done to eliminate a variable that had previously been demonstrated to be of profound significance (5, 10). Furthermore, on all occasions where seedling height determinations were made, the seeds were grown in petri dishes in a controlled environment room. Measurements were made to the nearest millimeter after either 6 or 7 days of growth, depending on the objectives of the particular experiment.

To determine the frequency of interchanges in the irradiated generation, cytological analyses were made at the first meiotic metaphase of X_1 plants that were grown to maturity (7). This laborious procedure was used because it avoided the controversy about the accuracy of using root tip cells for determining interchange frequencies (4, 30). Seeding chlorophyll mutation data were obtained on X_2 populations using the method of Stadler (27).

Experimental Results

Relation of seedling height to genetic injury

After treatment with 5,000 r of X-rays, a large sample of seeds was divided into four lots. Two lots were immediately hydrated for 45 minutes, one in the presence of oxygen and the other in the presence of nitrogen. The remaining two lots were stored over phosphorus pentoxide for 8 days at room temperature. After this period one lot was hydrated aerobically and the other lot anaerobically.

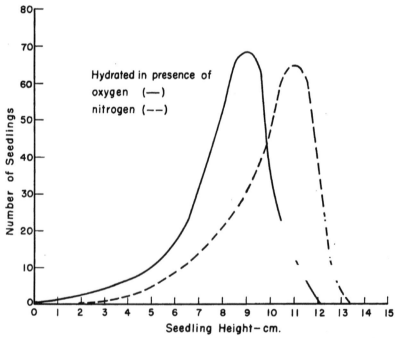

FIGURE 1.—*Distribution of seedling heights at 7 days from seeds given 5,000 r of X-rays and then immediately hydrated in the presence of either oxygen or nitrogen.*

Immediately after hydration the seeds were plated out in petri dishes and grown for 7 days. At that time individual seedling heights were determined (compare the frequency distributions in Figures 1 and 2) and the seedlings were placed in one of three height classes, *viz.*, 0 to 5 cm, 5.1 to 9 cm, and taller than 9.1 cm. The seedlings in each height class were than transplanted and grown to maturity to obtain interchange and mutation data (Figures 3 and 4, Table 1).

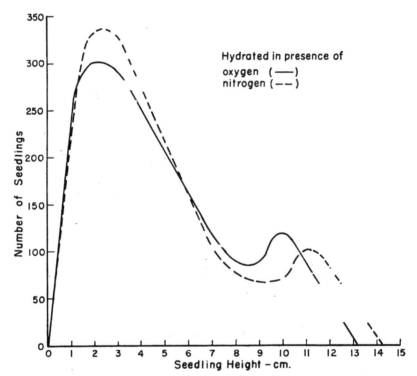

FIGURE 2.—*Distribution of seedling heights at 7 days from seeds given 5,000 r of X-rays and then stored for 8 days before hydration in the presence of either oxygen or nitrogen.*

It is significant that for those seeds hydrated immediately after X-irradiation (Table 1) the data are not presented in terms of genetic damage in different height classes but rather for the population as a whole. The reason for this is that the distribution of seedling heights

TABLE 1.—HEIGHT OF SEEDLINGS AT 7 DAYS AFTER 5,000 R IRRADIATION IN RELATION TO
SURVIVAL, INTERCHANGE FREQUENCY, AND MUTATION FREQUENCY. *

Condition of hydration	Height classes, cm	Seedlings transplanted, No.	Survival to maturity, %	Sporocytes analyzed, No.	Interchanges observed/ 100 spikes	X_2 seedlings analyzed, No.	Mutations observed/ 100 seedlings
		Hydrated Immediately After Irradiation					
O_2	—	239	96.2	153	16.9	4,514	1.92
N_2	—	237	89.8	143	9.1	4,408	_0 65
		Hydrated 8 Days After Irradiation					
O_2	0–5	810	74.5	283	30.3	11,796	2.03
	5.1–9	329	96.6	165	13.3	4,322	2.19
	9.1	192	97.3	132	6.8	3,160	0.98
Total		1,331	83.3	580	20.2	19,278	1:90
N_2	0–5	705	74.3	299	25.0	11,483	1.63
	5.1–9	250	86.8	115	10.4	3,656	1.20
	9.1	250	94.4	150	6.0	4,798	0.43
Total		1,205	81.0	564	16.9	19,937	1.26

*Nonirradiated control populations evidenced no interchanges in 383 X_1 spikes and an X_2 chlorophyll mutation frequency of 0.001 mutant seedling per 100 X_2 plants.

about the mean was close to normal (Figure 1). Accordingly, too few seedlings fell in either the tall or short height classes to give a reliable estimate of the injury in those classes in comparison with those which fell within plus or minus one standard deviation of the mean.

From the data presented it is apparent that there is a good correlation between the extent of injury, as measured by seedling height, and the degree of genetic injury as measured both in terms of mutations and interchanges. It is also apparent (Table 1) that populations of individuals can be selected from seeds that have been X-rayed and stored which evidence no increase in genetic injury resulting from storage.

Relation of oxygen to manifestation of genetic injury

The demonstration that post-irradiation hydration in the presence of oxygen resulted in greater seedling injury to dormant seeds than post-irradiation hydration in the presence of nitrogen (5, 10) necessitated a critical analysis of the relation between dose of X-rays

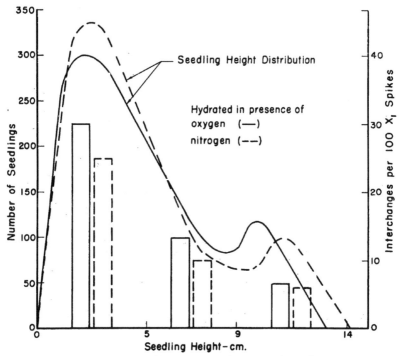

FIGURE 3.—*Relation of seedling height at 7 days to interchange frequency at microsporogenesis. (Plants grown from dormant barley seeds subjected to 5,000 r of X-rays and stored for 8 days before hydration in the presence of either oxygen or nitrogen.)*

and gnenetic injury under the two conditions of hydration. In this connection, a study was set up wherein seeds were subjected to one of a wide range of doses of X-rays and then immediately hydrated, either aerobically or anaerobically, and planted in the field. Immature inflorescences were collected from the X_1 plants and microsporocytes were examined to determine the frequency of interchanges induced in the material (Figure 5). At maturity up to five heads were removed from each plant and the seeds grown to determine the X_2 chlorophyll mutation frequency (Figure 6).

These genetic data demonstrate that when seeds are hydrated aerobically after X-irradiation there is an exponential relation

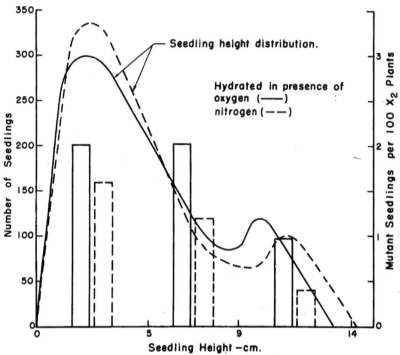

FIGURE 4.—*Relation of X_1 seedling height at 7 days to X_2 seedling muta-tion frequency. (X_1 plants grown from dormant barley seeds subjected to 5,000 r of X-rays and stored for 8 days before hydration in the presence of either oxygen or nitrogen.)*

between dose and interchange frequency, but that when they are hydrated anaerobically the relation is linear. Furthermore, they show that the mutation frequency is linear under both conditions of hydration.

The data suggest a relation between ion density and the role of oxygen in the production of both one- and two-hit events that will be presented in the discussion. Also, they indicate the likely reason for the conflicting views that have been expressed by different cytogenet-icists relevant to the nature of the regression of interchanges on dose (4, 30).

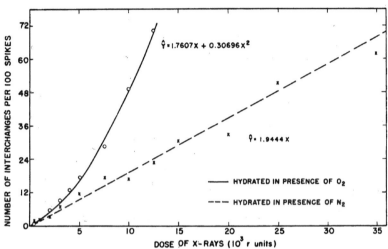

FIGURE 5.—*Relation of interchange frequency to dose when seeds are hydrated in the presence of either oxygen or nitrogen immediately following X-irradiation.*

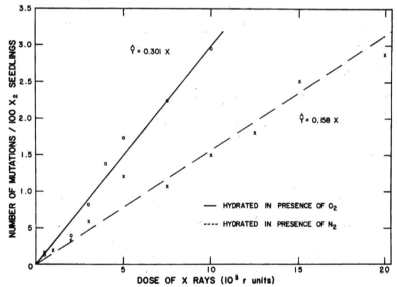

FIGURE 6.—*Relation of mutation frequency to dose when seeds are hydrated in the presence of either oxygen or nitrogen immediately following X-irradiation.*

Relation of pre- and post-irradiation temperature to manifestation of injury

Pre-irradiation temperature and the oxygen effect.—In previous investigations, prior to the demonstration that, following irradiation, seeds were injured by aerobic hydration and protected by anaerobic hydration, it was shown that protection from X-rays was also afforded to barley seeds when they were treated with a barely sub-lethal temperature immediately before irradiation (6, 24, 26). Present studies have proved that a pre-irradiation heat treatment of 75° C or 85° C gives effective protection when the treatment is for a duration of 24 hours. Furthermore, the protective effect is not necessarily associated with a detectable water loss and cannot be attributed to this factor. It has also been demonstrated that two months can elapse between the temperature treatment and irradiation without loss of protection, provided the seeds are not hydrated in the interim.

Because protection from X-irradiation was obtained by both pre-irradiation temperature treatment and by anaerobic post-irradiation hydration, it was deemed essential to determine whether or not the two types of protection were complementary. To determine this point seeds that had been heated for 24 hours at a temperature of 75° C and then X-rayed with a wide range of doses were immediately hydrated, either aerobically or anaerobically, and compared with seeds that were not subjected to the temperature treatment but otherwise handled in an identical manner (Figure 7).

These data show that the protective effect of the heat treatment was obtained under both conditions of hydration. Because other data (Figure 3) suggest a casual relation between genetic damage and seedling injury, and because it has been demonstrated that a pre-irradiation treatment with heat reduces the frequency of interchanges in X-irradiated seed (26), it is apparent that the protective effect afforded by post-irradiation hydration in the presence of nitrogen and pre-irradiation treatment with heat must be complementary; in the sense, that is, that they are both influencing the manifestation of the same kinds of genetic injury.

Pre-irradiation temperature and storage effect.—As previously indicated (5, 12), earlier work has shown that the injury to barley seeds from a given dose of X-rays can be enhanced by post-irradiation storage. In addition, data have been presented which show that a pre-

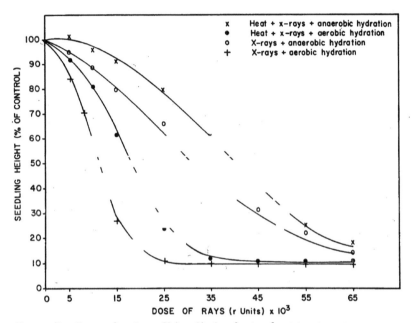

FIGURE 7.—*Protection from X-irradiation by pre-heat treatment.*

irradiation heat treatment results in a reduction in the radiosensitivity of seeds. The logical next step was to determine whether or not
pre-irradiation heat had any influence on post-irradiation sensitivity
to storage. To determine this the following study was conducted.
Seeds were subjected to a temperature of 75° C for 24 hours and then
irradiated with a dose of 15,000 r. After irradiation, different segments
of the population were stored at room temperature for periods ranging from 0 to 48 hours before being hydrated either aerobically or
anaerobically. Six days after hydration seedling height data were
obtained and compared with data obtained from populations treated
in precisely the same way except that they were not subjected to the
pre-irradiation temperature treatment. The data presented compare
only the two extremes of the post-irradiation storage period, 0 and
48 hours, because they illustrate the points of concern (Figures
8, 9, 10).

The following conclusions can be drawn from the data in this
experiment. First, that under conditions of either aerobic hydration

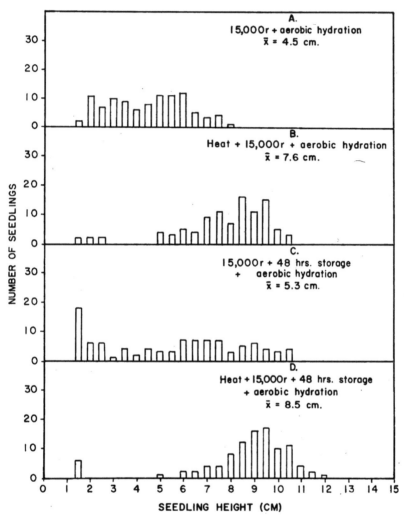

FIGURE 8.—*The effect of a 75° C pre-irradiation heat treatment on the injury to seeds that are hydrated aerobically before and after storage.*

(Figure 8) or anaerobic hydration (Figure 9) the distribution of seedling heights was more nearly normal in the populations that received heat treatment than in the population that did not. Furthermore, the most skewed distribution occurred in the populations that were not

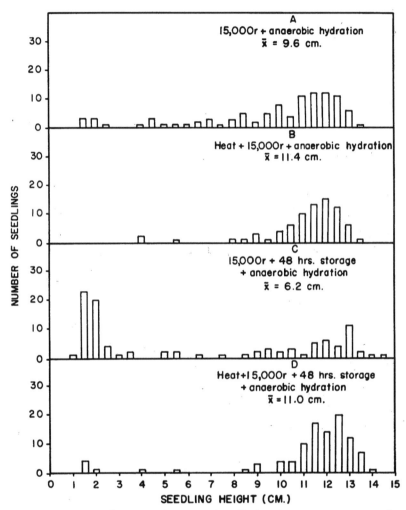

FIGURE 9.—*The effect of a 75°C pre-irradiation heat treatment on the injury to seeds that are hydrated anaerobically before and after storage.*

heated but in which germination was delayed for 48 hours (Figures 8 and 9, part C). Second, pre-irradiation heat treatment completely eliminated the increase in injury that usually occurs when seeds are stored after irradiation (Figures 8 and 9, compare parts A and C with

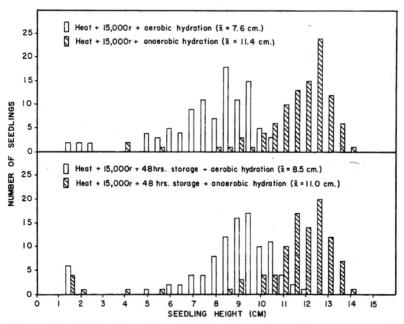

FIGURE 10.—*Elimination of sensitivity to storage but not to oxygen result-ing from a pre-irradiation temperature treatment at 75°C.*

parts B and D). Third, while pre-irradiation heat treatment elimi-nated storage injury, it did not modify the sensitivity of the seeds to aerobic hydration, even when they were stored for 48 hours before the initiation of hydration (Figure 10). This particular observation is considered to be of especial significance and will be dealt with in the discussion.

Post-irradiation temperature extreme and oxygen and storage effects.—The demonstration that the injury to seeds could be increased by storing them at room temperature after X-irradiation emphasized the need to determine to what extent the injury progres-sion, as a function of time, was temperature dependent. To elucidate this problem, an extensive two phase study was set up. In the first phase seeds were irradiated and then immediately stored at one of three temperatures (−78° C, 20° C, and 85° C) for different periods of time before they were hydrated either aerobically or anaerobically (Figure 11A). In the second phase, after X-irradiation, the seeds were stored for 120 hours at −78° C. They were then removed from this temperature and subjected to additional storage periods, at either 20° C or 85° C, before being hydrated (Figure 11B).

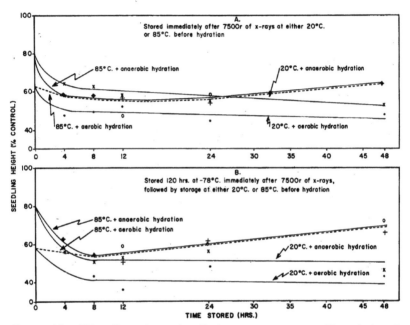

FIGURE 11.—*The effect of post-irradiation temperature on X-ray-induced storage injury and the suppression of injury progression by low temperature.*

These data clearly show that, for at least the first few hours, at both 20° C and 85° C injury increases progressively as a function of time stored, with the most rapid increase occurring at the highest temperature. Furthermore, they show that sensitivity to hydration in the presence of oxygen is eliminated first by the highest temperature. In addition, they demonstrate that after about 48 hours storage at 85° C there is a partial recovery from injury. Essentially the same result was obtained when seeds were stored at 75° C for longer periods of time (Figure 12).

The data also demonstrate (16) that storing seeds at −78° C for 120 hours essentially prevents the progression of post-irradiation

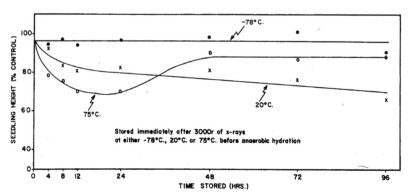

FIGURE 12.—*Initial thermal enhancement of and ultimate thermal recovery from X-ray-induced storage injury.*

injury. Apparently the radiation-induced damage is maintained in relatively labile state by the low temperature because, on removal from this temperature to temperatures of either 20° C or 85° C, the progression of injury, as a function of time, is similar to that obtained when seeds are stored at these temperatures immediately after irradiation (Figure 11A and B).

Discussion

Seedling injury (growth retardation during a finite period) following germination of irradiated seeds has been used commonly as an indication of the genetic injury that has been induced in the cells of the meristem. The validity of using this estimate is reasonable on *a priori* grounds. In addition, it has been repeatedly demonstrated that there is a good correlation between seedling injury and genetic injury over a wide range of doses with X-rays and other irradiations. Despite this, the problem of the relation between seedling height and genetic injury has long been of concern because the distribution of seedling heights about the mean was often more skewed in X-rayed material than in neutron-irradiated seed (9). The problem took on added significance when studies were undertaken on the effects of post-irradiation storage on seedling injury because the variability in seedling heights in the stored populations became extreme; actually, it often approached the bimodal.

Clearly, there was a need to determine if the suspected relationship between seedling height and genetic injury held true when seedlings in populations of X-rayed seed, which had been subjected to post-irradiation storage, were subdivided into different height classes and analzed for frequencies of interchanges and mutations. Such a study was conducted and the data presented show beyond any reasonable doubt that there is a good correlation between seedling injury and genetic injury (Figures 3 and 4, Table 1). Indeed, they warrant the conclusion that the relationship is one of cause and effect.

At the present time there is no obvious explanation as to why seeds in a population undergo different degrees of storage injury (Figure 2), although it seems likely that it is associated with either an environmental variable that has not been elucidated, to permeability of the cell, or to a different physical state of radiosensitive sites within the nuclei of different seeds. These possibilities will be explored later.

Basing his arguments on the first root tip cell divisions in the germinating seed, Wolff (29) came to the opposite conclusion to that presented here, *viz.*, that changes in seedling height during storage were not correlated with radiation-induced chromosomal damage. He suggested that, "the pattern of damage can be correlated to the behavior under similar regimes of long-lived radiation-induced radicals". Presumably the injury resulting from such radicals was considered to be nongenetic in nature because it did not become manifest in the form of chromosome breaks. It is unlikely that the difference between Wolff's results and those presented here is due to the method of analyzing for genetic injury, because root-tip studies similar to Wolff's that have been conducted in this laboratory confirm the interchange and mutation data that are presented. There must be some other explanation for the difference and this problem should be re-examined.

Early work demonstrated that the relation between interchange frequency and dose was exponential when Tradescantia microspores were subjected to X-irradiation (22) and linear when they were bombarded with neutrons (16). From these data it was concluded that, when biological material is treated with neutrons, the dense clusters of ionization that occur usually break two chromosomes simultaneously which then have a good probability of exchange; thus, the one to one relation between interchange frequency and dose. In contrast with this, it was postulated that the free electrons resulting from absorption of X-rays, because they can produce a range of ion densities within a sensitive volume, often only break one chromosome at a time. Accordingly, the probability of having two breaks simultaneously present for exchange would increase at a greater rate than the first power of the dose, with the result that the dose interchange relationship would be exponential. It has since been demonstrated (8) that, with barley seeds, the relation between dose and interchange frequency is linear when the seeds are subjected to both X-rays and thermal neutrons. It is now known that the linear relation obtained with X-rays only occurs in seeds with a moisture content above 10 per cent at the time of irradiation or in seeds with water contents below that level when they are hydrated anaerobically immediately after irradiation (see following discussion). From these earlier results (8) it was concluded that only the densely ionizing tails of the X-ray-induced free electrons cause immediate chromosome breakage and that then two chromosomes are usually broken "simultaneously".

This conclusion was disputed by Wolff and Luippold (30) on the grounds that the scoring of dicentric bridges in root-tip cells was a poor means of determining interchange frequencies. They pointed out that in their analyses there was actually an exponential relation between interchange frequency and dose when decentrics were scored at metaphase and a linear relation when they were scored at anaphase. This particular consideration ignored the fact that the interchange frequency as detected at the first metaphase division of microsporogenesis was also linear (8). A more pertinent and convincing argument should have included data on neutron bombardment, the water content of the seeds at the time of irradiation, the conditions under which the seeds were hydrated after irradiation, and the frequencies of interchanges at microsporogenesis.

What appears to be a more likely answer to the problem is that the two workers were probably using test materials that had non-uniform water contents and, therefore, differential sensitivities to post-irradiation modification. To develop the significance of this consideration, it will be necessary to diverge briefly to examine one possible explanation as to why post-irradiation effects were observed with X-rays and not with neutrons (8, 12).

It has been postulated that post-irradiation sensitivity to oxygen might depend upon the distribution of energy within critical sites. As stated earlier, with X-rays there would be some sensitive volumes that were sparsely ionized and some that were densely ionized. If post-irradiation oxygen treatment influenced only sites that received one or a few ionizations, effecting a single chromosome break, the frequency of such sites, and their corresponding breaks, would increase linearly with the dose while the frequencies of interchanges between such breaks would increase at a power greater than unity.

To test this consideration seeds were treated with one of a wide range of doses of X-rays and then immediately hydrated either aerobically or anaerobically. The seeds were then permitted to develop into mature plants and the frequency of interchanges was determined at microsporogenesis.

The data obtained (Figure 5) are in agreement with the hypothesis presented, for the frequency of interchanges in this material increased exponentially when the seeds were hydrated in the presence of oxygen and linearly when they were hydrated anaerobically.

These data have greater significance than merely in apparently resolving the confusion that has heretofore existed concerning the interchange-dose relationship. They suggest that many radiation-induced breaks are in fact "potential" breaks or "lesions", that either become manifest as actual breaks or are eliminated, depending upon the presence of oxygen in the system.

In regard to these lesions, it is worth considering whether or not it would be correct to use the term "restitution" in referring to the elimination of breaks under conditions of anaerobic hydration. Wolff and I have discussed this matter in personal communications and are agreed that, if the relationship between dose and interchange frequency is exponential for seeds hydrated in the presence of oxygen, which it is, then it is reasonable to conclude that "restitution" (as the term is defined in radiobiological literature) occurred in the material hydrated in the presence of nitrogen.

The effect of temperature treatments before, during, and after irradiation on X-ray-induced injury in seeds has been studied over a period of 30 years (25). Unfortunately, in the early work, the profound effects of slight changes in the water content of seeds on their senstitvity to pre- and post-irradiation treatments was not appreciated. Consequently, there was often little correspondence between differ-·

ent workers' findings and the data were very largely ignored. Added to this was the fact that, at the times these studies were conducted, the test materials used in most radiobiological studies were not suitable for demonstrating the influence of environment on radiosensitivity and there was general skepticism concerning reports that the damage to biological systems could continue after the cessation of irradiation. Indeed, this particular opinion was expressed very recently, as far as a post-irradiation effect of oxygen is concerned (3).

These two reasons undoubtedly account for the fact that little consideration was given to the work of Kempton and Maxwell (17), who demonstrated that maize seeds X-rayed at either the temperature of liquid air or at 61° C evidenced less radiosensitivity than seeds X-rayed at room temperature.

Now that some of the environmental parameters which affect radiosensitivity can be controlled, and it is established beyond any reasonable doubt that this applies to damage to the chromosome, speculation is warranted concerning the biophyiscal mechanisms that are involved.

Of the possibilities that might be examined, three seem worthy of mention. First, that all damage to the chromosome results from the direct absorption of energy. This does not impugn the possible significance of intramolecular energy transfer mechanisms. Second, that all damage to the chromosome results from absorption of energy by the ambient layer, the effect on the chromosome being initiated by active radicals or some other mechanism for transferring energy. The following discussion would appear to rule out the need for further consideration of this possibility. Third, that damage to the chromosome results from a combination of the first two kinds of events.

For the reasons indicated *(loc. cit.),* it seems reasonable to conclude that the energy released in regions of high specific ionization is more than sufficient to break a chromosome at the initial center of absorption. This particular type of event would not likely be modifiable by the changes in the physical structure of the chromosome that could conceivably result from high temperature treatment or the other conditions that have been imposed on seeds during the course of this work. Data supporting this statement have been obtained in the study relating to the nature of the interchange-dose relationship (Figure 5) when seeds were hydrated in the presence of oxygen and nitrogen. They show that the linear components in both curves are indistinguishable from one another. This is rather good evidence that oxygen, at least, has no discernible effect on breaks resulting from dense clusters of ionization.

From these considerations, it seems reasonable to suggest that most, if not all, environmental factors that effect radiosensitivity either mitigate or enhance the probability of sparsely ionized sites becoming biologically detectable. On the assumption that the detectable events are a manifestation of genetic injury, one could conceive of both metastable (reversible) states within the chromosome and in the ambient layer as initial stages. Distinguishing between these two possibilities would be difficult, if not impossible, in studies with cells. Possibly studies with virus particles should be expanded in an attempt to resolve the problem.

At this point it should be emphasized that, because of the sigmoidal nature of the curve expressing seedling injury as a function of dose (Figure 7), it is often necessary to choose specific doses to demonstrate the significance of particular kinds of pre- and post-irradiation treatments. This accounts for the fact that 15,000 r was used to show the protective effect of a pre-irradiation temperature of 75° C (Figures 8, 9, 10), whereas 7,500 r was used to show the effect of post-irradiation temperature (Figure 11) on the progression of injury from storage.

It is now appropriate to look at the interrelations between hydration, storage, and oxygen availability in connection with the manner in which they influence the manifestation of radiation-incited changes as seedling injury. It can be stated with certainty that the addition of water to the seed after irradiation, irrespective of other pre- and post-irradiation treatment conditions, has a quenching influence on those radiation-induced changes which might otherwise progress for a considerable period of time. Furthermore, absorption of water is equally effective in eliminating injury progression under both anaerobic and aerobic conditions, although more total injury may occur in the material subjected to aerobic hydration under specific treatment regimes.

Most previous investigators have suggested that the increase in injury which occurs during storage includes the oxygen-sensitive component (5, 12). It was demonstrated (5) that as post-irradiation storage time increased, sensitivity to oxygen decreased. This particular relation appears to hold under all but one set of conditions. When seeds are subjected to a temperature of 75° C for 24 hours before they are irradiated, they subsequently show no sensitivity to storage. However, their sensitivity to oxygen immediately after irirradiation is as great as in non-heat treated seeds and, furthermore, it persists for at least 48 hours (Figure 10). These last data suggest that there is no relation between the effects of oxygen and storage on seeds. This problem will be discussed later.

There appear to be two reasonable explanations as to why a pre-irradiation treatment with heat prevents the dissipation of sensitivity to oxygen. First, that the cell membranes are modfied by the temperature treatment so that they become impermeable to oxygen. Second, that radiosensitive molecules within the nucleus are so re-oriented by high temperature that the reactive sites are enfolded within them and are not exposed to oxygen until hydration occurs. Both of these explanations suggest that heat-treated seeds would retain a radiation-induced labile state more or less indefinitely under appropriate conditions. The data presently available support this suggestion.

In their recent report Powers, et al. (21) have demonstrated that, in bacteria, maximum protection from X-irradiation by high temperature (80° C) only occurs when the temperature treatment is administered during or immediately after irradiation. Furthermore, maximal thermorestoration or thermal annealment, as they have chosen to call this phenomenon, takes place only in the absence of oxygen. In contrast with the spore work, for the first few hours following irradiation there is a more rapid increase in the injury to X-rayed seeds when they are maintained at 85° C (Figures 11 and 12). Interestingly enough, however, seeds maintained at the higher temperatures for periods of 48 hours or more showed some thermorestoration of injury. One other significant fact is that sensitivity to oxygen was completely eliminated after a 15-minute post-irradiation heat treatment at either 75° C or 85° C, whereas it persisted for considerably longer periods of time at 20° C.

Speculation seems warranted concerning the mode of action of high temperatures on critical molecules that results in the elimination of injury from storage, but not from oxygen, when the temperature treatment is given before irradiation and the reverse when it is given after irradiation. Let us assume that the injury occurring during storage results from the transfer of energy from the site of initial absorption (primary event) to a second more sensitive site (secondary event), whereas injury from oxygen results directly from oxidation of the primary event. Let us further assume that it is a matter of chance whether or not oxidation occurs at the primary site before the energy associated with it is transmitted to a secondary site. If such was the case, it is conceivable that pre-treatment with heat could modify the structure of the critical molecule in such a way that the energy associated with the primary event could not be transferred or otherwise dissipated. Under such a circumstance the oxygen-sensitive primary site should persist indefinitely, which is precisely what the data indicate (Figure 10).

Using the same model, the effect of a post-irradiation heat treatment, which rapidly eliminates sensitivity to oxygen, could be explained by assuming that the thermal energies involved were sufficient to enhance the rate at which energy was transferred away from the site of primary absorption.

Summary

Dormant seeds of barley with an embryo water content of 4 per cent were used for biophysical studies concerned with the influence of pre- and post-irradiation treatment conditions on the manifestation of seedling and genetic injury. In the course of the investigations, designed to examine some of the environmental parameters that were known to affect radiosensitivity, the following observations were made:

1. Seeds germinated immediately after X-irradiation and then grown for 8 days showed a near normal distribution of seedling heights about the mean. However, when seeds were stored at room temperature before germination, the distribution of seedling heights about the mean became progressively skewed, as a function of time, for a period of at least 8 days. Determination of the extent of genetic injury in different seedling height classes in material that was stored for 8 days showed that there was an inverse relation between seedling height and genetic injury. On the average, seeds that fell into the shortest height class evidenced between two and four times as many interchanges and mutations as seeds that fell into the tallest height class.

2. The interchange frequency increases exponentially with dose when seeds are hydrated aerobically immediately after X-irradiation but linearly with dose when they are hydrated anaerobically. For both conditions of hydration the mutation frequency increases linearly with dose. However, aerobic hydration results in the production of more mutations per roentgen than does anaerobic hydration.

3. Pre-irradiation temperature treatments for 24 hours at 75° C, showed a protective effect over a wide range of doses of X-rays. The protective effect of the treatment was manifest when the seeds were hydrated either aerobically or anaerobically and was not associated with a detectable change in water content. Furthermore, the percentage protection obtained, per unit of injury, under both conditions of hydration was similar. It was also demonstrated that the protective effect of the heat treatment against a subsequent dose of X-rays persisted for at least 2 months between heating and irradiation provided the humidity of the storage atmosphere was not changed. In addition, a pre-irradiation temperature of 75° C eliminated the increase in injury that usually accompanies post-irradiation storage. However, such a pre-treatment did not modify the sensitivity of the seeds to post-irradiation hydration in the presence of oxygen.

4. In contrast with the effects of a pre-irradiation heat treatment, a short post-irradiation storage temperature of either 75° C or 85° C generally enhanced the rate of increase in injury that was observed during post-irradiation storage at room temperature. However, 48 hours storage at both of these temperatures resulted in a distinct thermorestoration of injury. Of particular significance was the demonstration that treatment of the seeds at either 75° C or 85° C for as little as 15 minutes after they were irradiated eliminated their sensitivity to post-irradiation hydration in the presence of oxygen.

Part II. Theoretical Considerations Relating to Acute Irradiation of Successive Generations

Actually since 1930, but particularly since the conclusion of the Second World War, ionizing radiations have been used by plant geneticists in efforts to induce heritable changes that will be of economic significance. By and large these studies have involved attempts to induce so-called point mutations. The organisms most commonly studied have been the small-grained cereals, peanuts, and corn. Data obtained from studies with these species have been discussed in other sections of this Symposium. The intent of the present report is to mention only theoretical considerations that may be of relevance to the plant breeder and which have received very little attention by most workers.

In using ionizing radiations in plant improvement, one must take into consideration the degree of ploidy and the breeding behavior or mode of reproduction of the species with which he intends to work. In addition, one must be cognizant of the fact that ionizing radiations not only induce point mutations but also all manner of structural chromosomal anomalies, such as reciprocal translocations, inversions, duplications, and deletions. Furthermore, he must be aware that the frequency with which these types of events are produced, and their ease of detection, will be dependent upon the species with which he is working and the ontogenetic stage of the plant to which the irradiation is given.

To reduce this complex of variables in the present report, unless otherwise stated, we shall consider only what may possibly be achieved through seed irradiation of the naturally self-fertilizing cereals which occur in nature as diploids, tetraploids and hexaploids.

Mutations From a Single Dose of Radiation

On *a priori* grounds, in the X_2 from populations receiving one dose of radiation it would be expected that plants expressing mutant phenotypes for qualitative characters would occur commonly in the diploid species, with a lower frequency in tetraploid species, and only rarely in hexaploid species. The reason for this is simply that most radiation-induced mutations are recessives and the simultaneous mutation of genes which influence the expression of the same character, but are located on homeologous chromosomes, would be a relatively rare event, even under optimal conditions of treatment. The essential validity of this consideration has been borne out by chlorophyll mutation studies on tetraploid and hexaploid oats and wheat.

In addition to the above consideration, there is another important factor which makes it difficult to recover a large number of mutant phenotypes following administration of a single dose of radiation to an organism. This is because ionizing radiations induce chromosomal aberrations as well as mutations. Such aberrations are often responsible for cell death and it is apparently a simple matter of chance whether or not either a mutational event, a chromosomal anomaly, or both, are induced in a cell following the absorption of photons or ionizing particles. Accordingly, as the dose to which the population is subjected is increased, the number of cells in the population which contain both aberrations and mutations also increases. Because only one or a very few aberrations can result in cell death, it is apparent that the probability is very low of inducing mutations at more than a few loci before a lethal aberration is induced.

From these considerations, it should be apparent that similar complications prevail when attempts are made to obtain mutations for quantitative traits. As many of the most important characters with which the breeder works are under polygenic control, the desirability of using ionizing radiations in other than a "one-shot" approach in breeding programs should be obvious.

Mutation From a Dose Administered to a Sequence of Generations

In diploids

It seems obvious that a method that would increase the relative frequency of mutations to lethal or semi-lethal aberrations in a population, before it is subjected to the expensive process of screening for economic mutations, would be of distinct value to the geneticist. At this time there appears to be only one simple way in which this can be achieved. That is to irradiate successive seed generations of the material under test, always selecting for re-irradiation seed from those spikes or panicles which do not evidence structural chromosomal anomalies. This is a particularly simple procedure in diploid species, such as barley, because during the development of the plant from irradiated seed all cells which are cytologically deficient in their chromosome complement are incapable of competing with more normal cells. The end result is that, effectively, only reciprocal interchanges and inversions are detected in the analysis of sporocytes arising from irradiated diploid seed. Reciprocal interchanges invariably induce 25 to 50 per cent sterility. Thus, when selecting seed for re-irradiation one can select it only from fertile heads. Because mutations and chromosomal aberrations tend to be independently induced events (15), seed from fertile heads will carry as many mutations as seed from semi-sterile heads. While seed selected in this way may incidentally carry an inversion it would likely be of little consequence to the breeder and could be easily eliminated if the need was indicated.

In suggesting the use of recurrent irradiation in diploid species for inducing maximum genetic variability for both quantitative and qualitative traits, it is recognized that gene mutations of both a deleterious and beneficial nature will be induced in the same cells and that the method does not provide for their separation between generations. From this it follows that the immediate products from recurrent irradiation of a diploid could not likely be used as commercial varieties. However, after screening for mutants of agronomic value in a population that had been subjected to recurrent irradiation, it would be easy to produce hybrids between the original progenitor and the irradiated material to place the character, or characters, of value in what was an otherwise desirable genetic background. In so far as the author is aware, this particular approach has received little consideration by plant breeders.

In polyploids

When attempting the induction of detectable mutations in the self-fertilizing polyploids among the small grains, a special set of circumstances confronts the geneticist. For every pair of genes that influence the expression of a character in the diploid, both quantitative and qualitative, there presumably usually exist two and three times as many genes in the tetraploid and hexaploid, respectively.

If considering mutation only for qualitative characters, theoretical assumptions suggest that continued re-irradiation of the seed generation of the polyploids, in which the genes governing the same order of biochemical function are in the homozygous dominant condition on homeologous chromosomes, will ultimately result in the appearance of a high frequency of chlorophyll mutations in the population. When such mutations appear, F_2 segregation for the mutant trait in crosses with the original parental type should be 15:1 in the case of the tetraploid and 63:1 in the case of the hexaploid. Among the normal F_2 phenotypes from these crosses there should exist genotypes which contain the normal gene in the homozygous dominant condition on one pair of homologs and are homozygous recessive on the other pair or pairs of homeolgous chromosomes. In plants of this genetic constitution crosses with the recessive phenotype should give a 3:1 segregation for the mutant character. In effect, then, it can be said that the polyploid has been "diploidized". It follows that the recurrent re-irradiation of any polyploid that shows bivalent pairing should eventually result in the material used in such studies being indistinguishable, by simple Mendelian tests, from naturally occurring diploids.

It might well be asked if there exist any factual data to support these considerations. The answer is simply that the incidence of chlorophyll mutations in hexaploid oats has increased each generation through six successive generations of re-irradiation. Mutant types have been perpetuated, but crosses with the parental type have not been made.

In addition to the hexaploid oat work, which will be mentioned again in the next section, recurrent irradiation studies are also underway at the University of Minnesota with diploid and tetraploid oats and tetraploid and hexaploid wheat. The incidence of chlorophyll mutations in the wheat material after irradiation of two successive seed generations supports the general considerations.

It seems pertinent to ask the following question: If diploidization of a polyploid is indeed possible, is it likely that it will have any practical significance? The answer, based on theory and existing recurrently irradiated plant material, is yes. To give substance to this assertation it is again necessary to resort to theoretical considerations and the findings of other workers. Heterosis in corn is commonly attributed to differences in alleles which govern expression of the same characters, i.e., heterozygosity at several loci, or to epistasis. Similarly, it has been demonstrated by Wallace and Vetukhiv (28) in Drosophila that genes which are recessive and deleterious when in the homozygous condition may benefit the survival and growth of a fly in which they are in the heterozygous state.

What has this to do with diploidization of a self-fertilizing polyploid, specifically, for ease of presentation, hexaploid wheat? As has been indicated, in the hexaploid often there are presumably six genes, one on each of three pairs of homeologous chromosomes, which influence the expression of a given character. Furthermore, apparently these genes are often in the homozygous dominant condition on more than one pair of the homeologous chromosomes. Now, if in diploids a heterotic effect is obtained when differences exist between alleles on a pair of homologs, it seems possible that in a polyploid differences between genes which govern the expression of the same character, but are on homeologous chromosomes, may also confer increased vigor to plants in which the condition exists. If such is the case, the self-fertilizing polyploid would have an advantage over the diploid in that the "heterozygous" state (between genes on homeologous chromosomes) could be perpetuated by selfing and thus provide permanent heterosis by merely maintaining the alleles on one pair of homologous chromosomes in the homozygous dominant condition.

In regard to the possible usefulness of this approach in plant breeding, it seems relevant to mention a promising study that is in progress, using recurrent irradiation on hexaploid oats. In this program the following procedure has been used. Seed from a single plant from three different oat varieties, Park, Missouri 0–205, and Clintland, was increased and irradiated. In the X_1, 1,000 fertile panicles were chosen from different plants in each of the three varieties. The seed from these panicles was re-irradiated and planted in panicle rows. From these panicle rows one fertile panicle was chosen from one agronomically suitable-looking plant to form the seed source for the next cycle of radiation. This procedure has been continued throughout the entire program, which has now run through six cycles and is presently in the seventh.

Some interesting general observations on this material are worth reporting, although at this time there are no grounds for assuming that a state of permanent heterosis exists in any of the material. First, and most significant, is the fact that variability in the material for height, panicle type, and maturity date was not in evidence until after the second cycle of radiation. Subsequent to this there has been a continued increase in variability for these characters. At the end of the fifth cycle of radiation the variability was so extreme in all of the material, with a considerable number of types that evidenced characteristics of agronomic interest, that a number of plant selections were made for testing with the parental type. These selections were increased in 1959 and the promising ones were turned over to experienced plant breeders who will compare them in yield tests with the parental types in 1960. It is important to emphasize that after this long series of recurrent irradiation, where selection was made each generation for desirable plant types, most of the material in the nursery is agronomicaly undesirable. This is particularly true of the material derived from Clintland and to a lesser extent to that derived from Missouri 0–205 and Park, in that order.

Another point that warrants consideration is the fact no attempt was made to isolate the irradiated material from outcrossing. However, as indicated, each year seed for re-irradiation was taken only from completely fertile panicles. Accordingly, there should have been no more outcrossing in the material than the breeder usually finds in his nursery plots. As the incidence of outcrossing should be quite low, it seems unlikely that this can account for the extreme variability that has been observed or for the apparently desirable types that have been selected for further study. However, this possibility cannot be ignored and studies are presently under way with polyploid series in oats and wheat, where isolation is being practiced, to resolve this question.

The Use of Intraspecific Chromosome Structural Changes
Duplications

Among the complex types of chromosomal aberrations induced by ionizing radiations is a class known as the "duplication". As the name implies, this means that a duplication of genetic material governing specific gene functions exists in the chromosome complement. Duplications may arise in two simple ways which do not simultaneously result in a loss of genetic material in the cell. The first involves translocations between corresponding arms of homologous chromosomes that were broken at dissimilar points. The second involves breakage and reunion of sister chromatids in which the break points were not at corresponding loci.

From the preceding discussion on polyploids it should be evident that they have built into them a system of duplications of what can amount to whole sets of genes in the case of a true autopolyploid. It would appear, then, that radiation-induced duplications would likely be of greatest practical use in diploid species. Accordingly, the discussion will center around their possible use in self-fertilizing diploids.

The previously mentioned work of Wallace and Vetukhiv (28) with Drosophila, which has demonstrated that flies heterozygous for one or a few loci may be better adapted to their environment than when they are either homozygous dominant or recessive for these loci, is information that could be of import to the plant breeder using diploid species if he could devise some method of obtaining lines that would breed true for the heterozygous condition. There seems to be only one possible method for achieving this end, within the confines of the diploid, and that is to obtain a duplication of the first type mentioned so that the normal and mutant alleles are located on the same chromosome. This would necessitate irradiating F_1 seeds that were heterozygous for the allele under consideration.

One might well ask three questions concerning the feasibility of using the method: (A) What is the probability of obtaining a translocation at dissimilar points on homologous chromosomes? (B) What is the probability that the duplicated segment of the interchange will involve the gene in question? (C) What is the likelihood of recognizing the duplication once it is produced?

The answer to the first question is that translocations between opposite arms of homologous chromosomes occur with a frequency that is apparently due to chance (19). It can be assumed, therefore, that translocations between corresponding arms would occur with the same frequency. From this it follows that the lower the number of chromosomes in the species under study, the more readily will such interchanges arise. Actually, in diploid barley and maize, cytogenetic studies have shown that translocations between opposite arms of homeologous chromosomes are a relatively common occurrence (4, 6, 20).

The second question cannot be answered except to point out that, based on purely physical assumptions, a gene in the median position of an arm should be involved in duplications more commonly than one close to the centromere or on the distal end of an arm.

The answer to the third question is simply that it depends on whether or not the heterozygous state produces a recognizable phenotype. If it does, plants carrying a duplication involving a dominant and recessive allele should be readily detected because they will either not segregate for the character in question or give aberrant segregation ratios.

Duplications arising in the second-mentioned manner, *viz.*, by joining of sister chromatids that were broken at dissimilar points, are exceedingly common, as based on cytological observations. However, they maye be extremely difficult to detect phenotypically and would appear to have little practical value except in cases where additive gene action prevailed in the production of a particular trait, such as disease resistance, pigmentation, etc.

Production of fertile tetraploids

Since the discovery that colchicine was an effective polyploidizing agent, plant breeders have doubled the chromosome complement of numerous species with which they work. In only a few instances have the induced polyploids been useful economically. One immediate reason for this is that meiosis in an autopolyploid is disturbed by multivalent formation which, along with some undetermined physiological imbalance, apparently results in sterility and aneuploid types arising in the progenies. If these problems could be overcome, adequate tests of polyploids, originating from common genetic backgrounds, could be undertaken and the genetic diversity within a species that was available to the breeder could be increased accordingly.

One possible method of overcoming multivalent associations would be to reorganize structurally the chromosomes of a species to the point that very little or no homology remained between the structurally modified lines and the progenitor. At this stage F_1's between the modified lines and the progenitor could be doubled to produce a fertile tetraploid that would be similar to artificially and naturally occurring amphidiploids.

At attempt to produce such structural modification experimentally in barley is underway, using ionizing radiations to induce translocations, inversions, duplications, and deficiencies. For the past six generations, the procedure has been to use irradiation to induce at least one interchange per generation in each of four lines originally selected from four different varieties. Incidental to the addition of the interchanges, but equally effective in inducing structural differentiation, are the addition of the other anomalies mentioned.

Hybrids between some of the material, in which structural differentiation of the chromosomes is being attempted, and the parental genotype have just been achieved. They will be examined cytologically in the spring of 1961 and it is to be hoped that some of the F_1's will show the kind of reduced pairing that is common in two of the asynaptic mutants that have arisen from the material.

Use of Interspecific and Intergeneric Chromosome Interchanges

With few exceptions, cereal breeders are limited in their improvement programs to using the phenotypes they can obtain by recombining the genes that exist in the species with which they are working. The reason for this is simply that when wide crosses can be effected, there is frequently little homology between chromosomes from the two parents and correspondingly little recombination takes place. The consequence is that there is often either divergence toward parental types or the progenies from such crosses contain additions of whole chromosomes to the basic complement. The lines in which whole chromosomes have been added to a basic complement have invariably proved too inferior for commercial usage.

In most instances, what the breeder is seeking from intergeneric and inter-specific crosses is one or a few characters, such as disease or insect resistance, that he can incorporate in a commercial strain without impairment to that strain. Recently ionizing radiations have been effectively used to achieve this where conventional methods failed. In this regard, Sears (23) has reported the successful translocation of a segment of chromosome carrying a gene for leaf rust resistance from *Aeiglops umbellulata* to *Triticum vulgare,* and Elliott (14) has had comparable success in transposing a piece of chromosome carrying a gene for stem rust resistance from *Agropyron elongatum* to *T. vulgare.*

Because of these successes, there would appear to be considerable merit to the suggestion that ionizing radiations should be used extensively in hybrids involving Triticum, Avena, and their related genera, to obtain recombinant types that cannot be obtained by conventional methods. Using the method outlined, it should be possible to derive lines that contained all the 42 chromosomes of *T. vulgare* plus one pair from a related genus. Theoretically, where there is no homology between chromosomes involved in an intergeneric cross, it should be possible to derive as many strains containing a different pair of chromosomes from the related genus as in the "n" number of chromosomes in that genus. By screening the derived strains, it could be determined whether or not the added pair of chromosomes carried a factor or factors of economic significance. In cases where some economic worth was apparent, irradiation studies could be undertaken to translocate a region of the chromosome carrying the gene or genes of significance into the *T. vulgare* background of chromosomes. Once this had been achieved it would be a simple matter to eliminate the extra pair of chromosomes from the related genus by backcrossing to *T. vulgare* and selecting only those progenies with a complement of 42 chromosomes.

Concluding Remarks

In recent years a great deal of attention has been given to the use of one acute dose of ionizing radiation for inducing mutations of a beneficial nature. At the same time, relatively little consideration has been directed to other possible approaches, involving both mutations and chromosome structural modifications, which would appear to have great potential promise. The present report is an attempt to emphasize proved and theoretical considerations which suggest that effectively utilized ionizing radiations will likely become of increasing significance to the applied geneticist.

References

1. Adams, J. D., and Nilan, R. A. 1958. After effects of ionizing radiation in barley: II. Modification by storage of X-irradiated seeds in different concentrations of oxygen. *Radiation Res.,* **8:** *111–122.*

2. ————, ————, and Gunthardt, H. 1955. After effects of ionizing radiation in barley: I. Modification by storage of X-rayed seeds in oxygen and nitrogen. *Northwest Sci.,* **29:** *101–108.*

3. Alper, T. 1956. The modification of damage caused by primary ionization of biological targets. *Radiation Res.,* **5:** *573–586.*

4. Caldecott, R. S. 1955. The effects of X-rays, 2 Mev electrons, thermal neutrons, and fast neutrons on dormant seeds of barley. *Ann. New York Acad. Sci.,* **59:** *514–535.*

5. ————. 1958. Post-irradiation modification of injury in barley—its basic and applied significance. *Proc. 2nd Intern. Conf. Peaceful Uses of Atomic Energy. New York: United Nations,* **27:** *260–269.*

6. ———— and Smith, L. 1952. The influence of heat treatments on the injury and cytogenetic effects of X-rays on barley. *Genetics,* **37:** *136–157.*

7. ————, ————. 1952. A study of X-ray-induced chromosomal aberrations in barley. *Cytologia*, 17: *224–242.*

8. ————, Beard, B. H., and Gardner, C. O. 1954. Cytogenetic effects of X-ray and thermal neutron irradiation on dormant seeds of barley. *Genetics,* 39: *240–259.*

9. ————, Frolik, E. F., and Morris, R. 1952. A comparison of the effects of X-rays and thermal neutrons on dormant seeds of barley. *Proc. Nat. Acad. Sci.,* 38: *804–809.*

10. ————, Johnson, E. F., North, D. T., and Konzak, C. F. 1957. Modification of radiation-induced injury by post-treatment with oxygen. *Proc. Nat. Acad. Sci.,* 43: *975–983.*

11. Conger, A. D., and Randolph, M. L. 1959. Magnetic centers (free radicals) produced in cereal embryos by ionizing radiation. *Radiation Res.,* 11: *54–66.*

12. Curtis, H. J., Delhias, N., Caldecott, R. S., and Konzak, C. F. 1958. Modification of radiation damage in dormant seeds by storage. *Radiation Res.,* 8: *526–534.*

13. Ehrenberg, A., and Ehrenberg, L. 1958. The decay of X-ray-induced free radicals in plant seeds and starch. *Arkiv Fysik,* 14: *133–141.*

14. Elliott, F. C. 1957. X-ray-induced translocation of Agropyron stem rust resistance to common wheat. *Jour. Hered.,* 48: *77–81.*

15. Gaul, H. 1958. Present aspects of induced mutations in plant breeding. *Euphytica,* 7: *275–289.*

16. Giles, N. H. 1940. Induction of chromosome aberrations by neutrons in Tradescantia microspores. *Proc. Nat. Acad. Sci.,* 26: *567–575.*

17. Kempton, J. H., and Maxwell, L. R. 1941. Effect of temperature during irradiation on the X-ray sensitivity of maize seeds. *Jour. Agr. Res.,* 62: *603–618.*

18. Konzak, C. F., Caldecott, R. S., Delhias, N., and Curtis, H. J. 1957. The modification of radiation damage in dormant seeds. *Radiation Res.,* 7: *326.*

19. Koo, F. K. S. 1959. Expectations on random occurrence of structural interchanges between homologous and between non-homologous chromosomes. *Amer. Nat.,* 43: *193–199.*

20. Morris, R. 1955. Induced reciprocal translocations involving homologous chromosomes in maize. *Amer. Jour. Bot.,* 42: *546–550.*

21. Powers, R. L., Webb, R. B., and Ehret, C. F. 1961. Storage transfer and utilization of energy from X-rays in dry bacterial spores. *Bioenergetics symposium. Radiation Res.,* **Suppl. 2.** In press.

22. Sax, K. 1941. Types and frequencies of chromosal aberrations induced by X-rays. *Cold Spr. Harb. Symp. Quant. Biol.,* 9: *93–101.*

23. Sears, E. R. 1956. The transfer of leaf-rust resistance from *Aegilops umbellulata* to wheat. *Brookhaven Symp. in Biol.,* **9:** *1–22.*

24. Smith, L. 1946. A comparison of the effects of heat and X-rays on dormant seeds of cereals with special reference to polyploidy. *Jour. Agr. Res.,* **73:** *137–158.*

25. ————. 1951. Cytology and genetics of barley. *Bot. Rev.,* **17:** *1–355.*

26. ———— and Caldecott, R. S. 1948. Modification of X-ray effects on barley seeds by pre- and post-treatment with heat. *Jour. Hered.,* **34:** *173–176.*

27. Stadler, L. J. 1930. Some genetic effects of X-rays in plants. *Jour. Hered.,* **21:** *3–19.*

28. Wallace, B., and Vetukhiv, M. 1955. Adaptive organization of the gene pools of Drosophila populations. *Cold Spr. Harb. Symp. Quant. Biol.,* **20:** *303–310.*

29. Wolff, S. 1960. Post-irradiation storage and the growth of barley seedlings. *Radiation Res.,* **12:** *484.*

30. ———— and Luippold, H. E. 1957. Inaccuracy of anaphase bridges as a measure of radiation-induced nuclear damage. *Nature,* **179:** *208–209.*

31. Zimmer, K. G., Ehrenberg, L., and Ehrenberg, A. 1957. Nachweis langlebiger magnetischer Zentren in bestrahlten biologischen Medien und deren Bedeutung fur die Strahlenbiologic. *Strahlentherapie,* **103:** *1–15.*

Comments

ROBINSON: I question here, and in Doctor MacKey's presentation, the assumption of the superiority of the heterozygote, i.e., the importance of overdominance, particularly in quantitative characters. Reference has been made to work in Drosophila for the importance of epistasis. I cite the work of Vethukiv *(Evol., 1954)* and by Vethukiv and Beardmore *(Genetics, 1959)* where the later results do not support the conclusion of important epistasis given in 1954.

The importance of super- or overdominance in quantitative characters has, to my knowledge, still to be convincingly demonstrated.

Heterosis occurs in variety crosses of maize, but neither overdominance nor epistasis may be necessary for an adequate explanation of the heterosis. Genetic diversity resulting from different alleles in different parents could account for the heterosis observed.

CALDECOTT: I would make three points because I believe they reflect our differing views on the motivation for doing the research reported and the nature of the genetic systems involved.

1. To test the hypothesis, or "assumption" to use your word, that the heterozygote is superior to the homozygote is not, in itself, bad. One could have set up the opposite hypothesis which would have been subject to the same kind of criticism that you render but which, I sense, you would condone.

2. The fact that overdominance has not been "convincingly" demonstrated is the crux of the whole issue. If it had been clearly demonstrated, or proved not to exist, both in diploids and polyploids (and I emphasize the latter), I would have no interest in conducting the studies outlined as they would contribute nothing new to our knowledge. In this regard, in my opinion, it is dangerous to generalize from what are admittedly inadequate studies with two species, if you accept the negative data in Drosophila, both of which are diploids, and neither of which is naturally inbred or exists in nature at the polyploid level.

3. You make a distinction between genetic diversity and epistasis. These two concepts need not be, indeed they cannot realistically be, treated independently. Therefore, it seems inconsistent to me that, on the one hand, you postulate that heterosis may result from genetic diversity but, on the other hand, not from epistasis. If you recall, in the text it is specifically stated that the procedures being used should be of significance if heterosis is due either to the differences between genes governing the same order of biochemical function or to epistasis.

MacKEY: Wheat is neither strictly autoploid nor alloploid. It varies with the characteristic. My own investigations indicate this rather clearly. Characteristics essential for the ability to complete life and to reproduce show a high degree of autoploidy, while other characteristics not decisive to life or death but determining morphology, adaptation, etc., may follow a more alloploid pattern. The reason for this difference is that the former category will have a more conservative differentiation on the diploid level, while the latter may evolve rather freely as evident from the morphology of *Triticum monococcum*, *Aegilops speltoides*, and *A. squarrosa*, the parental genomes of 6x wheat.

First, when diploids are added to form polyploids, the essential genes can mutate more readily without severe deleterious effects, but also here, at least, one pair has to be kept intact. The others are, however, now free to mutate in directions that may have been impossible on the diploid level. The degree of diploidization revealed in my experiments with *dicoccum* favors the idea that this process has selective advantage.

It may be a kind of genomic heterosis effect. In addition, the dosage of every gene is very unlikely always optimal at the level reached by the addition of whole genomes. For phylogenetical reasons, I am just now studying the degree of diploidization in different 4x wheats, since it is still not settled whether they should be traced to one or more processes of tetraploidization.

CALDECOTT: What you have stated is precisely the basis on which the studies outlined here were conducted.

GRUN: There are two aspects on which I have questions to ask. The first concerns the plan described to see whether you can establish permanent heterosis in self-pollinated hexaploid cereals. As I understand it, you plan to irradiate plants repeatedly over a number of generations, then cross back onto the original parental lines and attempt to select vigorous lines from the progeny. While you may well increase vigor by this procedure, I wonder whether you would be safe in assuming that such vigorous derivatives are a result of a heterotic interaction between homozygous dominant and homozygous recessive alleles of the same gene on homologous chromosomes. Might not such vigorous derivatives result instead from new epistatic interactions resulting from the induced mutations?

The second question concerns your use of the word "diploidization". I am assuming that you are implying by this a progressive change from autoploidy toward alloploidy. I wonder whether mutations from dominant to recessive alleles of polyploid plants, as you have described, would necessarily lead the plant more towards alloploidy. Might not such changes be better termed as simple gene mutations than adding the implications involved in calling them diploidizations?

CALDECOTT: In reference to your first question, the answer is yes, and this fact is mentioned in the text.

Regarding the second question, the term diploidization was chosen simply because after recurrent irradiation one should be able to make crosses that would give a wide range of genes that would segregate in a 3:1 ratio. (See text.)

DERMEN: If mutation was induced in the genotype AA AA AA, in a hexaploid wheat, and it was changed successively to Aa Aa Aa, would all three mutations be alike?

CALDECOTT: They could be, but it is very unlikely.

DERMEN: Does AA AA AA constitution represent an autoploid or an amphiploid condition?

CALDECOTT: It could represent either, providing one was dealing with bivalent pairing.

GARBER: On what basis are autopolyploids and allopolyploids to be distinguished when individual loci are involved? When the term "diploid-ization" is used, is it restricted to converting replicate loci (autoploids) to only two loci for alleles or may it be used for converting any poly-ploid to a strictly bivalent type?

CALDECOTT: As I am working with allopolyploids I have not concerned myself with this problem. The distinction will, of course, be related to whether or not there is bivalent or quadrivalent pairing. The use of the term diploidization should be generally taken to imply the conver-sion of replicate genes, those which govern the same order of biochem-ical function, to one of their allelic forms so that, after selecting appro-priate genotypes out of F_2 hybrids between the mutated stock and the progenitor, genotypes can be obtained which, when crossed with one another, will give 3:1 ratios.

GABELMAN: Seedlings produced from seeds irradiated with low levels of X-radiation (less than 1,500 r) occasionally show stimulation in growth. What is the nature of this stimulation? How do you rationalize this potential stimulatory effect in evaluating and interpreting the response of seedlings to higher levels of irradiation as used in your experiments on radiosensitivity?

CALDECOTT: I am not aware of any reproducible studies with barley seeds that show that low doses of ionizing radiations cause a stimulation in growth, albeit this may be true in other species. I do not know what the nature of the stimulation is in other species and, because it does not exist in barley, have not had the problem of rationalizing the effect with the data presented.

OSBORNE: As Doctor Caldecott has mentioned, the Brookhaven coop-erative Program has for some years included preradiation stabilization of seeds at controlled relative humidity which, I believe, is 65 per cent.

I would merely add that, since the beginning of the UT-AEC Cooperative Program in 1956, we have also brought all seeds to constant moisture
at 65 per cent relative humidity. They are then sealed in plastic, irradiated, and returned to the cooperator. There seems little doubt that
this procedure keeps sensitivity (in terms of toxicity), storage effects, and
oxygen effects at a minimum and the experiments are highly repeat-
able.

Discussion of Session IV

I. J. JOHNSON

Caladino Farm Seeds, Inc., Wheaton, Illinois

Progress in plant improvement may be dependent upon the incorporation into a new genotype of either one or many gene pairs. Several examples can be given in which but a single gene with major effects has meant the difference between success or failure in the economic culture of a crop plant. The best examples of the importance of single (or few) gene effects are the addition of disease resistance lacking in otherwise agronomically desirable genotypes in wheat, oats, barley, flax, and a number of other crops. Often, also, a simple gene mechanism may control plant height, as in sorghum, and a relatively few genes may control photoperiodic response or time of maturity and thereby adaptation to a new environment. All favorable simple gene affects need not be due to the action of dominant genes. Many cases can be cited where the recessive is the "desirable" type and the dominant the "undesirable" type. Hence, in the total realm of plant improvement, one must not confine his thinking only to quantitative characters for which expression usually is conditioned by many genes, each with small effects and with action varying from additive to epistatic.

Throughout the long evolutionary processes in the development of plants there has been accumulated a large number of mutant types. These have been partially catalogued in several crops and maintained as world collections. These mutant types provide a "gene pool" and are the basic working stocks in many of our economic food plants.

The mere existence of a known source of supply of genes to meet specific needs does not imply that *all* of the genes useful in plant breeding are now in existence. When a search for desired genes fails to uncover sources of new characters desired in plant improvement, speeding the process of natural mutation through artificial means surely becomes an important part of plant breeding. During the progress of this symposium, several examples (dwarf cotton, early maturing Hybiscus, and rust-resistant Mcrion bluegrass) of new

405

genes useful in plant improvement and arising through the use of mutagenic agents have been reported.

Many studies reported in the literature have shown that irrespective of the mechanism involved or the kind of mutagenic agent employed there is no problem in *producing* mutant types. MacKey in his paper strongly emphasized the importance of devising more adequate screening procedures to isolate those kinds of mutants from the many arising from irradiation that may be useful in higher plants. Nelson also has shown the complexity of screening for useful mutants (as well as natural variants) in microorganisms.

Nearly every study in mutation breeding has shown that newly formed mutants generally are expected to be associated with chromosome aberrations that may be deleterious to the organism as a whole. This is not surprising. Present techniques to produce mutations are not selective. For these reasons it would be indeed surprising if a single gene mutant in higher plants would produce a line useful *per se.* This does not limit the usefulness of mutation breeding, but only serves to emphasize the necessity of transferring a newly formed desirable mutant gene into a more usable gene background through hybridization. The example given by Langham for nondehiscent sesame is an example of the difficulty that can occur in separating the undesirable plieotropic "side affects" from the major gene or genes.

The discussion to this point largely has been concerned with single or simple gene effects in naturally self-pollinated crops. Perhaps it would be appropriate to discuss next the more complex problems associated with mutation breeding for modification of characteristics in a desired direction when character expression is dependent upon multi-gene effects. In seeking to improve quantitative characters by irradiation the problem of identification becomes most difficult. General experience has shown that the heritability of quantitative characters is low and hence only limited progress can be expected from selection on a single plant basis. Evaluation of mutant progenies in replicated trials (as in the normal procedure of testing variability derived by hybridization) consequently becomes a necessity.

In any approach to plant improvement the breeder must compare the probability of obtaining improvement by alternative approaches. The long history of plant improvement has clearly shown the advantages to be gained by first defining the *needs* for improvement, then carefully surveying the parental material available which together may contain the desired genes for recombination, followed by hybridization to create the pool of variability upon which subsequent selection and evaluation can be practiced. To me, one of the most serious shortcomings of mutation breeding for characters whose expression is conditioned by many genes is the transfer from the philosophy of "creating planned variability" to creating "chance variability". At best, the desired recombination of many genes has a relatively low probability, but it is advantageous to know something about the odds as based on previous experiments.

During the progress of this symposium examples have been cited of gains made by following the mutation breeding route. Other examples have been given by Burton and Caldecott showing lack of significant progress. These demonstrations through experimentation that gains or lack of gains have been made neither prove nor disprove the value of mutation breeding. The critical test is the comparative progress that could have been made with *similar expenditure of time and resources* to achieve a particular goal. Modern plant breeding approaches with the customary limitations of resources at research institutions require that progress be equated in terms of gains per unit of investment.

Throughout the reports in the literature on mutation breeding for improvement in quantitative characters there seems to be an urge on the part of the breeder to achieve progress faster than he has a right to expect. As previously stated, presently used mutagenic agents are not selective. It would be expected that as many (or more) undesirable chromosome alterations would be derived as favorable ones. The evolutionary development of favorable internal and relational balance in our present economic plants has been a slow one. Evidence presented in the literature[1] has shown that after many generations following hybridization in barley important measurable changes have occurred in such hybrid-derived populations. Is it not likely that much more potential gains could be achieved if populations derived from induced mutation were allowed to be subjected to a considerable period of time for "restoration" to chromosome balance before

---

[1]Bal, B. S., Suneson, C. A., and Romage, R. T. Genetic shift during 30 generations of natural selection in barley. *Agron. Jour., 51: 555–557. 1959.*

attempts were made to select individuals from them for progeny evaluation? Selection of plants from a heterogeneous population within the limits of sampling that are imposed at best represent only a limited portion of the potential population range. It would seem to me that the probability of excluding the most unproductive end of the population distribution would be greatly enhanced if selection pressure for survival of individuals were allowed to operate on such mutation-derived variants.

During the progress of this session of the symposium major emphasis has been directed to the use of mutagenic agents to produce new variants in self-pollinated plants. The literature does not record many irradiation experiments with the large group of economic plants in which cross pollination is the rule. Obviously, it would be difficult to distinguish between natural and induced mutations in populations that predominantly are heterozygous as a consequence of random cross pollination. The real question, however, is whether or not there is need to produce new variability within a crop in which abundant natural variability already exists. It is a common belief of plant breeders that we have sampled and evaluated only a small portion of the total gene supply in such species and hence adding to this reserve, for the most part, is not likely to be an important breeding procedure. It should not be inferred, however, that among established inbred lines (as in maize) all desired characters have been fixed by inbreeding and selection. The use of mutagenic agents to produce desired variants among existing inbred lines is in reality a problem comparable to their use in naturally self-pollinated crops.

In summary, to this discussant, one should view the usefulness of induced mutations not as a sole means to attain a desired objective, but rather as a supplement to our present well-established procedures that have adequately demonstrated their value in plant breeding. This is not a "negative" point of view. If a desired character cannot be found within the existing pool of germ plasm in a species or in its related species, there is ample justification to create new variability through induced mutations even though our present techniques are comparatively crude in respect to gene selectivity. Finally, the use of mutation breeding techniques rarely can be expected to immediately produce agronomically desirable variants. The undesirable and uncontrolled side effects due to chromosome rearrangements may upset the balance of gene complexes that characterize the total genetic complex needed in our present-day highly specialized crop varieties.

Session V

Possibilities for the Future

H. F. Robinson, *Chairman*
North Carolina State College,
Raleigh, N. C.

Mutagenic Specificity and Directed Mutation[1]

HAROLD H. SMITH

Brookhaven National Laboratory,
Upton, L. I., N. Y.

IRECTED control of induced mutation is an important objective in both theoretical genetics and its practical application. The means of achieving and improving this control is through better understanding of mutagenic specificity. Evidence of specificity has been reported for mutations among genes (inter-locus) and among alleles (inter-allelic and/or intra-locus), for localization of chromosomal breakpoints, and for relative frequencies of gene mutations *vs.* chromosome breaks.

In the first part of this paper, experimental evidence will be summarized for a nonrandomness, differential effect or specificity of induced mutation in organisms ranging from higher plants and animals to bacteriophage. The selectivity may be expressed in broad spectrum experiments of "forward" mutations and chromosome breakage, or, more specifically, in differences of back mutation response of particular alleles to a series of mutagenic agents. This discussion will be followed by considerations relating interpretation of mutagenic specificity to mechanisms of action and structure of the genetic material.

Evidence for Mutagenic Specificity

Higher Plants

The evidence for mutagenic specificity in higher plants is confined for the most part to "forward" (impaired function) mutations between loci (inter-locus specificity) and gene mutations *vs.* chromosome breaks. The Swedish group of investigators have been most active in this field and have reported results on different spectra of mutations produced in barley following treatment with different mutagens (28, 29, 71).[2] Chlorophyll mutations of the *viridis* type are

---
[1]Research carried out at Brookhaven National Laboratory under the auspices of the U. S. Atomic Energy Commission.
[2]See References, page 430.

induced with increasing frequency and of the *albina* type with comparative decreasing frequency through the series of mutagenic agents: neutrons, X-rays, ethylene oxide, myleran, nitrogen mustard, ethylene imine, di(β–chloroethyl) phenyl alanine, and nebularine (purine–9–d–riboside). The spectrum of spontaneous chlorophyll mutations lies between that of myleran (di–methane–sulfonyloxy–butane) and nitrogen mustard. There is also an indication of different frequencies in mutations for mildew resistance and susceptibility compared to chlorophyll mutations when the spectrum of spontaneous, X-ray-induced, and chemically induced types are compared (22).

Differential inter-locus response following exposure to ionizing radiations of different ion densities has been reported for 69 analyzed radiation-induced erectoid mutants located at 22 different loci in barley (29). Comparisons among mutants at five of the loci are particularly informative. Locus *a* (11 to 12 mutants) has not given mutations with neutrons, locus *b* (4 mutants) gave mutations in only one variety (Golden) of the three tested, locus *c* (17 mutants) gave mutations from exposure to sparsely as well as densely ionizing radiations and these were commonly associated with detectable chromosome breakage, and loci *d* (10 mutants) and *m* (7 mutants) also gave mutations from both types of irradiation but in only one case was a detectable chromosome break involved. These results are indicative of differential mutagenic effects of different radiation sources, though there is still a regretable paucity in numbers.

Localization or nonrandomness of chromosome breakage is another manifestation of differential control of the mutation process in its broader sense. It has been known for about a decade that radiomimetic compounds have the ability to break chromosomes preferentially. In contrast, most evidence indicates that radiations are nonspecific, or at least less specific, in effecting breakage between and within chromosomes. The highly reactive compounds, namely, nitrogen mustards, di (2:3–epoxypropyl) ether, and beta–propiolactone, induce breaks selectively in heterochromatic regions in the middle of the short chromosomes of the *Vicia faba* complement. The more weakly reactive radiomimetic compounds that have been tested extensively, namely, maleic hydrazide and 8–ethoxycaffeine, produce a concentration of breaks in the long satellited chromosome of *Vicia faba*.

With the former, a structural isomer of the pyrimidine base uracil, breaks are concentrated near the centromere; with the latter, a purine derivative, breaks are mainly in the region of the nucleolar constriction (47, 64, 65). The effects of maleic hydrazide and 8–ethoxycaffeine may conceivably be related to their structural resemblance to nucleic acid bases. KCN treatments produce an apparent random distribution of breaks among Vicia chromosomes which are, however, concentrated within chromosomes in heterochromatic regions (36). Analysis of exchanges between localized and randomized chromatid breaks may be used to gain information on chromosome structure in specific regions (46).

Specificity of the mutagenic agent itself is clearly shown by the wide variation in the ratio of gene mutations to chromosomal aberrations. For some chemicals, as maleic hydrazide, the ratio is essentially zero, for others, as nebularine, it is large if not infinite; while certain alkylating agents and ionizing radiations fall in between these extremes (17). Diethyl sulphate has been found to combine the qualities of inducing a high frequency of mutations in barley with few accompanying chromosomal structural aberrations (31). There is, also, some indication of a different spectrum of chlorophyll mutations produced by diethyl sulphate compared to gamma radiation. The highly efficient mutagenic alkanesulphonic esters, as ethyl methane sulphonate (33) appear to cause little chromosome breakage. On the other hand, myleran is efficient in producing localized chromosome breaks in Hordeum and Vicia (48). We have found that ethyl methane sulphonate is ineffective at 0.1 M in producing loss of endosperm marker genes in the short arm of chromosome 9 in treated pollen of maize. By this same test (39) diepoxybutane was found to induce breaks approximately at random along this chromosomal arm, a region in which there is no conspicuous localization of heterochromatin. The number of anaphase bridges induced by maleic hydrazide and certain of its derivatives was observed by Graf (27) to be correlated with the number of heterochromatic knobs in the variety of maize treated.

The clearest evidence of specificity in induced chromosome breakage is, then, the preference shown by certain chemicals for affecting particular heterochromatic regions. No definitive explanation is yet available; but the phenomenon may be due to different spacial relations of chromosomal strands in eu- vs. hetero-chromatic regions, or to different placement with reference to the nuclear membrane. Significant information may be forthcoming from experiments underway utilizing radioactive-labeled chemical mutagens. Such experiments should provide answers to questions regarding incorporation and possible localization of the mutagen in the chromosome and whether the mutagenic agent acts directly or indirectly with the chromosomal constituents. Callaghan and Grun (10), using C^{14} labeled maleic hydrazide, observed that this compound is incorporated into chromatin material, is localized in the nucleoli, and is uniformly distributed in eu- and hetero-chromatin.

An example of induced mutation leading to directional genetic change in populations is found in investigations on radiation-induced polygenic variation. Experiments with Drosophila will serve as an illustration. Utilizing lines that no longer responded to selection for number of sternopleural hairs, Scossiroli (59, 60) and Clayton and Robertson (11) were able to induce, with radiation, new genetic variability. Selection was practiced for both higher and lower number of hairs, but response was realized only in the high lines. In our laboratory, Daly (12) induced genetic variability for flowering time in *Arabidopsis thaliana* with γ-radiation and selected for early and late flowering. In the R_3 generation, 80 lines were developed from selected R_2 plants. Up to 50 per cent of the lines selected for earliness were found to be significantly earlier than nonirradiated selected control populations. None of the late-flowering lines differed significantly from the corresponding controls. We have had similar experiences with a radiation-induced quantitative character in *Nicotiana tabacum*. Although it cannot be stated unequivocally that the mutations *per se* are biased in one direction, it is nevertheless clear that the ultimate result of induced variability in quantitative characters may frequently, and perhaps characteristically, be a directed change in the population. Further information on this phenomenon is urgently needed for clarification of its genetic basis as well as its evolutionary and practical significance. It is not evident that the "easy direction" is necessarily a detrimental one, as e.g., earliness in Arabidopsis; but rather it appears to be characteristic of the genotype to respond to a mutagen by a greater yield of genetic variability in one direction from the parental mean compared to another.

Animals

The most extensive series of experiments on differential muta-genicity reported with metazoan material are those of O. G. and M. J. Fahmy on *Drosophila melanogaster*. As in the experiments on higher plants, these involve selectivity of mutagenic agents among loci, and are concerned with "forward" mutations which are frequently associated with deficiencies. In earlier publications the Fahmy's reported the following differences in mutants recovered after treatment with certain alkylating agents compared to X-rays: a higher proportion of visible to lethal mutations, new visible mutations not previously produced spontaneously or after irradiation, and a different distribution of breakpoints and lethals in the X chromosome. These claims have been critically reviewed recently by Auerbach (4) and Altenburg and Altenburg (3) and will not be discussed further here.

Subsequently, additional papers by Fahmy and Fahmy (19, 20, 21) have been published giving further experimental evidence of differential induction of mutations in Drosophila. In one of these (20) was reported differential response of different regions of the X chromosome in Drosophila to the same mutagenic compound (phenylalanine mustard). In the region proximal to *f* they found an excess of visibles in the chromosome segment *f-car* and an excess of lethals in the segment from *car* to the centromere. The Fahmy's contend that their data indicate a qualitative difference in the physical nature of visibles and lethals. They suggest that visibles may result from molecular re-orientation which alters the transmission of the genetic code, whereas lethals can be attributed to elimination or inactivation of the gene. Obviously, it is not presently possible to test this hypothesis. The Fahmy's have recently shown (21) that S–2–chloroethylcysteine is virtually specific to a particular cell stage of the male germ line in *Drosophila melanogaster*. This compound, they reported, is ineffective on mature sperm, but is very effective on the early spermatogonia.

In the silkworm Nakao, Tazima, and Sakurai (51) found differences in the relative frequencies of two egg color mutants *(pe* and *re)* following treatment of the males with two structurally different mutagenic mustards, nitromine and alanine "mustard". With the former the ratio of *pe* to *re* was over 1 (1.06–1.58), with the latter it was 0.6. Most of the mutations were due to deficiencies so that tentatively there appears to be a specific relation between the chemical agent and the position of breakage of the chromosome.

Microorganisms

The most productive and illuminating of recent studies on mutagenic specificity are those carried out with microorganisms: fungi, bacteria, and bacteriophages. The outstanding advantage of the experimental methods used with these materials is the extraordinarily high resolving power which permits localization of genetic events with sufficient accuracy to approach a correlation with macromolecular substructure. These investigations differ from those on higher plants and animals in that inter-allelic or intra-locus rather than inter-locus specificity is tested, and also, back-mutation analysis is used almost exclusively.

In Neurospora mutagenic specificity has been reported for two particular alleles at two loci, one adenineless, the other inositolless, which have been combined in a double mutant. These alleles have mutated at different relative rates under the action of eight different agents tested, i.e., six chemical mutagens of the alkylating type, ultraviolet light, and X-rays (38, 69, 70). The proportion of $ad^+ : inos^+$ mutants recovered ranged from 1.5 (ethyl methane sulphonate) to 3,800 (bromoethyl methane sulphonate) among the chemical mutagens. With ultraviolet light the proportion was 0.5, for X-rays 16, and for spontaneous mutations 10. The widely different results with the two methane sulphonates, ethyl and bromoethyl, demonstrate the pronounced specificity which related chemicals may have for the genetic material. A further indication of differential mutagenic effect of these two alleles is that, although ultraviolet induced only about half as many reversions to ad^+ as $inos^+$, if the UV exposure is preceded by a 20-minute treatment with formaldehyde then about four times as many ad^+ as $inos^+$ mutants are recovered (45).

Preliminary evidence of gene-specific effects, i.e., inter-locus specificity, for 8–ethoxycaffeine, dimethyl sulfate, and ultraviolet light, has been obtained recently in the ascomycete Ophiostoma multiannulatum (75) and in Schizosaccharomyces pombe (32).

In bacteria early evidence indicating that not only different loci but also different alleles of the same locus may react differently to the same mutagen was reported by Demerec (13) in 1953. Thirty-five

nutritional deficiencies in *Escherichia coli* involving 11 amino acids
were tested for frequency of spontaneous reversions and those induced
by MnCl₂, ultraviolet, and beta–propiolactone. The material included
a number of alleles governing deficiencies for tryptophane, methionine, histidine, and leucine. These studies were extended by Glover
(26) to additional alleles and mutagens. He showed that in *E. coli* the
frequency of induced mutations at one locus can be influenced by the
type of allele present at another locus.

Extensive evidence of inter-allelic specificity of induced mutation has now been demonstrated in *Salmonella typhimurium*. About
60 per cent of auxotrophic and fermentation alleles in this species are
"mutagen stable", i.e., although they mutate spontaneously, their
mutation rates are not increased by treatment with any of a large
number of mutagens which are effective on other alleles (14). Back-
mutation tests have revealed that different "mutagen-labile" alleles
at a locus may differ widely in rates of induced mutability. The most
extensive analysis reported to date is that of Kirchner (37) with a
series of alleles at the histidine locus in two strains of *S. typhimurium*.
The histidine mutants were either of spontaneous origin or induced
by either ultraviolet irradiation or 2–aminopurine. The mutagenic
agents used to revert the mutant alleles to prototrophy were 2–aminopurine (2AP), 5–bromodeoxyuridine (BDU), sodium nitrite (NO₂),
t–butyl hydroperoxide (TBP), and beta–propionactone (BPL). Of 54
mutants tested in one strain 23 were mutagen-stable. Data on the 31
mutagen-labile mutants are summarized in condensed form from
Kirchner's results in Table 1.

In the table + signifies an increased back-mutation frequency, −
means that there was no increase. The highly reactive polymerizing
agents, BPL and TBP, are grouped, as are the base analogues, BDU
and 2AP. The mutant alleles show considerable specificity of response
to the mutagens, and this is further emphasized if the relative amount
of increased mutation is also taken into consideration. Seventeen of
the alleles responded to the mutagenic action only of the highly
reactive compounds, nine to both groups, four only to the analogue
2AP, and only one to all three categories of mutagenic agents. Eleven
of the 12 auxotrophs that were obtained originally after treatment
with 2AP showed an increased reversion frequency with BDU,
whereas the ultraviolet-induced auxotrophs showed no particular

TABLE 1.—INDUCTION OF REVERSE MUTATIONS WITH CHEMICAL MUTAGENS IN 31
MUTAGEN-LABILE ALLELES AT THE HISTIDINE LOCUS OF *Salmonella typhimurium.* *

Highly reactive polymerizers		Base analogues		Deaminizer	No. of
(BPL)	(TBP)	(2AP)	(BDU)	(NO₂)	alleles
+	+	−	+	+	1
+	+	+	−	−	6
+	−	+	+		1
+	−	+	−		2
+	+	−			6
+	−				9
−	+	−	−	−	2
−	−	+	−	−	4

*After Kirchner (37). See text for meaning of symbols.

mutational pattern. The interpreted significance of these results will
be discussed below under mechanisms of mutagenic action.

Patterns of reversion of 22 mutants at two tryptophan loci (try C
and try D) in *Salmonella typhimurium* have been investigated following treatment with four chemical mutagens (15, 56). The compounds
used were the base analogues 5–bromouracil (5BU) and 2–aminopurine (2AP), and the alkylating agents diethyl sulphate (DES) and
beta–propiolactone (BPL). Many of the alleles responded differently
to the treatments, as shown in Table 2, where + indicates that reversions were induced and − that there was no increase in frequency of
reversions. As with alleles at the histidine locus, those tested at the
tryptophan loci also showed marked specificity to chemical mutagens.
DES induced reversions in all 18 of the mutants that reverted, of

TABLE 2.—INDUCTION OF REVERSE MUTATIONS WITH CHEMICAL MUTAGENS IN 22 ALLELES
AT TWO TRYPTOPHAN LOCI OF *Salmonella typhimurium.* *

Alkylating agents		Base analogues		No. of alleles
DES	BPL	2AP	5BU	
+	+	+	+	9
+	+	+	−	1
+	−	+	+	5
+	+	−	−	2
+	−	−	−	1
−	−	−	−	4

*After Balbinder in Demerec, *et al.* (15). See text for meaning of symbols.

which 15 also responded to the base analogues. Mutants that were affected by one base analogue were also, with one exception, affected by the other.

Zamenhof (73) has reported that heating (135° to 155° C) of vegetative cells *(E. coli)* or spores *(B. subtilis)* in the dried state in vacuum produced mutations at frequencies of 1 to 10 per cent, certain mutants appearing more frequently than others. It was suggested that this difference may be attributed to a differential susceptibility to alteration by heat of different loci.

The first evidence that the base analogue 5–bromouracil induces mutations more frequently at specific sites, distinctly different from those most mutable spontaneously ("hot spots"), was afforded by the experimental results of Benzer and Freese (8) with rII alleles of bacteriophage T4. These experiments were carried out consequent to evidence that 5–bromouracil is incorporated in place of thymine in the DNA of phage and bacteria (16, 74) and that this incorporation causes mutation (42). Subsequently, Brenner, Benzer, and Barnett (9) demonstrated the induction by proflavine of another series of mutations which occur preponderantly at sites of the rII locus distinctly different from those characteristic of the spontaneous and 5–bromouracil spectra. The mutant alleles can be mapped by techniques of high resolution which permit the positioning of mutant sites in regions of the order of magnitude of a few nucleotide pairs in a DNA molecule.

Freese (23, 24, 25) has extended these investigations with rII alleles to include mutagenic patterns of additional base analogues to test the reversion frequencies of induced and spontaneous mutations, and particularly to provide a theoretical basis for better understanding of the mutation process and mutagenic specificity. Of the additional analogues investigated, 2–aminopurine, 2,6–diaminopurine, and 5–bromodeoxyuridine were found to be mutagenic. The mutability spectrum of 5–bromodeoxyuridine was essentially the same as that of 5–bromouracil but different from that of 2–aminopurine. Each differed from the spectrum of spontaneous mutants. A striking difference was shown in the pattern of reversion of rII alleles that had been induced originally by different mutagens. Of the mutants induced by base analogues, 95 to 98 per cent were reversible by base analogues; whereas only 14 per cent of spontaneous mutants and 2 per cent of the proflavine mutants were reversible by treatment with base analogues. Furthermore, mutants induced by 5–bromouracil are more readily reverted by 2–aminopurine than by 5–bromouracil and, *vice versa,* for those originally induced by 2–aminopurine. In summary, mutagenic specificity has been demonstrated in rII alleles both for forward mutations, i.e. different spectra of "hot spots", and reverse mutations. The latter show different reversion rates both in general, according to the origin of the mutant, as well as individually, according to the specific site.

Mechanisms for Mutagenic Specificity

Most of the evidence for mutagenic specificity, as can be seen from the preceding review, comes from experiments in which nucleic acid base analogues or alkylating agents were used as the mutagenic agents. It is unlikely that mutations are induced by a common mechanism in all organisms or for all mutagens. However, the best working hypothesis so far proposed is that hereditary information is determined by the base pair sequence in DNA and that true intragenic mutations involve fundamentally an alteration in this sequence.

Freese (23, 24, 25) postulated that certain base analogues through "mistakes" in incorporation and replication of DNA would cause errors in base pairing and an ultimate change in sequence of the normal bases. Mutations are considered to result from these "transitions", i.e., replacements of purine by purine or pyrimidine by pyrimidine. For example, diagrammatically:

1. Substitution of a pyrimidine analogue, 5–bromouracil (5BU)
    A–T (5BU) → A–5BU → G–5BU → G–C
    G–C 5(BU) → G–5BU → A–5BU → A–T

2. Substitution of a purine analogue, 2–aminopurine (2AP)
    T–A (2AP) → T–2AP → C–2AP → C–G
    C–G (2AP) → C–2AP → T–2AP → T–A

A and G represent the natural DNA purine bases adenine and guanine, respectively; T and C, the pyrimidine bases thymine and cytosine. The new stable base pair, i.e., transition from one normal purine-pyrimidine pair to another, at the mutant site would be expected to be established after two replications of DNA, subsequent to the initial incorporation of the analogue. Rudner (55) confirmed that mutations induced by 5–bromodeoxyuridine and 2–aminopurine in *Salmonella typhimurium* become established as mutant clones only after two DNA replications (50).

The importance to the problem of mutagenic specificity of these recent concepts and experiments is that mutagenic activity can be related to genetic molecular structure. The mapping of sites with different mutation frequencies under the influence of different mutagenic agents can be localized by use of Benzer's high-resolving power technique (6, 7), with an accuracy that approaches the nucleotide level. If it is assumed that only two kinds of nucleotide pairs are present in DNA, the appearance of different "hot spots" would seem to indicate that the mutability of a particular nucleotide pair is dependent upon its position in the genome, perhaps the specific sequence in a particular area of the macromolecule.

The reversion of base analogue-induced mutants in both phage (24, 25) and bacteria (37) by subsequent treatment with base analogues is confirmatory evidence that the original mutations were caused by transitions for which change in both directions would be expected. Reversion by a particular mutagen of specific mutants, which involve a restoration of function owing (in theory) to re-establishment of a nucleotide sequence, can be considered as specificity in the strictest sense. This is the basis of the back-mutation test used in Neurospora and bacteria and for reverting rII-type mutants in phage.

The infrequency of base analogue inducible reversions with mutants of spontaneous or proflavine origin led Freese to postulate that most of these arose by a process different from "transition", namely, "transversion", which involves a replacement of a purine by a pyrimidine, or *vice versa* in a given DNA chain.

Nitrous acid will produce mutations in tobacco mosaic virus after *in vitro* treatment of isolated RNA (49, 57). Since de-amination with nitrous acid converts cytosine to uracil, adenine to hypoxanthine, and guanine to xanthine (58), it is considered that these new bases produced *in situ* will have altered pairing properties and cause the change of a nucleotide pair in subsequent replication of DNA (25). The mutagenic effect would depend upon "transitions", as with the base analogues. Nitrous acid has been found to be highly mutagenic for *E. coli* (34, 35) and to produce a spectrum of mutations which differs significantly in some respects from the spontaneous and ultraviolet-induced spectra, but not from that caused by disintegration of incorporated phosphorus–32. Nitrous-acid-induced rII mutants in T4 phage are, for the most part (87 per cent), reverted by base analogues (25).

In contrast to the sound theoretical basis (actually as yet unproved by experiment) for the mutagenicity, and by plausible extension the mutagenic specificity, of base analogues and nitrous acid, there is little known at present which would explain mutagenic specificity of alkylating agents. These compounds react with proteins, nucleic acids, and nucleoproteins; chiefly with sulphhydryl (−SH), amino (−NH₂), and acid (−COOH) groups (54, 61). Reactions with the phosphate groups, amino and imino groups, and ring nitrogens of the purines and pyrimidines of DNA or precursors are, in all probability, of importance in mutagenicity. It is not known which of these reactions is primarily responsible for the ultimate genetic changes. Based on *in vitro* reactions some investigators have been inclined to emphasize esterification of the phosphate groups (1, 2), which results in the formation of unstable triesters; others, the direct alkylation of the purine or pyrimidine ring (40, 41). Ethylation in phage DNA may be particularly significant and may lead to faulty replication (43). Reiner and Zamenhof (53) reported differential sensitivity of DNA bases of calf thymus when treated *in vitro* with the alkylating agents dimethyl sulphate, diethyl sulphate, and nitrogen mustard. These agents attach their alkyl group to the purines, but not the pyrimidines, of DNA on the nitrogen in position 7 (72).

The alkylated DNA cannot itself be the mutant hereditary determinant since such a structure could not be replicated by metabolic cell processes. A preferred interpretation is that alkylation of a DNA group interferes with synthesis so as to increase the chance of errors in copying (44).

Szybalski (66) treated *Escherichia coli* cells with the alkylating agent triethylenemelamine (TEM) and investigated the mechanism by which mutants for streptomycin independence are produced. TEM reacted extensively with pyrimidines and with their nucleosides and deoxynucleosides, but not with the purines adenine and guanine. Szybalski postulated several sequential reaction steps to explain the effect of TEM: (a) semireversible absorption of TEM by resting cells; (b) chemical reaction between the mutagen and pyrimidine deoxynucleotides with the production of fraudulent DNA precursors; (c) incorporation of these unnatural analogues into DNA during synthesis of the latter; and (d) occurrence of permanent but nonlethal errors (mutations) in the base sequence during subsequent replication of the modified DNA. It is conceivable that if nitrogen mustards alkylate purines (Zamenhof) and ethylene imines pyrimidines (Szybalski), a basis may be provided for mutagenic specificity between these two classes of alkylating agents.

An ultimate step in demonstrating the mechanism of mutagenic specificity would be to show that a hereditary change induced by altering the nucleic acid has, in turn, produced a change in amino acid sequence of associated protein. Tsugita and Fraenkel–Conrat (67) have reported that the protein of a differential host mutant isolated after nitrous acid treatment and reconstitution of tobacco mosaic virus (TMV) RNA differs from that of the parent strain in that three amino acid residues are replaced by three others (proline, aspartic acid, and threonine by leucine, alanine, and serine).

Mutagenic Specificity and
Directed Mutation in Higher Organisms

Experiments on microorganisms, fungi, bacteria, and bacteriophage, have provided the unequivocal evidence for mutagenic specificity. These analyses have, for the most part, made use of techniques developed to explore the genetic fine structure of complex loci (14). It has been shown that specificity is characteristic of alleles at numerous sites in the gene locus rather than of the locus in general. This intra-locus or inter-allelic specificity has been demonstrated in Neurospora and bacteria only by the back-mutation test; in phage by both forward ("hot spot" spectra) and reverse mutations. Auerbach and Westergaard (5) suggest that forward mutations involve damage or malfunction by diverse mechanisms and are therefore less strictly specific; whereas, repair of a damaged allele would require a specific mechanism. Two mutagens which show an allele-specific effect in a back-mutation test may not show any striking specificity in a broad spectrum forward-mutation experiment.

A question of primary interest is, to what extent can the results with microorganisms be related or applied to higher organisms? There are differences in the organization of the heritable material from phage to the chromosomes of higher forms in strand composition, chemical composition, and possibly in the organization of genetic fine structure. In higher plants and animals there is now evidence that the chromosome consists in cross section of 32 DNA strands before synthesis and 64 after DNA synthesis in prophase through metaphase (63). In contrast, from T2 phage to bacteria the number of DNA strands is considered to be usually 2, with exceptions. Furthermore, in phage, and apparently in bacteria, there is no close association of protein with the DNA as in the nucleoprotein composition of chromosomes in higher forms. It is not yet clear that gene loci of higher forms are subdivisible into units of mutation, recombination, and function in the same way as complex loci in microorganisms; but the presence of pseudo-alleles in Drosophila and maize suggests that a basic similarity of organization may exist. Westergaard (70) has discussed further differences between genetic mechanisms in microorganisms and higher forms which may be significant in specificity.

The evidence for mutagenic specificity in higher plants and animals rests primarily on (a) the different spectra of gene mutants produced by different mutagens in barley and in Drosophila; and (b) the differential effect of certain mutagens on chromosome breakage both with respect to localization of breaks (usually in heterochromatic regions) and in the ratio of gene mutations to chromosomal aberrations. A common cause for many of these results may be that there is regional (particularly heterochromatic) sensitivity of the chromosome to damage by different mutagens and that the expressed change may vary in degree with the mutagen to give gross chromosomal breaks, small deficiencies (lethals), or "visible" gene mutations. (For more detailed discussion, see 4, 5, 17). It has been suggested by Westergaard that the higher proportion of visibles over lethals in Drosophila and the higher proportion of less drastic *viridis* over more extreme *albina* chlorophyll mutations in barley may reflect in each case a less drastic effect in the mutation event of certain chemical mutagens compared to ionizing radiations.

A large gap remains to be bridged between the intra-locus or site specificity observed in microorganisms and the inter-locus or chromosomal region specificity observed in higher plants and animals. The former implies a correlation between hereditary function and structure within DNA molecules; the latter may depend more on the chemical bonds that hold the DNA molecules together and also conceiv-

ably the binding with protein. Attachment of DNA molecules longi-
tudinally as well as DNA strands laterally may be through bonding of
diester phosphates with a divalent metal (62). These sites, as well as
attachment points of DNA through diester phosphate groups to
amino groups of proteins, may be susceptible to attack by alkylating
agents which are known specifically to react with phosphate groups
of DNA (54). What relation may exist between such reactions and
DNA genetic coding is unknown, but conceivably alteration of base
sequence between DNA "species" could be as significant as that within
a single DNA molecule.

In order to formulate a direct analogy between current explanations for inter-allelic specificity for reverse (restoration of function)
mutations in microorganisms in terms of DNA base sequence and
inter-locus specificity for forward (loss of function) mutations in
higher forms, it would appear necessary to postulate a qualitative or
quantitative difference in reaction sites affecting DNA code within
the functional gene locus involved. That is, either that the gene loci
differ qualitatively in susceptibility to change into viable or lethal
mutant sequences by the action of one mutagen compared to another,
or that there are different numbers of ways (numerically more or
fewer susceptible sites) among gene loci to change to heritable loss of
function with one mutagen compared to another. Parenthetically, to
those who work with cultivated plants forward and reverse mutations
may often lack evident distinction.

Close analogues of normal DNA bases have not, to date, been
reported as mutagenic in higher plants and animals (nebularine may,
however, be considered in this category). Owing to interest in the use
of these analogues in cancer chemotherapy much is known about their
incorporation and antimetabolic activity in mammalian cells (30).
Less is known about their effects on plants (52). We have
recently obtained preliminary evidence from experiments with
tritiated 2–aminopurine that this analogue is incorporated
in the DNA of nuclei of root tip cells of *Vicia faba* (Figure
1). Root tips were placed in a solution of tritiated 2–aminopurine (31×10^{-8} M, 3.7 µc/ml specific activity) for 8 hours
at 23° C. They were then grown in Hoagland's solution at 20° C for
24- and 48-hour periods, respectively; placed for 2 hours in 0.05 per
cent colchicine; fixed; and stained by the standard Feulgen smear

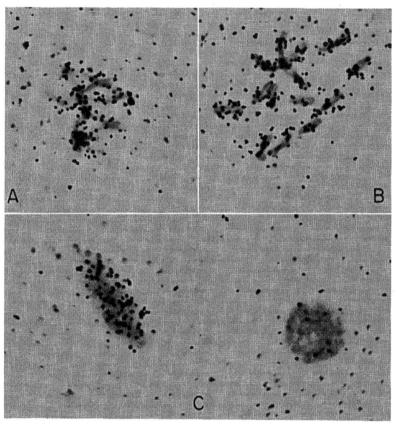

FIGURE 1.—*Nuclei of root tip cells of* Vicia faba *that have been grown in tritiated 2–aminopurine for 8 hours. A and B show labeled metaphase chromosomes 24 hours after removal from the base analogue solution. C shows labeled (left) and unlabeled (right) resting nuclei after 48 hours of recovery.*

technique. The slides were then dipped in photographic emulsion and exposed for 12 days. The 24-hour recovery period slides showed evidence of tritium decay in a number of metaphase chromosomes (Figure 1, A and B) and some resting nuclei. After 48 hours of recovery few metaphase chromosomes appeared labeled, but many resting nuclei were (Figure 1, C). Some of the label is observed in the cytoplasm. These observations are tentatively interpreted to indicate that 2–aminopurine, or at least that part with which the tritium label remains, is incorporated into the chromosomes and cytoplasm. It has been reported recently (68) that only a small quantity of unchanged 2–aminopurine is incorporated into DNA and RNA of *E. coli,* and that a large proportion is transformed into adenine and guanine.

Striking differences in growth of *Arabidopsis thaliana* can be induced by incorporation of 2–aminopurine and/or 5–iododeoxyuridine in the medium on which this plant can be cultured. Threshold concentrations which barely permit the plants to mature are 0.0008 M 2–aminopurine, 0.00005 M 5–iododeoxyuridine, and a combination of 0.0004 M of 2–aminopurine with 0.00005 M of 5–iododeoxyuridine. There has not yet been time to test for induced mutations in the progeny of plants so treated.

In a symposium on "Mutation and Plant Breeding" it is appropriate to make reference to the use of directed mutation. This has been an objective of plant breeders since artificial induction of mutations became possible. Today we are able, by a choice of mutagenic agents, to exercise some selectivity in differentially inducing chromosome breaks compared to gene mutations. This accomplishment is recent and there are many ramifications to explore in theory and to exploit in practice. Applications can be made in spite of our lack of understanding of the phenomena involved.

An ultimate objective would seem to be able to induce one particular mutation to the exclusion of all others. However, a more legitimate objective on the basis of present working hypotheses about the structures that store and transmit genetic information, i.e., sequences of a four (or fewer) word code, is limited control of the spectra of mutations rather than a complete "all-or-none" direction of specific mutations. As we have seen, there is limited evidence that with higher plants this can be accomplished in broad spectra experiments by using a variety of mutagens. The theoretical basis for specificity in forward mutations among gene loci remains obscure, and it will be difficult to devise definite experiments of sufficient resolving power with higher plants. The best material for experimentation would appear at present to be Neurospora where refined tests can be applied to a plant with a hereditary apparatus of apparently the same level of organization as green plants. Experiments using base analogues as mutagens should be particularly informative, if indeed, these compounds are mutagenic in higher forms.

References

1. Alexander, P. 1960. Radiation-imitating chemicals. *Sci. Amer.*, **202:** *No. 1, 99–108.*

2. ———— and Stacey, K. A. 1957. Comparison of the changes produced by ionizing radiations and by the alkylating agents: Evidence for a similar mechanism at the molecular level. *Ann. N. Y. Acad. Sci.*, **68:** *1225–1237.*

3. Altenburg, L. S., and Altenburg, E. 1959. Ratio of visible to lethal mutations produced in Drosophila by phenylalanine mustard. *Nature*, **183:** *699–700.*

4. Auerbach, C. 1960. Chemical mutagenesis in animals. *Chemische Mutagenese (H. Stubbe, editor), Berlin: Akad. Verlag, 1–13.*

5. ———— and Westergaard, M. 1960. A discussion on mutagenic specificity. *Chemische Mutagenese (H. Stubbe, editor), Berlin Akad. Verlag, 116–123.*

6. Benzer, S. 1957. The elementary units of heredity. In *A Symposium on the Chemical Basis of Heredity (W. D. McElroy and B. Glass, editors) Baltimore: Johns Hopkins Press, 70–93.*

7. ————. 1959. On the topology of the genetic fine structure. *Proc. Nat. Acad. Sci.*, **45:** *1607–1620.*

8. ———— and Freese, E. 1958. Induction of specific mutations with 5-bromouracil. *Proc. Nat. Acad. Sci.*, **44:** *112–119.*

9. Brenner, S., Benzer, S., and Barnett, L. 1958. Distribution of proflavin-induced mutations in the genetic fine structure. *Nature*, **182:** *983–985.*

10. Callaghan, J. J., and Grun, P. Incorporation of C^{14} labeled maleic hydrazide into the root tip cells of *Allium cepa, Vicia faba*, and *Tradescantia paludosa*. *Jour. Biophys. and Biochem. Cytol.* In press.

11. Clayton, G., and Robertson, A. 1955. Mutation and quantitative variation. *Amer. Nat.*, **89:** *151–158.*

12. Daly, K. 1960. The induction of quantitative variability by γ radiation in *Arabidopsis thaliana*. Abstract in *Genetics, 45: 983.*

13. Demerec, M. 1953. Reaction of genes of *Escherichia coli* to certain mutagens. *Symp. Soc. Expt. Biol.*, **7:** *43–54.*

14. ———— and Hartman, P. E. 1959. Complex loci in microorganisms. *Ann. Rev. Microbiol.*, **13:** *377–406.*

15. ————, Lahr, E. L., Balbinder, E., Miyake, T., Ishidsu, J.,

Mizobuchi, K., and Mahler, Brigitte. 1960. Bacterial genetics. *Carnegie Inst. Yearbook*, 59: *426–441*.

16. Dunn, D. B., and Smith, J. D. 1954. Incorporation of halogenated pyrimidines into the deoxyribonucleic acids of *Bacterium coli* and its bacteriophages. *Nature*, 174: *305–306*.

17. Ehrenberg, L. 1960. Chemical mutagenesis: Biochemical and chemical points of view on mechanisms of action. *Chemische Mutagenese (H. Stubbe, editor) Berlin: Akad. Verlag*, *124–136*.

18. ————, Gustafsson Å., and Lundqvist, U. 1959. The mutagenic effects of ionizing radiations and reactive ethylene derivatives in barley. *Hereditas*, 45: *351–368*.

19. Fahmy, O. G., and Fahmy, M. J. 1959. Differential gene response to mutagens in *Drosophila melanogaster. Genetics*, 44: *1149–1171*.

20. ————, ————. 1960. Cytogenetic analysis of the action of carcinogens and tumor inhibitors in *Drosophila melanogaster:* VII. Differential induction of visible to lethal mutations by related nitrogen mustards. *Genetics,* 45: *419–438*.

21. ————, ————. 1960. Cytogenetic analysis of the action of carcinogens and tumor inhibitors in *Drosophila melanogaster:* VIII. Selective mutagenic activity of S-2-chloroethylcysteine on the spermatogonial stages. *Genetics*, 45: *1191–1203*.

22. Favret, E. A. 1960. Spontaneous and induced mutations of barley for the reaction to mildew. *Hereditas*, 46: *20–28*.

23. Freese, E. 1959. The specific mutagenic effect of base analogues on phage T4. *Jour. Mol. Biol.,* 1: *87–105*.

24. ————. 1959. The difference between spontaneous and base-analogue induced mutations of phage T4. *Proc. Nat. Acad. Sci.,* 45: *622–633*.

25. ————. 1959. On the molecular explanation of spontaneous and induced mutations. *Brookhaven Symp. in Biol.,* No. 12: *63–73*.

26. Glover, S. W. 1956. A comparative study of induced reversions in *Escherichia coli. (Demerec, M., et al.) Genetic Studies with Bacteria, Carnegie Inst. Wash. Pub.,* 612, *121–136*.

27. Graf, G. E. 1957. Chromosome breakage induced by X-rays, maleic hydrazide and its derivatives. *Jour. Hered.,* 48: *155–159*.

28. Gustafsson, Å. 1960. Chemical mutagenesis in higher plants. *Chemische Mutagenese (H. Stubbe, editor) Berlin: Akad. Verlag*, *14–29*.

29. Hagberg, A., Gustafsson, Å., and Ehrenberg, L. 1958. Sparsely contra densely ionizing radiations and the origin of erectoid mutations in barley. *Hereditas,* 44: *523–530*.

30. Handschumacher, R. E., and Welch, A. D. 1960. Agents which influence nucleic acid metabolism. In *The Nucleic Acids (E. Chargaff and J. N. Davidson, editors), New York: Academic Press Vol. 3, 453–526.*

31. Heiner, R. E., Konzak, C. F., Nilan, R. A., and Legault, R. 1960. Diverse ratios of mutations to chromosome aberrations in barley treated with diethyl sulfate and gamma rays. *Proc. Nat. Acad. Sci., 46: 1215–1221.*

32. Heslot, H. 1960. *Schizosaccharomyces pombe:* un nouvel organisms pour l'etude de la mutagenese chimique. *Chemische Mutagenese (H. Stubbe, editor) Berlin: Akad. Verlag, 98–105.*

33. ———, Ferrary, R., Levy, R., and Monard, C. 1959. Recherches sur les substances mutagenes (halogeno-2 ethyl) amines, derives oxygenes du sulfure de bis-(chloro-2 ethyle), esters sulfoniques et sulfuriques. *Compt. Rend. Acad. Sci., 248: 729–732.*

34. Kaudewitz, F. 1959. Production of bacterial mutants with nitrous acid. *Nature, 183: 1829–1830.*

35. ———. 1959. Inaktivierende und mutagene Wirkung salpetriger Saure auf Zellen von *Escherichia coli. Zeits. Naturforsch., 14b: 528–537.*

36. Kihlman, B. A. 1957. Experimentally induced chromosome aberrations in plants: I. The production of chromosome aberrations by cyanide and other heavy metal complexing agents. *Jour. Biophys. and Biochem. Cytol., 3: 363–380.*

37. Kirchner, C. E. J. 1960. The effects of the mutator gene on molecular changes and mutation in *Salmonella typhimurium. Jour. Mol. Biol., 2: 331–338.*

38. Kölmark, G. 1953. Differential response to mutagens as studied by the Neurospora reserve mutation test. *Hereditas, 39: 270–276.*

39. Kreizinger, J. D. 1960. Diepoxybutane as a chemical mutagen in *Zea mays. Genetics, 45: 143–154.*

40. Lawley, P. D. 1957. The hydrolysis of methylated deoxyguanylic acid at pH 7 to yield 7-methylguanine. *Proc. Chem. Soc., Oct.: 290–291.*

41. ——— and Wallick, C. A. 1957. The action of alkylating agents on deoxyribonucleic acid and guanylic acid. *Chem. and Ind. (London), May: 633.*

42. Litman, R. M., and Pardee, A. B. 1956. Production of bacteriophage mutants by a disturbance of deoxyribonucleic acid metabolism. *Nature, 178: 529–531.*

43. Loveless, A. 1959. The influence of radiomimetic substances on deoxyribonucleic acid synthesis and function studied in *Escherichia coli* phage systems: III. Mutation of *T2* bacteriophage as a consequence of alkylation *in vitro:* the uniqueness of ethylation. *Proc. Royal Soc. London, Ser. B,* **150:** *497–508.*

44. ———. 1960. Some observations on the interaction between alkylating agents and phage and their relevance to problems of chemical mutagenesis. *Chemische Mutagenese (H. Stubbe, editor) Berlin: Akad. Verlag, 71–75.*

45. Malling, H., Miltenburger, H., Westergaard, M., and Zimmer, K. G. 1959. Differential response of a double mutant—adenineless, inositollness—in *Neurospora crassa* to combined treatment by ultra-violet radiation and chemicals. *Intern. Jour. Radiation Biol.,* **1:** *328–343.*

46. Merz, T., Cohen, N. S., and Swanson, C. P. 1959. The interaction of localized and randomized chromatid breaks produced by treatment with X-rays and radiomimetic compounds. Abstract in *Genetics,* **44:** *527.*

47. Moutschen-Dahmen, J. and M. 1958. Sur l'evolution des lesions causees par la 8-ethoxycaféine chez *Hordeum sativum* et chez *Vicia faba. Hereditas,* **44:** *18–36.*

48. ———. 1958. L'action du Myleran (di-methane-sulfonyloxy-butane) sur les chromosomes chez *Hordeum sativum* et chez *Vicia faba. Hereditas,* **44:** *415–446.*

49. Mundry, K. W., and Gierer, A. 1958. Die Erzeugung von Mutationen des Tabakmosaikvirus durch chemische Behandlung seiner Nucleinsäure *in vitro. Zeits. Vererbl.,* **89:** *614–630.*

50. Nakada, D., Strelzoff, E., Rudner, R., and Ryan, F. J. 1960. Is DNA replication a necessary condition for mutation? *Zeits. Vererbl.,* **91:** *210–213.*

51. Nakao, Y., Tazima, Y., and Sakurai, Y. 1958. Specificity of interactions between the individual gene locus and the structure of chemical mutagens. *Zeits. Vererbl.,* **89:** *216–220.*

52. Nickell, L. G. 1955. Effects of antigrowth substances in normal and atypical plant growth. *Antimetabolites and Cancer, AAAS, 129–151.*

53. Reiner, B., and Zamenhof, S. 1957. Studies on the chemically reactive groups of deoxyribonucleic acids. *Jour. Biol. Chem.,* **228:** *475–486.*

54. Ross, W. C. J. 1957. *In vitro* reactions of biological alkylating agents. *Ann. N. Y. Acad. Sci.,* **68:** *669–681.*

55. Rudner, R. 1960. Mutation as an error in base pairing. *Biochem. Biophys. Res. Comm.,* 3: *275–280.*

56. ———— and Balbinder, E. 1960. Reversions induced by base analogues in *Salmonella typhimurium. Nature,* 186: *180.*

57. Schuster, H., Gierer, A. and Mundry, K. W. 1960. Inaktivierende und Mutagene Wirkung der Chemischen Veränderung von Nucleotiden in Virus-Nucleinsäure. *Chemische Mutagenese (H. Stubbe, editor) Berlin: Akad. Verlag, 76–85.*

58. ———— and Schramm, G. 1958. Bestimmung der biologisch wirksamen Einheit in der Ribosenucleinsäure des Tabakmosaik virus auf chemischem Wege. *Zeits. Naturforsch.,* 13B: *697–704.*

59. Scossiroli, R. E. 1954. Artificial selection of a quantitative trait in *Drosophila melanogaster* under increased mutation rate. *Proc. 9th Intern. Congr. Genet. Carologia,* 6 *(vol. suppl. pt. II): 861–864.*

60. ————. 1959. Selezione artificiale per un carattere quantitativo in popolazioni di *Drosophila melanogaster* irradiate coll raggi X. *Com. Naz. Ricerche Nucleari, Div. Biol.,* CNB-4: *1–132.*

61. Stacey, K. A., Cobb, M., Cousens, S. F., and Alexander, P. 1957. The reactions of the "radiomimetic" alkylating agents with macromolecules *in vitro. Ann. N. Y. Acad. Sci.,* 68: *682–701.*

62. Steffensen, D. 1957. Effects of various cation imbalances on the frequency of X-ray-induced chromosomal aberrations in *Tradescantia. Genetics,* 42: *239–252.*

63. ————. 1959. A comparative review of the chromosome. *Brookhaven Symp. in Biol.,* 12: *103–118.*

64. Swanson, C. P. 1957. Cytology and Cytogenetics. *Englewood Cliffs, N. J.: Prentice-Hall, Inc., 391–399.*

65. ———— and Merz, T. 1959. Factors influencing the effect of β-propiolactone on chromosomes of *Vicia faba. Science,* 129: *1364– 1365.*

66. Szybalski, W. 1960. The mechanism of chemical mutagenesis with special reference to triethylene melamine action. *Developments in Industrial Microbiology (B. M. Miller, editor), New York: Plenum Press, 231–241.*

67. Tsugita, A., and Fraenkel-Conrat, H. 1960. The amino acid composition and C-terminal sequence of a chemically evoked mutant of TMV. *Proc. Nat. Acad. Sci.,* 46: *636–642.*

68. Wacker, A., Kirschfeld, S., and Träger, L. 1960. Über den Einbau Purinanaloger Verbindungen in die Bakterien Nukleinsäure. *Jour. Mol. Biol.,* 2: *241–242.*

69. Westergaard, M. 1957. Chemical mutagenesis in relation to the concept of the gene. *Experientia,* 13: *224–234.*

70. ————. 1960. Chemical mutagenesis as a tool in macromolecular genetics. *Chemische Mutagenese (H. Stubbe, editor) Berlin: Akad. Verlag, 30–44.*

71. Wettstein, D. von, Gustafsson, Å., and Ehrenberg L. 1959. Mutationsforschung und Züchtung. *Arbeitsgemeinschaft f. Forsch. des Landes Nordrhein-Westfalen,* 73: *7–60.*

72. Zamenhof, S. 1959. The chemistry of heredity. *Springfield, Ill.: C. C. Thomas.*

73. ————. 1960. Effects of heating dry bacteria and spores on their phenotype and genotype. *Proc. Nat. Acad. Sci.,* 46: *101–105.*

74. ———— and Griboff, G. 1954. *E. coli* containing 5-bromouracil in its deoxyribonucleic acid. *Nature,* 174: *307–308.*

75. Zetterberg, G. 1960. The mutagenic effect of 8-ethoxycaffein, caffein, and dimethylsulphate in the Ophiostoma back-mutation test. *Hereditas,* 46: *279–311.*

Comments

AUERBACH: I should like to throw in a word of caution about the so-called "true specificity" of Doctor Smith. I think one has to distinguish between the result, the mutation, and the process by which it is obtained. If one accepts the DNA model of the gene and the base-change model of mutation, then the result—a change of one allele into another—certainly is specific; but this does not mean that the action of the chemical on the DNA is specific. In fact, there is so far no proof that this is so in any case, although there is strong presumptive evidence for it in the case of bacteriophage. In other cases, reported for bacteria and fungi, mutagen specificity might reside at any one of the steps by which the primary chemical event is transformed into a detectable mutant. Doctor Haas has dealt with some of these steps, and there may be more of them. In some cases it could, in fact, be shown that the detected mutagen specificity depended on residual genotype, plating medium, or different kinds of pretreatment. This skeptical attitude to the theoretical interpretations of mutagen specificity does *not* imply a corresponding skeptical attitude to the possibility of obtaining specific types of desirable mutants by chemical treatments. I think, however, that the best hope for success in this field lies in attempts to find specificity at some of the later steps in mutagenesis. I should recommend strongly to investigate the influence of treated stage and experimental conditions on the mutation spectrum produced by a given chemical mutagen.

PETERSON: In terms of specificity (and, although it may be a special case or class of cases), one might consider the "microspecificity" that is involved in mutable gene systems in corn. The specificity involves the mutable systems concerned with a specific locus—example a_1. Among the different strains of maize there are present at this locus five or six different systems which have been analyzed with their own controllers of mutability. Each of these controllers has their own specific mutator. There is no interaction between members of different systems. Here is a case where the controller is responsive to its own specific activator. Note that the specificity does not only involve the locus, but the specific controlling elements.

COE: Although I do not work with mutability factors, I feel compelled to point out some aspects of these systems in relation to mutational specificity. First, these are broad-spectrum mutagens. In the work by McClintock originally, and since then in studies by Kramer, Nuffer, and others, these systems have been used to induce a variety of changes. Still, there is an interesting quantitative specificity involving the chromosome on which the factor is located, according to Brink. Second, these are highly specific mutagens, as shown in the original case reported by Rhoades for Dt and A in maize, in further study of A by Nuffer, and in the studies of wx by Sprague. In these cases, a mutable system, once induced, permits many variations in the expression of the affected factor. There is little doubt that employment of variants induced originally by mutability factors and further affected by subsequent mutations can provide specificity at least as great as that provided by present mutagens known to affect higher plants.

Increasing the Efficiency of Mutation Induction[1]

R. A. NILAN AND C. F. KONZAK

Washington State University, Pullman, Washington

G REAT STRIDES have been made in the study and use of induced mutations for the improvement of agricultural crops. This success has spurred researchers to accelerate their pace toward obtaining greater efficiency of mutation induction for plant improvement programs of the future.

A principal approach to this endeavor is through investigations of the induced mutation process. The specific objectives of these investigations are (a) to increase the total induced mutation yield through increasing dose tolerance of tissues and through altering the ratios of mutations to chromosome aberrations, and (b) to control and direct the induced mutation process for the production of desired mutations.

These objectives are being pursued through the control and manipulation of secondary factors which alter the response of tissues to radiation through the use of certain chemical mutagens that induce effects different from radiation and through the transfer of plant-cell compounds from radio-resistant to radio-sensitive species.

Several other approaches to increasing mutation yield and to increasing the economic feasibility of artificial mutagenesis in future plant breeding programs have been investigated. These include (a) the use of pollen and embryos in mutation induction, (b) diploidization of loci in polyploids by mutagen treatment, (c) alteration of reproductive mechanisms (especially self-incompatibility and apomixis) by mutagen treatment, and (d) development of efficient recurrent irradiation techniques. Such approaches are concerned with the appearance, detection, and selection of induced mutations rather than with the induced mutation process *per se*. However, they are so intimately related to the problem of increasing the efficiency of mutation induction that they must be considered in this discussion.

---

[1]Scientific Paper No. 2065 Washington Agricultural Experiment Stations, Pullman, Wash. Research supported by U. S. Atomic Energy Commission Contract AT (45-1)-353 and U. S. Public Health Service Grant A-2184.

The results already obtained from these investigations concerned with increasing the efficiency of mutation induction as well as indications of necessary future research will be the subject of this paper.

Investigations Related to the Induced Mutation Process

The manipulation of the induced mutation process appears to be a promising way for increasing the efficiency of mutation induction in plant improvement programs. Here the prime objective is to increase total mutation yield. Eventually we may obtain sufficient control and direction over the induced mutation process so that specific mutations may be produced.

Some progress toward both objectives has been achieved through the control and manipulation of secondary factors that influence the response of plant tissues to ionizing radiation. Progress has been achieved also through the use of certain chemical mutagens, such as diethyl sulfate and ethyl methane sulfonate, which induce a spectrum of effects different from radiation.

Plant radiobiological studies have shown that there are many factors that alter the response of tissues to the sparsely ionizing radiations such as X and gamma rays. Of these, the factors that have been more intensively investigated (21, 29)[2] are genotype, age, stage of cellular development, chromosome number and size, nuclear volume, moisture, temperature, atmosphere (oxygen, nitrogen, carbon dioxide, hydrogen sulfide, etc.), infra red radiation, and chemicals (colchicine, cysteamine, etc.) (21, 29).

Most modifying factors, however, do not seem to change the response of plant tissues to neutrons and other densely ionizing types of radiation (21, 29). Therefore, in mutation breeding, neutrons give repeated and predictable results regardless of the physiological condition and environment of the plant tissue. In spite of this advantage, neutron treatments cannot be experimentally modified for increasing total mutation yield and altering the induced mutation process. In this section, therefore, reference will be made only to the investigations involving X and gamma radiation.

In the investigations of the influence of secondary factors in irradiated tissues, the seed has been the chief experimental material. This

---

[2] See References, page 455.

is because the seed has several unique properties not possessed by the more actively metabolizing tissues, such as the vegetative bud or fresh pollen. One important characteristic of the seed in these radiobiological studies is its adaptability. It can be irradiated under a range of conditions that greatly alter the cellular environment. For instance, the seed can be desiccated, soaked, or frozen. It can be maintained under a vacuum, almost free of oxygen, or under high pressures of oxygen or other gases for extended periods. When dry it is resting, almost biologically inert, and the severe environmental treatments apparently cause little or no biological damage. After such treatments and controlled rehydration, the irradiated seed can be measured for damage using several biological criteria.

Rigid controls of environmental conditions before, during, and after exposure to radiation provide a means for learning about the specific modifying factors that influence the degree of radiation damage. Controls are also necessary for revealing the basic physical and chemical processes responsible for the cell damage incited by radiation (23, 32).

For other reasons, such as ease of handling during irradiation, chemical treatment, and culture, the seed has been the most widely used plant organ for the induction of mutations in plant breeding. Thus, the fundamental studies of the induced mutation process in the seed can provide pertinent information which may have an immediate and direct application for increasing the efficiency of mutation induction in crop plants.

Among the seeds used for radiobiological and chemical mutagen studies, the barley seed *(Hordeum vulgare* or *Hordeum distichum)* has been by far the most useful (29). The chief advantage in using this seed in these studies is that the effects caused by radiation and by chemical mutagens can be measured in terms of several criteria. These include the linear rate of M_1 seedling growth over a finite period; survival of M_1 plants following treatment and frequencies of seedling leaf flecking and cholorophyll-deficient chimeras in the M_1 plants; the number of spikes per M_1 plant and number of seedlings produced in the M_1 plant and spike; frequencies of chromosome bridges and fragments in the shoot-tips of mutagen-treated seeds; chromosome translocation and inversions at meiosis in the M_1 plants; and the frequency and proportion of chlorophyll-deficient seedling mutations

among the M_2 progenies. These criteria present a broad base for interpreting the effects of the radiation and the interaction between radiation effects and secondary factors and for comparative analysis of the
action of chemical mutagens on plant cells. By measuring both chromosomal and genetic damage induced by the same treatment, one
may obtain a greater understanding of the nature and control of the
induced mutation process.

Increasing Induced Mutation Yield

An increased total of induced mutations has been sought
through techniques that increase the radiation dose tolerance of tissues and reduce the amount of chromosome damage without an associated reduction of induced mutation frequencies in the treated cells.
These techniques have involved the control and manipulation of
secondary factors in irradiated seeds, the transfer of plant-cell compounds from radio-resistant to radio-sensitive species, and the use of
the chemical mutagens diethyl sulfate and ethyl methane sulfonate
which induce high mutation frequencies but relatively few chromosome aberrations.

Increasing the radiation dose tolerance

Under normal conditions, plant tissues have distinct levels of
tolerance to doses of ionizing radiation. Doses of radiation above these
levels lead to such low cell survival that most mutations are lost within the plant. These levels, which vary from species to species, are
governed to a large extent by numerous secondary factors, only a few
of which are understood. Nevertheless, through proper manipulation of a few of them in irradiated seeds, it has been possible to
increase the radiation dose tolerance, and, hence, mutation yield.

Extensive studies have shown that after-effects, as they are related
to time, oxygen, moisture, and temperature, are a most important
influence on the degree of damage in X- and gamma-rayed seeds
(3, 5, 6, 9, 10, 21, 23, 29, 30, 31, 32). Through proper control and
manipulation of oxygen, moisture, and temperature before, during,
and after irradiation, the dose tolerance of the seed has been greatly
altered. Because of the several recent and extensive reviews on this
subject, only a brief summary will be presented here.

Oxygen-effects and after-effects in irradiated barley seeds increase
radiation-induced damage, which in turn causes low cell and plant

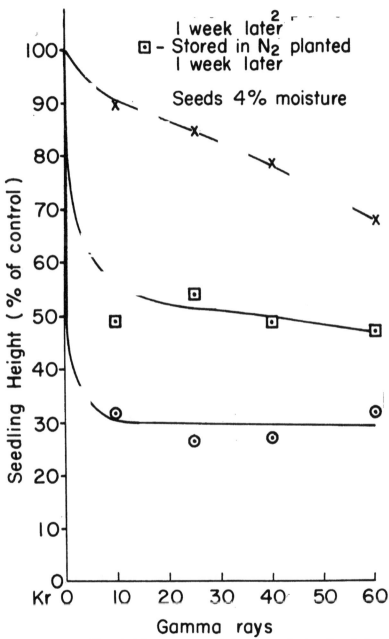

FIGURE 1.—*Seedling height response of barley seeds irradiated from 10 to 60 Kr and then stored in N_2 and O_2 for 1 week before planting.*

survival. Typical oxygen- and after-effect data are presented in Figure
1. Seeds at 4 per cent moisture were gamma-rayed from 10 Kr to 60 Kr
and then either planted immediately or stored in oxygen or nitrogen
for 1 week b$_e$fo$_{re}$ planting. Severe after-effects occurred in the oxygen-
stored seeds, while considerably fewer after-effects occurred in the
nitrogen-stored seeds. It is also interesting to note that a typical dose
response was exhibited by the seeds planted immediately after radiation, but no dose response occurred in the irradiated, stored seeds.

These oxygen- and after-effects now can be controlled (32). They
occur only slightly in seeds with over 12 per cent moisture, and are
even more reduced if these moist seeds are partially evacuated of
oxygen before irradiation and then irradiated at dry ice ($-78°$ C)
temperature. Controlling both the oxygen- and after-effects also
depends on the rehydration of the irradiated seeds in oxygen-free
water. The length of time necessary for rehydration depends upon
temperature of the water. Usually 1½ to 2 hours at $30°$ C is adequate.
Longer rehydration may be required for very dry seeds. Typical
results of these environmental controls on irradiated barley seeds are
shown in Figure 2.

By manipulating the oxygen and moisture content and controlling the after-effects, seeds can be made to tolerate and survive high

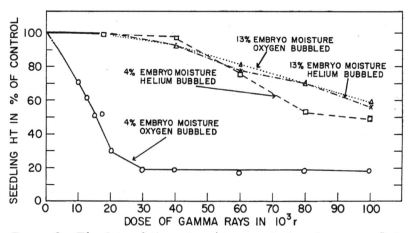

FIGURE 2.—*The interrelation of moisture content and oxygen effect
when seeds are frozen in dry ice ($-80°C$) prior to and during irradiation.*

radiation dosages. For instance, air-dried barley seeds, about 9 per cent moisture, given no proper pre- or post-treatment to control the oxygen- and after-effects can not survive 20 Kr of X-rays (even less if stored), whereas seeds given proper treatments to control after-effects can survive dosages of 80 to 100 Kr (23, 24). Mutation yields from these latter dosages are the highest yet recorded for irradiated barley seeds. Furthermore, with these treatments, more predictable and reproducible results have been obtained in seed irradiation experiments.

Another important means to alter the dose tolerance of plant tissues is the application of certain compounds that affect plant response to radiation. Only recently has this possibility been investigated (1, 2). Soaking pine seeds in extracts of mustard seeds increased their LD_{50}, where soaking mustard seeds in extracts of pine seeds reduced their LD_{50}. Protective effects were obtained after treatment with both acidic and basic fractions from mustard seeds. Furthermore, the phenolic and neutral fractions of the mustard seed extracts contained germination inhibitors. A study of the chemical fractions of pine seed extract revealed that the sensitizing action was centered in the organic acid fraction and was possibly a result of the presence of oleic and linoleic acids. It was suggested (1) that peroxides produced by irradiation of the unsaturated fatty acids may be the primary cause of the sensitization.

To date, the LD_{50} has been measured only in terms of seed germination. The influence of these radio-resistant or radio-sensitive fractions on the induced-mutation process has not been investigated. It appears, however, that this line of research may open up a whole new approach to the manipulation and possible control of the radiation tolerance of plants for increased mutation induction.

Increasing the ratio of induced mututations to chromosome aberrations

Possibly one of the chief causes of cell death in tissues irradiated at high dosages are chromosome aberration. Caldecott (3) has shown that chromosome aberrations in barley increase exponentially, whereas the mutations increase linearly with dose. This means that at high doses, chromosome aberrations concurrently induced limit the mutation yield. Therefore, for radiation to become more efficient for inducing mutations in plant breeding, the frequency of chromosome aberrations in irradiated cells must be decreased. Such a decrease will lead to a more favorable ratio of mutations to chromosome aberrations.

Some progress has already been achieved in decreasing frequencies of chromosome aberration in irradiated cells. Several reports (7, 12, 13, 17, 28) have indicated that manipulation of various factors of seed environment may decrease X-ray-induced chromosome aberration frequencies and maintain or increase the M_1 plant survival and M_2 mutation frequency.

A convincing demonstration of a possible control over the ratio of radiation-induced mutations to aberrations has been heat-shock post-treatments in our laboratory (23, 24).[3] Seeds were frozen in dry ice at about $-78°$ C, exposed to 80 Kr and 100 Kr, then immediately plunged into water at 60° C for 1 minute, and hydrated in distilled oxygen-free water at 32° C for $1\frac{1}{2}$ hours. Compared with the non-shocked treatments, there was an appreciable increase in survival of the M_1 plants from the heat-shocked seeds. Mutation yield from these treatments was higher than any previously recorded for irradiated barley seeds. It was determined that the improved survival rate after heat-shock was in part due to a reduction of chromosome aberrations (Table 1). We have more recently found that certain growth-regulator treatments after X-radiation also reduce chromosome aberration frequencies without decreasing mutation frequencies.

The mechanism conditioning this more favorable ratio of mutations to chromosome aberrations is yet unknown. Apparently it must affect processes that either cause mutations and gross chromosome aberrations at different frequencies or that result in restitution or repair of chromosome damage.

Considerable support for the possibility of obtaining more favorable ratios of mutations to chromosome aberrations in irradiated seeds has come from experiments involving certain chemical mutagens. It must be recalled here that most chemical mutagens are considered to be radiomimetic since their effects and even possibly their basic mechanisms of action are similar to those of ionizing radiations. Diethyl sulfate and ethyl methane sulfonate have produced exceptionally high frequencies of mutations in barley (11, 15, 18, 19, 20,

---

[3]More recent studies have established that the heat shock response, at least in part, depends on oxygen and moisture in the pre-radiation storage environment of the seeds.

TABLE 1.—RESPONSE OF IRRADIATED BARLEY SEEDS TO POST-RADIATION
HEAT-SHOCK TREATMENT.

Aberrations	80 Kr.				100 Kr.			
	Dry seeds		Moist seeds		Dry seeds		Moist seeds	
	Shock	No shock	Shock	No shock	Shock	No shock	Shock	No shock
Cells scored	688	694	302	301	198	194	382	393
Rods/cell	2.62	3.16	1.49	3.81	4.34	6.55	2.34	3.18
Dots/cell	2.12	2.48	0.97	2.23	3.28	3.96	1.69	2.27
Total fragments/cell	4.74	5.64	2.46	6.04	7.62	10.52	4.03	5.45
Bridges/cell	0.52	0.80	0.53	0.45	0.50	0.74	0.84	1.25
Seedling response	54.6	56.8	61.6	63.3	—†	—†	45.4	—†
Survival*	74.5	47.2	61.0	66.0	28.0	9.0	67.0	15.0
Mutations: Per cent/ M₁ plant	29.7	16.7	28.2	27.6	32.3	14.1	26.1	15.6
Per cent/ M₁ spike	12.7	8.5	11.3	12.9	16.9	9.0	13.5	10.7
Per cent/ M₂ seedling	2.6	1.8	2.6	3.2	4.1	1.0	2.7	4.0

*Percentage of control.
†Not recorded.

24). (See Figure 3.) The important new finding, however, is that these chemicals produce very few gross chromosome aberrations of the interchange type. (18, 24). (See Table 2.)

Recently, it has been found that chromosome bridges and, less frequently, fragments and bridges occur at anaphase I in M_1 plants

TABLE 2.—FREQUENCIES OF CHROMOSOME INTERCHANGES IN SHOOT-TIP CELLS OF TREATED
SEEDS AND IN POLLEN MOTHER CELLS OF M₁ PLANTS FOLLOWING TREATMENT OF
BARLEY SEEDS WITH DIETHYL SULFATE AND GAMMA RAYS.

Treatment	Field survival*	Cells scored	Fragments per cell	Bridges per cell	Spikes scored	Translocations observed
Gamma radiation 60 Kr.	75.8	300	3.40	0.52	175	46
Diethyl sulfate (1) 1½ hr. 30°C	75.1	300	0.00	0.00	—	—
Diethyl sulfate (2) 1½ hr. 30°C†	—	591	0.11	0.02	175	4

*Percentage of control.
†More severe treatment with new, more purified chemical.

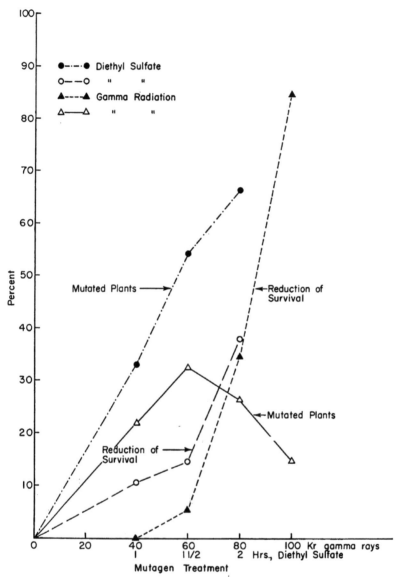

FIGURE 3.—*Comparison of effects of diethyl sulfate and gamma rays on survival of M_1 plants and frequencies of chlorophyll-deficient mutations in M_2 barley seedlings.*

from the chemical treatments. These aberrations indicate that inversions have been induced by the chemicals. This is of considerable interest since inversions are relatively rare in plants from irradiated seed. Furthermore, the finding that these chemicals induce a higher frequency of intra-chromosomal compared to inter-chromosomal type of aberration than radiation suggests a more subtle and delayed effect of the chemicals on the chromosomes.

It is already apparent that the effectiveness and efficiency of the chemical agents may be greatly influenced by modifying factors. For example, the uniformity in the growth of seedlings from seeds treated with diethyl sulfate seems to be measurably increased by the addition of certain divalent cations to the treatment solution. These results suggest that the hydrolysis product, ethyl sulfuric acid, in the treatment solution may chelate divalent cations in the cell membranes and facilitate the penetration of the active agent into the cells of the seed.

It is also interesting to speculate that the hydrolysis products from both diethyl sulfate and ethyl methane sulfonate may have important roles in the mutagenic and cytogenetic activity of these agents. Alcohol might be expected to act as a surface-active agent aiding penetration of the alkylating agents. The ethyl sulfuric and methane sulfonic acids, weak chelating agents, may affect cell membrane permeability. Also, by their chelating action on the divalent cations bonding the DNA, they may cause an opening-up of the multi-stranded chromosome, permitting the alkylation of additional as well as internal sites, as suggested by the Steffensen model (38). It is now already indicated, however, from the results of Strauss (39) that in bacteriophage at least (which seems to have a double or single-stranded chromosome), the chelating effect does not appreciably influence the frequency of induced mutations but reduces survival. Thus, the effect hypothesized may relate only to higher organisms.

It is also possible that chromosome aberrations of the type observed may result from the temporary chelation of the divalent ions bonding the DNA rather than from the alkylation of bases or phosphate groups. However, what is more important is that these possibilities can be subjected to experimental test, and future studies may provide even more exciting information on the nature of chemical reactions leading to chromosome aberrations and mutations.

It is obvious that more intensive studies on the nature and origin of chromosome aberrations and repair of chromosome damage are required before we can more effectively reduce or eliminate chromosome aberrations and increase mutations in irradiated cells. Investigations pursued by Wolff and his collaborators (42) have shed some light on the factors that govern the rejoining and restitution of chromosome breaks. However, recent papers by Revell (35, 36) suggest that radiation-induced chromosome reunions and exchanges or incomplete reunions and exchanges are derived from possible chemical bonding initiated by the ionizing radiation. If this suggested mechanism proves to be real, it could mean that radiation-induced changes in the chromosome may not result in aberrations until the late prophase of the irradiated cells. This suggests that a relatively long period during mitosis would be available for altering the course of radiation damage to the chromosomes. Experiments designed to test and then use this new hypothesis for the reduction or elimination of gross chromosome aberrations and for increasing the efficiency of mutation induction are now underway in our laboratory.

Directing the Induced Mutation Process

The greatest concentration of effort in directing the induced mutation process has been through the manipulation of secondary factors in irradiated seeds and the use of chemical mutagens. Here again, these studies are only possible because of the wide variety of conditions and controls that can be applied to seeds.

Indications that the spectrum of chlorophyll-deficient mutations in barley can be altered have been published by the Swedish group (41) in the past several years. Small shifts of the mutation spectrum have been caused by certain variations of the experimental conditions during X-irradiation. It also has been reported that the chlorophyll-deficient mutations produced by neutrons may show a different spectrum from those induced by X-rays. However, these results have not always been reproducible. It has not been determined whether the results are due to alterations in the sensitivities of individual genes or to unequal selection of these mutations during the ontogeny of the plant.

Probably some of the most convincing evidence that different mutagenic treatments can produce differential mutability of loci is from studies of the *erectoid* mutations of barley. Treatment with densely ionizing neutrons will induce more *erectoid* mutants than treatments with sparsely ionizing X and gamma rays. Furthermore, neutron treatments will affect only some individual *erectoid* loci; with X-ray, other loci will be affected (16). Whether these results are due to differential sensitivities of specific loci, to changes in modifying genes at other loci, to environmental or other modifying factors, or to differential selection of certain mutations, the practical importance of these findings still remains—a specific changed type can be more readily selected from one kind of treatment than from another.

Some of our investigations on the influence of moisture and heat shock on radiation damage also show a shift in the mutation spectrum among M_2 chlorophyll-deficient seedlings. From dry seeds but not from moist, the proportions of *albina* and *viridis* to total mutants decreased with increasing doses. On the other hand, the *albo-viridis* and *xantha* mutations from the dry seeds increased with increasing radiation dose. The mutation spectrum in the heat-shock experiments more closely resembled the spectrum obtained from seeds exposed to lower doses (24). There is some indication that these alterations in the proportion of mutation types are associated with the selective influence of induced chromosome aberrations. A previous report (10) has suggested that the proportion of *viridis* mutants was smaller and that of the *xantha* type larger at higher radiation doses. This alteration appeared to be associated with spike sterility which, in turn, may be a result of chromosome aberrations.

After treatments of barley seeds with several chemical mutagens, considerable variation in the proportion of M_2 chlorophyll-deficient seedling mutants has been observed (11, 15, 41). In our studies, it has been found that the gamma-radiation spectrum of induced-mutation types differs from that of diethyl sulfate (Table 3). This difference might be related to the fact that very few chromosome aberrations are produced by the chemical. In the plant from irradiated seed, aberrations would influence the selection and recovery of induced mutations. On the other hand, it is possible that in the DNA molecule some of these chemicals can bring about a specific change which is expressed as a specific type of mutation. Stronger evidence that there is a differential sensitivity of individual genes

TABLE 3.—COMPARISON OF MUTATION SPECTRA INDUCED BY DIETHYL SULFATE
AND GAMMA RADIATION.

Phenotypic categories	Number of mutations		Per cent of total mutations		99% confidence interval	
	Gamma radiation	Diethyl sulfate	Gamma radiation	Diethyl sulfate	Gamma radiation	Diethyl sulfate
Albina	272	262	48.6	30.3	43.2–54.8	24.8–35.5
Viridis	211	387	37.7	44.8	31.5–42.8	39.2–50.9
Xantha	25	88	4.5	10.2	2.2– 6.7	6.8–14.0
Tigrina	33	52	5.9	6.0	3.4– 9.0	3.6– 9.2
Striata	19	75	3.4	8.7	1.8– 6.1	6.1–12.8
Total	560	864				

to chemical treatment has been obtained from the experiments of Westergaard and his associates with back mutations with Neurospora (40).

Investigations Related to Appearance, Detection, and Selection of Induced Mutations

Several other approaches to greater efficiency and to increased economic feasibility of artificial mutagenesis in plant breeding should be explored more intensely. These include (a) the use of pollen, zygotes, and embryos in mutation induction; (b) diploidization of loci in polyploids by mutagen treatment; (c) alteration of reproductive mechanisms, especially self-incompatibility and apomixis, by mutagen treatment; and (d) development of efficient recurrent mutagen treatment techniques. Most of these are concerned with the appearance, detection, and selection of mutations rather than with the induction of mutations *per se*. However, they will be considered briefly here since progress along these lines can extend the usefulness of mutagen techniques.

Irradiation of Pollen, Zygotes, and Embryos

Most mutation breeding experiments have been initiated either with seeds of sexually propagated crops or with buds of cuttings in vegetatively propagated plants. However, both seeds and buds are multi-cellular, complex tissues presenting certain difficulties for induced mutation detection and selection. These difficulties relate to (a) the chimeral nature of mutated sectors; (b) a rapid loss of induced cytogenetic changes during subsequent mitoses and, in sexual crops, during meiosis of the M_1 plant; and (3) overgrowth of mutated cells by nonmutated cells (8, 33).

Greater efficiency might be obtained by irradiating a plant when half or all of its genes are located in an individual cell. This immediately suggests the application of mutagens to pollen or zygotes. However, pollen, zygotes, and embryos have been used for inducing mutations in very few physical or chemical mutagen experiments. Conger (8) has discussed in some detail a possible improved method for inducing mutation by using pollen instead of seeds, and some results have been obtained.

Much more work needs to be done on this technique, since its efficiency does not approach modern seed irradiation technics. From exploratory studies, Konzak (21) has noted that pollens of diploid maize and barley suffered such pronounced damage from radiation that high seedling mutation rates were not obtained. In one study (22), barley pollen was exposed to doses of X-rays, thermal neutrons, and ultraviolet radiation. Records were kept of the number of seeds set, number of aborted and shriveled seeds produced, the number and condition of the plants grown from the M_1 seed, and the number of second generation progenies showing mutations. M_1 plants carrying chromosome translocations or inversions and the frequencies of semi-sterility in M_2 and M_3 progenies were also recorded. Although the numbers studied, totalling approximately 1,200 seeds from irradiated pollen, were not large enough for firm conclusions, a disappointingly low number of mutations was recovered for the effort expended. Moreover, the ratio of mutations to chromosome translocations induced by X-rays and neutrons in pollen was considerably less than observed in seed. Several chromosome translocations were induced by the ultraviolet radiation treatments, but the frequency of mutations observed also was relatively low compared with that obtained from X-ray treatment of seeds.

A renewal of extensive work with pollen would seem to be warranted, however, if for no other reason than to take advantage of the fact that the mathematical analysis of induced mutation rates can be more critically estimated. Also, with irradiated pollen, an M_1 plant carrying a mutation is almost always completely heterozygous for the mutations rather than a sectorial chimera as is the case with irradiated seeds. Moreover, better comparisons would be possible with chemical and ultraviolet treatments of pollen. Before these studies can be practical, however, certain major difficulties must be surmounted. These include maintaining pollen viability in some species and adequate control over modifying factors.

As early as 1930, Stadler (37) compared the probable advantages of inducing mutations in zygotes or pro-embryos over those of seeds. However, because of several difficulties inherent in these materials, including the low dose tolerance, no further studies were conducted to explore the merit of zygotes or even immature embryos in higher plants as organs for the induction of mutations by use of radiation.

In the past two years, the use of these organs for mutation induction in barley has been investigated with some promising results (26, 27). The irradiation of pro-embryos has resulted in a significant increase in the size of the M_2 mutant population and in the frequencies of tillers of a single plant with the same mutation compared with that obtained from treated seed. These advantages indicate that irradiated zygotes or pro-embryos may have definite practical significance for the induction of mutations in certain self-fertilizing crops. However, the efficiency of this technique will need considerable improvement to approach the results now possible with seed irradiation.

Diploidization of Polyploids

Another means of facilitating the detection and selection of mutations is by the diploidization of loci in polyploids through mutagen treatment. Many of our crop species are polyploids; thus the duplication, triplication, etc., of loci, restrict the segregation and appearance of induced mutations. It may be possible to alter these loci through mutagen treatment so that they behave as in a diploid. Their alleles would then segregate in normal Mendelian ratios, and induced changes at these loci would be easier to detect. Some success with this technique has already been achieved and reported at this symposium (6) and elsewhere (4).

Alteration of Reproductive Mechanisms

The reproductive system in cross-fertilized species influences and limits the segregation and recovery of induced mutations. For this reason, very little research has been conducted to determine the possibilities of mutation induction in these species. In the self-incompatible species, this limitation might be overcome by producing self-compatible or self-fertile strains of these plants through physical or chemical mutagen-induced changes of the self-incompatible (S) alleles. A precedent for this approach may be found in a paper by Lewis (25).

Recurrent Mutagen Treatments

Another means to produce higher mutation yields is mutagen treatment of successive seed generations prior to selection of mutations. In a self-fertile crop, such as barley, the procedure would be to treat the seeds from fertile inflorescences at each generation. As shown by Gaul (14), the frequency of point mutations or mutations not associated with gross aberrations is independent of fertility. Thus, through successive mutagen treatments of seeds from only fertile inflorescences, it should be possible to accumulate high frequencies of mutations in a line.

This approach, as outlined by Caldecott (4) in more detail for irradiation treatments, is essentially a mechanical means for removing induced chromosome aberrations while increasing the total accumulated radiation dose that can be applied. It is probably far less efficient than the controlled seed irradiation procedures described earlier, but since it can be applied independently from other controls or treatments, it has special merit.

Summary and Conclusions

Considerable progress has been made in recent years in increasing the efficiency of mutation induction by radiation in seeds. Greater precision, repeatability, and prediction in radiation experiments have been achieved and higher frequencies of induced mutations have been obtained through rigid control of after-effects, moisture, oxygen, and temperature. Furthermore, through control and manipulation of these secondary factors, certain deleterious effects of the radiation, such as low survival and high chromosome aberration frequencies can be reduced; and some control and direction of the induced mutation process may be realized.

Certain chemical mutagens, such as diethyl sulfate and ethyl methane sulfonate, may be more efficient than radiation in certain aspects of mutation induction. In barley seeds, these chemicals induce very high mutation frequencies without an appreciable frequency of gross chromosome aberrations, and apparently produce a slightly different mutation spectra from X or gamma rays.

It is obvious, however, that the required information for still greater progress in increased efficiency of mutation induction in this area is at present only fragmentary. Much more work should be conducted in radiobiology, biochemistry, genetics, and cytology to obtain more adequate information on induced mutations for plant breeders of the future.

In the final analysis, it is apparent that the greatest advances toward increasing the efficiency of mutation induction in plant tissues will come only when the basic mechanisms of radiation and chemical mutagen effects in cells are completely understood. Studies of the influence of secondary factors on radiation damage in seeds have extended our concept of the physical and chemical pathways of effects of ionizing radiation in cells. They have shown that much of the radiation damage in the chromosomes or genes is due not to the direct ionizing event but to intermediate agents (10, 32). Certain of these agents, free radicals, have been detected in seeds following irradiation. Furthermore, they seem to be influenced by the various secondary environmental factors to the same degree as the radiobiological effects. The apparent association of these radiation-induced radicals with biological damage was first demonstrated with barley seeds and now has been confirmed and extended with bacterial spores (34). Further studies on the nature of these intermediate agents are necessary to increase our knowledge of mechanisms that produce radiation damage in seeds.

It is unfortunate that much useful information obtained from the seed radiobiological research has not been generally applied by plant scientists, particularly plant breeders, in seed radiation studies for practical or theoretical purposes. The techniques for controlling secondary factors that have been described here and in more detail in other publications have been ignored to a large extent. And yet it should now be apparent to all who use seeds in radiation experiments that the biological effects of a given dose of radiation are meaningless unless the moisture content of the seed is known and unless oxygen, temperature, and after-effects have been adequately controlled. Certainly, information is now available that will provide greater precision, repetition, and prediction in seed radiation experiments. A discussion of the techniques and methods that should be followed in all seed radiation experiments will be summarized by the authors in an early issue of *Radiation Botany*.

As we consider the needs of future plant breeders in terms of genetic variability and of techniques for manipulating this variability, it is imperative that we keep abreast of modern advances in genetics and biochemistry. Many of these advances may well provide means of inducing mutations in plants with greater efficiency than ever before possible with physical or chemical mutagens. The gene mutation controlling elements of McClintock and Brink, and gene transduction through plant viruses may well be a future means for inducing high frequencies of directed mutations in crop plants. These and other similar possibilities, however, may be completely overshadowed when we consider that deoxyribonucleic acid (DNA), the basic genetic material, may soon be synthesized artificially. When this is accomplished and when methods for its incorporation into reduplicating chromosomes are found, man may have his most powerful means for controlling and directing the heredity of the plants and animals upon which he depends for his survival.

References

1. Bowen, H. J. M. 1961. Effects of seed extracts on radiosensitivity. In *Effects of Ionizing Radiations on Seeds and Their Significance for Crop Improvement. Proc. Sci. Symp. Sponsored by I.A.E.A. and F.A.O., Karlsruhe, Germany, Aug. 8–12, 1960.*

2. ———— and Thick, J. 1960. Factors from seed extracts that modify radiosensitivity. *Radiation Res.,* **13:** *234–241.*

3. Caldecott, R. S. 1959. Post-irradiation modification of injury in barley—its basic and applied significance. *Proc. 2nd Geneva Conf. on Peaceful Uses of Atomic Energy,* **27:** *260–269.*

4. ————. 1959. Irradiation and plant improvement. In *Germ Plasm Resources in Agriculture. Symp. Sponsored by Amer. Assoc. Adv. Sci. Dec. 1959, 1–12.*

5. ————. 1961. Seedling height, oxygen availability, storage, and temperature: Their relation to radiation-induced genetic and

seedling injury in barley. In *Effects of Ionizing Radiations on Seeds and Their Significance for Crop Improvement. Proc. Sci. Symp. Sponsored by I.A.E.A. and F.A.O., Karlsruhe, Germany, Aug. 8–12, 1960.*

6. ———— and North, D. T. 1961. Factors modifying the radio-sensitivity of seeds and the significance of the acute irradiation of successive generations. *This Symposium, 365–404.*

7. ———— and Smith, L. 1952. The influence of heat treatments on' the injury and cytogenetic effects of X-rays on barley. *Genetics,* **37:** *136–157.*

8. Conger, A. D. 1957. Some cytogenetic aspects of the effects of plant irradiations. *Proc. 9th Oak Ridge Reg. Symp., 59–62.*

9. Ehrenberg, L. 1959. Radiobiological mechanisms of genetic effects: A review of some current lines of research. *Radiation Res. Suppl.,* **1:** *102–123.*

10. ————. 1960. Induced.mutation in plants: Mechanisms and principles. *Genet. Agrar.,* **12.** In press.

11. ————. 1960. Chemical mutagenesis: Biochemical and chemical points of view on mechanisms of action. In *Chemische Mutagenese. Abhandl. dtsch. Akad. Wiss., Akademie Verl. Berlin,* **1:** *124–136.*

12. Favret, E. 1959. Mutation research on crop plants in Argentina. *Proc. 2nd Eucarpia Congress, 76–77.*

13. Gaul, H. 1957. Die Wirkung von Röntgenstrahlen in Verbindung mit CO_2, Colchicin, und Hitze auf Gerste. *Zeit. Pflanzenzücht.,* **38:** *397–429.*

14. ————. 1958. Present aspects of induced mutations in plant breeding. *Euphytica,* **7:** *275–289.*

15. Gustafsson, Å. 1960. Chemical mutagenesis in higher plants. In *Chemische Mutagenese. Abhandl. dtsch. Akad. Wiss., Akademie Verl. Berlin,* **1:** *14–29.*

16. Hagberg, A., Gustafsson, Å., and Ehrenberg, L. 1958. Sparsely contra densely ionizing radiations and the origin of erectoid mutations in barley. *Hereditas,* **44:** *523–530.*

17. Hayden, B., and Smith, L. 1949. The relation of atmosphere to biological effects of X-rays. *Genetics,* **34:** *26–43.*

18. Heiner, R. E., Konzak, C. F., Nilan, R. A., and Legault, R. R. 1960. Diverse ratios of mutations to chromosome aberrations in barley treated with diethyl sulfate and gamma rays. *Proc. Nat. Acad. Sci.,* **46:** *1215–1221.*

19. Heslot, H. 1960. Action D'agents Chimiques Mutagènes sur Quel-

ques Plantes Cultivées. In *Chemische Mutagenese. Abhandl. dtsch. Akad. Wiss., Akademie Verl. Berlin,* 1: *106–108.*

20. ————, Ferrary, R., Lévy, R., and Monard, Ch. 1961. Induction de Mutations Chez L'orge. Efficacite Relative Des Rayons, Du Sulfate D'Ethyle, Du Methane Sulfonate D'Ethyle et de Quelques Autrex Substances. In *Effects of Ionizing Radiations on Seeds and Their Significance for Crop Improvement. Proc. Sci. Symp. Sponsored by I.A.E.A. and F.A.O., Karlsruhe, Germany, Aug. 8–12, 1960.*

21. Konzak, C. F. 1957. Genetic effects of radiation on higher plants. *Quart. Rev. Biol.,* 32: *27–45.*

22. ————. Unpublished communication.

23. ————, Curtis, H. J., Delihas, N., and Nilan, R. A. 1960. Modification of radiation-induced damage in barley seeds by thermal energy. *Can. Jour. Genet. Cytol.,* 2: *129–141.*

24. ————, Nilan, R. A., Legault, R. R., and Heiner, R. E. 1961. Modification of induced genetic damage in seeds. In *Effects of Ionizing Radiations on Seeds and Their Significance for Crop Improvement. Proc. Sci. Symp. Sponsored by I.A.E.A. and F.A.O., Karlsruhe, Germany, Aug. 8–12.*

25. Lewis, D. 1946. Useful X-ray mutations in plants. *Nature,* 158: *519–520.*

26. Mericle, L. W. 1960. Irradiation of developing barley embryos at specific stages during ontogeny, its effects on subsequent development and some potential applications of this method to a breeding program. *Barley Newsletter,* 3: *21–22.*

27. ————, Sparrow, A. H., and Mericle, R. P. Unpublished communication.

28. Nilan, R. A. 1954. The relation of carbon dioxide, oxygen, and low temperature to the injury and cytogenetic effects of X-rays in barley. *Genetics,* 39: *943–953.*

29. ————. 1956. Factors governing plant radio-sensitivity. *U.S. A.E.C. Rpt. No. TID-7512. Proc. Conf. on Radioactive Isotopes in Agriculture., U. S. Govt. Printing Office, 151–162.*

30. ————. 1960. Factors that govern the response of plant tissues to ionizing radiation. *Genet. Agrar.,* 12: *283–296.*

31. ————. 1959. Radiation-induced mutation research in the United States of America. *Proc. 2nd Eucarpia Congress, 36–47.*

32. ————, Konzak, C. F., Legault, R. R., and Harle, J. R. 1961. The oxygen effect in X-rayed barley seeds. In *Effects of Ionizing Radiations on Seeds and Their Significance for Crop Improve-*

ment. Proc. Sci. Symp. Sponsored by I.A.E.A. and F.A.O., Karls-ruhe, Germany, Aug. 8–12.

33. Osborne, Thomas, S. 1957. Symposium on Radiation genetics: Mutation production by ionizing radiation. *Proc. Soil and Crop Sci. Fla., 91–107.*

34. Powers, E. L., Webb, R. B., and Ehret, C. F. 1960. Storage, transfer, and utilization of energy from X-rays in dry bacterial spores. *Radiation Res. Suppl.,* **2:** *94–121.*

35. Revell, S. H., 1959. The accurate estimation of chromatid breakage and its relevance to a new interpretation of chromatid aberrations induced by ionizing radiations. *Proc. Roy. Soc. (London) B,* **150:** *563–589.*

36. ————. 1960. Some implications of a new interpretation of chromatid aberrations induced by ionizing radiations and chemical agents. In *Chemische Mutagenese. Abhandl. dtsch. Akad. Wiss., Akademie Verl. Berlin,* **1:** *45–46.*

37. Stadler, L. J. 1930. Some genetic effects of X-rays in plants. *Jour. Heredity,* **21:** *3–19.*

38. Steffensen, D. 1957. Effects of various cation imbalances on the frequency of X-ray-induced chromosomal aberrations in Tradescantia. *Genetics,* **42:** *239–252.*

39. Strauss, B. Unpublished communication.

40. Westergaard, M. 1960. Chemical mutagenesis as a tool in macromolecular genetics. In *Chemische Mutagenese. Abhandl. dtsch. Akad. Wiss., Akademie Verl. Berlin,* **1:** *30–44.*

41. Wettstein, D. von., Gustafsson, Å., and Ehrenberg, L. 1959. Mutationsforschung und Züchtung. *Arbeitsgemeinschaft Forsch. des Landes Nordrhein - Westfallen,* **73:** *7–60.*

42. Wolff, Sheldon. 1960. Problems of energy transfer in radiation-induced chromosome damage. *Radiation Res. Suppl.,* **2:** *122–132.*

Comments

DAVIES: I wish to comment on acute irradiation. We have compared somatic mutation rates at the V^{by} locus in *Trifolium repens* with dose rates from 25 to 80,000 rads per hour, and find that we obtain about 10 times as many with the latter as the former.

LEWIS: If I understood the use of terms mutation yield and mutation spectrum correctly, would you not finally be more interested in spectrum than yield? In other words, the emphasis would be on different mutants rather than repeats of the same mutant. Possibly yield and spectrum are closely correlated.

NILAN: Yes, as we have discussed in the paper we are very much interested in finding ways of altering the mutation spectrum and of directing the induced mutation process toward specific desired changes. However, we think that increasing yield of many mutations in a genotype is also very important. I am afraid we do not have enough information yet to determine the relationship, if any, of yield and spectrum.

BRAWN: Albinos may be due to a variety of causes ranging from point mutations to deletions of some size. Beneficial mutations are not likely the result of deletions and so may differ in environment-treatment response. What is the likelihood that the treatments you have used may be specific for inducing only those mutants (albinos, etc.) which you used as the criteria for measuring mutation rate? Are beneficial mutants induced to the same degree and in the same way?

GROBMAN: Doctor Brawn's question has relevance in regard to resolution limits in the evaluation of different mutagens as to their effects in producing variable mutation spectra. I am particularly disturbed with the presentation in this paper and the preceding ones of chlorophyll mutation data to substantiate differences in mutation spectra due to different mutagens. The synthesis of chlorophylls in chloroplast grana is a very complex biochemical process, conditioned by at least 20 environmental factors, and in barley by at least 200 to 300 different genes, as it appears to be evidenced by Swedish mutation work (and Doctor MacKey may check me on this point). Scoring of chlorophyll mutations would, therefore, be a gross and inefficient manner of defining the limits of particular mutagen-induced mutation spectra. Characters conditioned in their phenotypic experssion by few genes, if selected, would give a much finer resolution of mutation spectra, in studies where differences between different mutagens were of primary interest.

NILAN: On the basis of the high chlorophyll mutation frequencies from diethyl sulfate, we have looked for morphological mutants in mature M_2 barley plants. Last summer high frequencies of these mutants were found. This indicates, I believe, that there is a good correlation between the chlorophyll and morphological mutants in barley.

MERICLE, R. P.: One factor which has not been given much consideration to date at this meeting is that of dose-rate. By this, I do not mean chronic versus acute, but rather, within the framework of acute or semiacute. In our proembryo irradiations of barley, for example, we have achieved more than a 6-fold increase in mutation rate to albinism by a 4-fold decrease in dose-rate, when total dose and irradiation conditions were identical. These mutations were in all cases ones in which the mutant-carrying sector encompassed two or more heads in the X-1 plants, and are probably of a similar nature since segregation ratios in the X-2 were the same. With the continued interest in increasing mutation rates, while at the same time decreasing the frequency of gross chromosomal aberrations, we suggest that, especially in instances when it is desirable to use large radiation doses, the usual dose-rates be reduced by $1/4$, $1/10$, or so. An optimum balance between total dose and rate of delivery should be sought, since lowering the dose-rate too much may result in loss of effective mutation frequencies.

The Efficacy of Mutation Breeding[1]

WALTON C. GREGORY

North Carolina Agricultural Experiment Station, Raleigh, N. C.

IT IS THE purpose of this paper to discuss some of the factors expected to affect the efficacy of mutation breeding and to bring forward data from which comparisons of efficacy may be drawn between hybridization and mutation as methods of generating genetic variance for purposes of selection. The paper will not be concerned with the efficacy of various selection and screening procedures, it being considered that these, important as they are, are by no means restricted to mutation breeding and, if improved, would benefit any method of breeding. The discussion has meaning in man's effort to create plant breeding capital in contrast to the accepted methods of liquidating natural mutational assets accumulated over evolutionary time.

The basic questions are (a) whether the power of present known mutagens is such as to produce the kinds of changes which natural mutation and selection have provided us, and (b) given this power, with what efficacy can it be used with plants of various breeding structure?

The Power of Mutagens

There is reason to believe that despite the different frequencies of similar mutations obtained with different mutagens (15, 14, 54),[2] the chief limiting factor in mutation production and mutant recovery is the genic constitution of the experimental organism and not the type of mutagen used. Thus, for the plant breeder, a knowledge of what might be called mutant expectation in his material may be more important than a resolution of the mechanism of mutational change at the submicroscopic level.

[1]Contribution from the Field Crops Department, North Carolina Agr. Exp. Sta., Raleigh, N. C. Published with the approval of the Director of Research as Paper No. 1253 of the Journal Series. This work was supported by the U. S. Atomic Energy Commission as part of Contract AT–(40–1)–1747.

[2]See References, page 482.

461

Useful mutant occurrences in many kinds of plants have been reported and a number of reviews have been published concerning their usefulness in plant breeding (2, 17, 31, 38, 40, 46, 48, 51, 55). Numerous characteristics, such as grain yield, straw stiffness, chlorophyll type, maturity date, grain weight, plant height, disease resistance, alkaloid content, etc., have been observed to vary significantly under the influence of mutagens. MacKey (31) concluded his paper on mutation breeding in Europe by saying that the evidence attests that *any* agronomic characteristic can be improved by induced mutations. The more recent reviews of Borg, *et al.* (2), Gaul (17), Prakken (40), Scholz (46), and Stubbe (51) have given generally optimistic accounts of the possibilities of mutation breeding.

In interpreting the results obtained with useful individual mutants it should be remembered that the statistical characteristics and therefore interpretations are considerably different in situations where useful deviates are recognized among a great number of variants in an artificially mutated population and in situations where a character is pre-chosen for study. Few *bona fide* cases of the latter can be identified with certainty in the literature on mutation plant breeding. Among these cases, Brock and Latter (6) pre-chose flowering date in their studies on subterranean clover, Oka, *et al.* (39) pre-chose heading date and plant height in studies with rice, and Gregory and Gregory (unpublished) pre-chose flowering response to day length in their study of Hibiscus.

Some other pre-chosen characteristics have not proved so successful as breeding ventures. For example, Gregory (unpublished) pre-chose the reversal of geotropism in the young peanut fruit. Following an individual observation of hundreds of thousands of X_2 and X_3 plants he has not, to date, observed what could be established as a diminution of the geotropic response in peanut fruits. Apple (unpublished) found in three different flue-cured varieties of tobacco, that two of the varieties were naturally very highly susceptible to "black shank", *Phytophthora parasitica nicotianae;* the third had a measurable tolerance to the disease. By making use of artificial inoculation in a combination of seedling flat and field techniques, Apple was able to screen extremely large numbers of individual X_2 plants. After several years of work he has been unable to discover any increase in tolerance to black shank in the two highly susceptible varieties. In the variety bearing some tolerance to the disease significant increases in tolerance have been selected among its progenies. Cooper and Gregory (9) observed an increase in the variability of tolerance to Cercospora leaf spot in peanuts, but this also was in a variety where tolerance was already at a relatively high level.

The conclusion appears inescapable that despite the large number of desirable mutations reported in the literature there have been sought highly desirable characteristics which have not been attained by artificial mutation after extensive trials. Furthermore, since many investigators do not report the negative results of breeding trials, there may be many more instances of inefficacy of mutation breeding than we suspect. It has to be admitted also that with a new character no *a priori* decision can be made as to (a) whether a certain desired change can be had through mutation, (b) how much mutagen should be applied, or (c) how large a population would be required. One might ask, for example, what it would require of mutation breeding to produce a peanut with tendrils. Miracles of mutation or even the more optimistic predictions of breeders simply cannot be realized because of points of no return attained already in the evolution of species.

These considerations appear to contradict the thesis that mutagenic agents simply increase the rate of natural mutation and the logical conclusion therefrom that since nearly every conceivable change has occurred in plants and animals, nearly every conceivable change should occur if sufficient number of mutations were produced. The fallacy lies in the failure to recognize the evolved relationships between variation at the gene level and the acceptance of that variation by the genome and by the organism.

Under the various conditions of nature various organisms have found the balances which worked for them in particular environments. They have opposed the conservatism of linkage to the opportunities of crossing over, the possession of a few large chromosomes to having a large number of small ones, self-fertilization to cross-fertilization, and have evolved many of the possible compromises of different levels of these characteristics.

Mutation, sometimes thought of erroneously as an extra-organismal force, is a characteristic of organisms related particularly to their breeding structures. If the hazard of mutation has been met by different breeding structures in different ways these adaptations affect the expected efficacy of mutation breeding.

Mutation and Breeding Structure

Crossbreds

By *breeding structure* I refer to the manner of reproduction of an organism and the consequent organization of its genome. It is the purpose of this section to point out that efficacy of mutation breeding cannot be divorced from the role of mutation in the evolution of breeding structure. That breeding structure is a result of natural selection was an early contribution of genetics and received its first general summary in Darlington's *Evolution of Genetic Systems* (11). This was followed by more explicit association of mating system with the polygenetic organization of the genome as brought out especially in a series of papers by Mather and given a general summary by him (32). The developments in population genetics during the last two decades (13), together with the indicated genomic control of mutation in plants (3, 4, 5, 16, 33, 34, 35, 42, 43), have suggested that the kind and abundance of mutation itself have evolved in conjunction with the evolution of breeding structure.

The concept of the crossbred population as a system of heterozygous genotypes maintained in frequencies optimal to the demands of environment has been generally established (8, 13, 23, 24, 28, 29, among others). Much work has been done that demonstrates the advantages of deleterious recessives in higher than base mutation-rate frequencies in crossbreds (28). The maintenance of heterozygosity by selection in crossbreds was further supported when Dempster (12) showed that mutation to deleterious recessive was inadequate to account for the variance observed. Recent tests of specific situations in known genetic material show that individual heterozygotes may be at a disadvantage in uniform environment (18, 26) with a consequent return to homozygosity (30). These facts suggest that a price in excellency may be required in a chosen environment in return for the opportunity to change in a changing environment. How generally these conclusions apply is still subject to experimental confirmation since the numerical relationships of lethals, sublethals, vitals, supervitals, etc., are only imperfectly known even in Drosophila (7, 37).

Numbers of loci which are carrying deleterious alleles are thought to be maintained in various degrees and frequencies in some sensitive equilibrium by breeding system as well as by external environment. The fact that such a genetically variable population presents, in the wild at least, a uniformity of wild type has been discussed by Lerner (28). The widely accepted explanation of the coincident maintenance in the crossbred of genetic diversity and phenotypic uniformity lies in the phenomenon of heterozygosis. Thoday (52, 53) was able to show that heterozygosity was essential for the preservation of bilateral symmetry in the fruit fly.

The conclusion that mutation itself is in equilibrium with requirements of the mating system derives from the simple genetic situation that given a mutant A, advantageous in heterozygous state, every other mutation which occurs affecting breeding system will be selected in terms of A's conferred benefits. Furthermore, that particular organization of the genome which permitted the mutability of A will be favored in the sense that not only will those forms possessing the A-mutating quality be favored in selection but also those forms that possessed qualities controlling the A-mutating frequency. The genome itself and its intrinsic characteristics are conceived to evolve with breeding system to give the breeding structure of the population. The cross-pollinator is thought to have a genotype suited not only to meeting the contingencies of its environment but to the controlling of the variability in its own heterozygous organization. A measurable supply of new mutation is steadily furnished the crossbred (47, 49). It is thought that the rate of supply is determined by natural selection in keeping with the demands of breeding system and environment. Such a dominance-dependent organization would tend to neglect the evolution of high thresholds to expressivity for individual alleles. Any radical change in breeding system, external environment, or mutation rate would result in immediate shifts in the genetic composition of the population.

The above considerations concern the efficacy of mutation breeding with normally outbred organisms where the chief protection against deleterious mutation resides in heterozygosity. If selection has already increased the load of mutations to the optimum for the exigencies of the environment for population size, for loss of mutant alleles through homozygosis, and for balanced mutation rate through modulators, any marked increase in mutation would likely result in a diminution of fitness. Hutchinson (27) and Muller (37) especially have maintained that little is to be expected of mutation breeding with this system of mating. These strictures do not preclude positive results from mutation breeding of crossbreds (38) in instances of small breeding populations or where the crossbred population is near or at the limit of its range, either case necessitating a past history of inbreeding. They do suggest that mutation breeding, in a panmictic population of large size and great complexity, unaccompanied by a radical departure from customary breeding procedure, would probably be ineffectual. The somewhat negative conclusion here should take into account the selection work reported on irradiated populations of Drosophila (45).

Lacking the necessary experimental data to prove or to disprove the above conclusion and unable to review the enormous literature allied to this subject, I would like to stop with the suggestion that the effectiveness of selection following mutagenic treatment vs. no treatment, with and without radical change of breeding procedure, be investigated in a panmictic crop plant—shall we say rye or corn.

Selfbreds

Mather has emphasized the contrasting organizations of the genetic systems in self- and cross-pollinators. In an extension of Mather's thesis, Gregory (19, 22) postulated that self-pollinators, faced with the alternatives of elimination or fixation of every mutation, have evolved genomic systems capable of absorbing relatively large numbers of mutations of small effect without their necessarily reaching the thresholds for phenotypic expression. According to this hypothesis genetic organization favoring resiliency of the phenotype of the homozygote in the self-pollinator would tend to become established by selection. The genetic factors controlling the tendency of mutations to occur in the direction of resiliency would likewise become ensconced in the genome. The genome would become laced with supporting modifier gene complexes with little other obvious or major effects than to provide variation in expressivity consistent with a changing environment. Each chromosome would come into a form of internal balance, as Mather has postulated, with respect to the amount of modulating material optimum for the genome to maintain in relation to external environment.

The relations of self-pollination, mutation, and expressivity permits a genetic explanation for Stebbins' (50) observation that self-pollinators possess, "a relatively high degree of phenotypic plasticity so that the individual is susceptible of tremendous modification in the face of adverse or extreme conditions".

The homozygote's capacity to adapt to great environmental change combined with the preservation of phenotype suggests that protection against expression of newly acquired mutation may be high (22). It is the possibility that the height of this barrier may permit a sufficient number of simultaneous mutations that makes the mutation breeding of selfbreds attractive. It should be said that not every breeder of self-fertilized crops shares this view (1, 27).

Comparative effects of Irradiation of Pure Lines and Their Hybrids

Notwithstanding the theoretical possibility that normally self-fertilizing species may utilize mutation as a resource of genetic adaptability in a manner comparable to the use of heterozygosity by cross-fertilizing species, it is still uncertain whether self-fertilizers are capable of absorbing to advantage very much larger mutation rates than those established in the species by natural selection. The solution of this problem requires knowledge of the effects of mutation on characters equatable to fitness in a number of different species. Interest in the effects of mutagenic treatment on quantitative characters, some of which are the most likely indicators of fitness, has been slow in its development. As a consequence of this it will be difficult to give a satisfactory assessment of the situation at the present time.

Papers which have laid the foundation for further experiments on quantitative characters have been published by Brock and Latter (6), Daly (10), Gregory (19), Mertens and Burdick (36), Oka (39), and Rawlings, et al. (41). For an assessment of the efficacy of mutation breeding, experiments designed to make comparisons between conventional breeding procedure and mutation breeding are required. Gregory (20) suggested that radiation-induced variance should be cumulative with that induced by hybridization. Preliminary results have been reported (21) on expressions of F_2 dominance in hybrids of mutant selections from the same pure line. In the absence of such work from other laboratories it is my purpose to devote the remainder of this paper to the description of such an experiment conducted with peanuts. (See, however, Krull, C. F., Agronomy Abstracts, 1960, page 50.)

In 1953 all possible hybrid combinations were made among six lines of peanuts. All of these lines had been reproduced from initial single plant selections followed by sufficient automatic self-fertilization to assure their relative homozygosity. Three of them were selections in X_5 generation of high-yielding mutants from the same pure line. The other three were selections in F_{11} generation from the hybridization and progeny testing program which, at that time, was conducted separately from the radiation experiments. A sufficient number of cross-pollinations was made to produce at least 100 F_1 seeds of each of the 15 possible F_1 hybrids. In each cross the 100 F_1 seeds were divided into two lots of 50 each. One 50-seed lot of each hybrid and a 125-seed sample of each parent were given a treatment of 15 Kr of X-rays.

The packages of peanuts were arranged all at a time upon a curved target surface in random order. At half time the packages were all removed, turned over, re-randomized, and placed upon the target surface for the remainder of the treatment. The radiation was delivered at the rate of 62 to 63r per minute at a distance of 1 meter from a 1,000-Kv tube equipped with beryllium window. It is thought that all received equal and uniform doses of X-rays. The seed had been stored dry for 6 months prior to irradiation and moisture content was uniform. (The author is indebted to Dr. William T. Ham, Jr. (25) for dosimetry measurements and the use of the X-ray tube, Biophysics, Medical College of Virginia, Richmond.) The treated seeds and their controls were planted in individual plots the day following X-ray treatment. In the fall of 1954 the plants were harvested individually. The treated hybrids were designated as F_1X_1. The F_2 and F_2X_2 generations were grown in 1955. At the end of the F_2 and F_2X_2 year, 12 of the 15 original crosses were available for experimentation. Nine of these were placed in a replicated experiment on the peanut testing station located in the central North Carolina coastal plain. This experiment was lost due to fall storms. Three of the crosses were placed on another station on the western edge of the coastal plain. These were harvested without mishap. They were Cross I (C12×A18), Cross II (C12×YT24), and Cross III (C12×YT13). C12 and A18 were F_4 selections from two different hybrids in F_{11} generation. YT24 and YT13 were X_3 selections in X_5 generation from the same pure line.

The F_3 and F_3X_3 experiment involving these three crosses was designed in the following manner. Ten F_2 generation plants were harvested individually from each of five F_1 generation families of the following treatments: P1, P2, P1X_2, and P2X_2. Ten plants were harvested individually from two sets of five F_1 generation families in the F_2 and F_2X_2. This provided an equal number of progenies of P, PX, F, and FX. The individual plants and the families from which they were chosen were taken at random except for the specification that enough seed be produced to conduct the experiment. The individual plant progenies of these selections were planted in the F_3 generation and arranged in the field according to the experimental design presented in Table 1. Dry weights of fruits were obtained and appropriate analyses conducted including analyses of variance of the individual F_1 and F_1X_1 families.

The overall effect of the irradiation by generation is shown in Table 2 where all values are presented in percentage of the mean of the entire experiment. Differences among treatments in blocks were highly significant in four of the six blocks and significant in the fifth. Thus the reduction in mean performance occasioned by X-ray treatment is a substantial factor in the breeding expectations from this material. The cross means are almost identical.

The genotypic standard deviation (s_G) of each F_1 family was determined as

$$\sqrt{\frac{V_p - V_e}{r}}$$

where V_p is the progeny mean square, V_e the individual F_1 family error mean square, and r the number of replications. The treatment means and the average s_G among

TABLE 1.—EXPERIMENTAL DESIGN AND MODEL ANALYSIS FOR COMPARING P1, P2, P1X, P2X, F$_2$, AND F$_2$X IN THREE PEANUT HYBRIDS.

Experimental design			Treatment		
Cross I...........	Block 1*	P1	P1X	F$_2$	F$_2$X
	2	P2	P2X	F$_2$	F$_2$X
Cross II..........	Block 1	P1	P1X	F$_2$	F$_2$X
	2	P2	P2X	F$_2$	F$_2$X
Cross III.........	Block 1	P1	P1X	F$_2$	F$_2$X
	2	P2	P2X	F$_2$	F$_2$X

Replications...	3	
Blocks in reps..	4	
Between treatments in blocks.....	6	
Between trts. in block 1.............................		3
Between trts. in block 2.............................		3
Between F$_1$ families in trts. in bl...............................	32	
Between F$_1$ fam. in trts. in bl. 1........................		16
Between F$_1$ fam. in trts. in bl. 2........................		16
Between F$_2$ progenies in F$_1$ fam. in trts. in bl................	360	
Reps. x trts. in bl..	18	
Reps. x fam. in trts...	96	
Reps. x F$_2$ prog. in F$_1$ fam................................	1,080	
Pooled error................	1,194	
Total...	1,599	

*Block 1 consisted of 10 F$_2$ progenies of 5 F$_1$ families each of treatments P1, P1X, F$_2$, and F$_2$X; Block 2 consisted of P2, P2X, F$_2$, and F$_2$X. PX and FX refer to P-irradiated and F-irradiated, respectively.

progenies by treatments are given in Table 3. From this table the effects of radiation *vs.* hybridization may be observed in the variation among the means as well as in the differences between treatment grand means. The average genotypic variance for P was exceedingly small, being zero for four cases out of six. Gregory (20) postulated that the variation induced by radiation might be cumulative with that of hybridization such that $\sigma^2_G P + \sigma^2_G FX = \sigma^2_G PX + \sigma^2_G F$. When calculated over all three crosses, this expectation was approached in the present experiment with $\sigma^2_G P + \sigma^2_G FX = 76.9\%$ of $\sigma^2_G PX + \sigma^2_G F$.

The above generalities are of importance, but the detailed behavior of the individual F$_1$ generation families holds the greater

TABLE 2.—EFFECT OF RADIATION UPON F₂ PROGENY MEANS IN PERCENTAGE OF THE MEAN OF ALL TREATMENTS AND CROSSES, FRUIT YIELD.

Cross	Treatments				
	P	PX	F	FX	Cross x̄
	%	%	%	%	%
Cross I:					
Bl. 1....................	102.0	98.8	101.4	96.2	99.6
Bl. 2......	101.7	95.0	98.3	103.3	
Cross II:					
Bl. 1....................	96.4	94.5	103.2	96.3	100.5
Bl. 2....................	106.9	97.7	108.1	101.5	
Cross III:					
Bl. 1....................	100.5	93.6	105.4	103.7	99.9
Bl. 2....................	102.5	89.5	106.4	97.5	
x̄.....	101.7	94.9	103.8	99.8	100.0

significance for the plant breeder. The F₂ genotypic standard deviations for individual families are presented for each treatment in

TABLE 3.—MEAN YIELD (x̄) OF F₂ PLANT PROGENIES AND GENOTYPIC STANDARD DEVIATIONS (s_G) IN POUNDS OF DRY FRUIT.

Cross	Treatment							
	P		PX		F		FX	
	x̄	s_G	x̄	s_G	x̄	s_G	x̄	s_G
Cross I:								
Block 1........	2.38	0.06	2.31	0.31	2.37	0.22	2.25	0.36
2........	2.37	0.00	2.22	0.40	2.29	0.19	2.41	0.24
Cross II:								
Block 1........	2.25	0.00	2.21	0.27	2.41	0.22	2.25	0.32
2........	2.50	0.00	2.28	0.19	2.52	0.16	2.37	0.39
Cross III:								
Block 1........	2.35	0.00	2.19	0.26	2.46	0.22	2.42	0.27
2........	2.39	0.04	2.09	0.22	2.48	0.23	2.28	0.27
Grand x̄.........	2.37		2.21		2.42		2.33	
Av. s_G..........		0.02		0.27		0.21		0.31

the three crosses in Figure 1. Here the data are presented as percentage of the mean of all treatments in each cross, i.e., as genotypic coefficients of variability. The center line labeled "cross mean" is equal to 100 per cent of all treatments in the cross. The bases of the bar graphs, shown slightly extended, represent the means of the F_1 generation families. The s_G arising from differences among progenies in each family is indicated by the height of the bar for each treatment shown at the bottom of the chart. Since interest is primarily in the variation in excess of the means, only the s_G above the mean is presented. The deviations of the family means from the cross mean are shown on the same scale as the genotypic standard deviations from the family means.

The mean family yield of PX was less than that of any P in 18 out of the 30 comparisons made in these two treatments. Of the 12 remaining comparisons s_G was nonsignificant in five PX families. This is to say that in the P $vs.$ PX comparisons 7 PX families out of 30 had equal or higher means in addition to higher variances than any of their P standards.

In the case of the F_2 $vs.$ PX, 17 out of 30 family means in the F_2 were higher than any PX family mean. In eight of these 17, s_G was small in magnitude, failing to attain significance. This is to say that in 9 out of 30 families of F_2 the F_1 family means were higher than any PX and, in addition, possessed a significant variance. In reverse, 13 of the 30 PX were lower than any F_2 and none higher.

In the case of the F_2 $vs.$ F_2X_2, 9 of 30 family means of the F_2X_2 were lower than any family mean of F_2. In reverse, seven F_2 were higher than any F_2X_2, while only two F_2X_2 were higher than any F_2. (This could happen since the comparisons were made cross by cross.)

It is obvious from the data that if outcrossing were too costly either in time or loss of collateral characters, progress from selection in PX could be expected. If the character here measured were the only one under consideration, it is likewise obvious that in two of the three crosses greater progress from selection would be expected in the F_2, while in one of the three greater progress would be expected in the F_2X_2. When this prognosis was tested (two locations, six replications, 100 plants per plot) with bulked F_2 family seed in the F_4 generation, the results confirmed the prediction

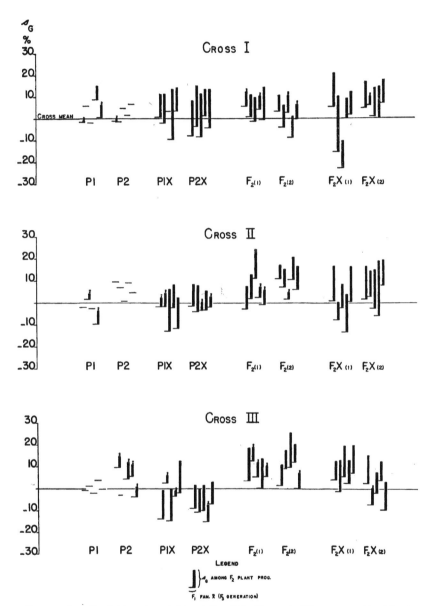

FIGURE 1.—*Variation in yield of* Γ, ΓX, Γ, *and* ΓX *among* F₂ *plant progenies from three crosses in the* F₃ *generation. The variation is shown in genotypic standard deviations from the* F₁ *family means in percentage of the mean of all treatments,* $\bar{x}$ *i.e. (genotypic C.V.* $= s_{G/\bar{x}} \times 100$).

in a highly satisfactory manner. The frequency distributions of the F_2 progenies for the three crosses are shown in Table 4. The F_4 frequency distributions of the highest yielding F_2 lines from the three crosses are shown in Table 5.

It is apparent from Table 3 and Figure 1 that radiation reduced the means and increased the variances. It appears also that the

TABLE 4.—FREQUENCY DISTRIBUTION OF THE F_2 PROGENY MEANS IN THE F_3 GENERATION, 1956.

Cross	Treat-ment	Class mark, lbs.									
		0.83	1.10	1.38	1.65	1.93	2.20	2.48	2.76	3.03	3.31
I	P					4	37	51	8		
	PX	1	1	5	6	11	21	45	10		
	F				4	11	33	40	12		
	FX		2		7	15	25	27	22	2	
II	P				2	7	29	56	6		
	PX	1	1		3	12	45	34	4		
	F				1	5	26	40	21	6	1
	FX	1	3		8	13	22	31	19	3	
III	P					2	41	48	9		
	PX			5	10	19	34	30	2		
	F				1	11	20	35	26	6	1
	FX				3	18	26	31	19	3	

TABLE 5.—FREQUENCY DISTRIBUTION OF THE F_4 MEANS OF F_2 PROGENIES SELECTED ON F_3 PERFORMANCE, 1957. *

Cross	Treatment	Class mark, lbs. †			
		14.08	15.69	17.30	18.91
I....................	F	1	5	2	
	FX	1	4	5	
II....................	F		1	4	
	FX		2	1	
III..............	F			2	
	FX		3	2	

*The variation in number of progenies per cross resulted from influence of market grade.
†Class interval equals LSD = 1.61 lbs.

larger variances were often associated with the smaller means. This inverse relation is brought out more sharply by the distribution of the standard deviations on the array of the F_1 generation means. The individual regressions which were found to be negative and highly significant for the P1X + P2X and significant for F_2X are shown in Figure 2. The regression was much more marked in PX than in FX, but the difference in slope was nonsignificant at the 0.05 level.

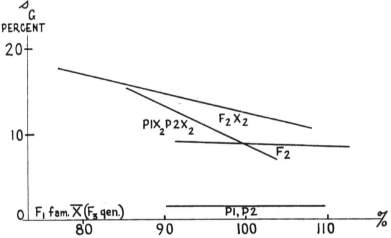

FIGURE 2.—*Regression of genotypic C.V. ($s_{G/\bar{x}} \times 100$) of F_2 progenies on F_1 family mean in percentage of the mean of all treatments, $\bar{x}$.*

In the light of the negative regression of s_G on F_1 generation family mean in the irradiated material, the question arises as to the relative magnitudes of the variances remaining in selected high and low F_2 generation families. Will negative regression of variance on mean have disappeared in the distribution of s_G for F_3 generation progenies on F_2 generation family mean? The tests to evaluate the means and variances of the high selections of F_2 and F_2X_2 as F_7 progenies in F_9 generation will not be conducted until 1961. However, four crosses from the experiment lost to the fall storms of 1956 and involving some of the same parental material as the crosses shown above have furnished material for making such tests.

Bulked F_2 generation progenies by F_1 generation families of the four crosses A (YT32 × YT13), B (A18 × YT32), C (C12 × YT32), and D (C12 × B35) were tested in a replicated yield trial in F_4 generation (1957). From each treatment in each of the crosses the highest and the lowest yielding F_2 generation families were chosen from the F_1 generation family showing the maximum range in yield of dry fruits. Table 6 shows the comparative values of the treatment means

TABLE 6.—YIELDS IN POUNDS OF THE F_2 PROGENIES FROM THE F_1 FAMILY SHOWING MAXIMUM RANGE IN YIELD IN F_4, 1957.*

Cross	F_2 progeny	Treatment					
		P1	P1X	P2	P2X	F_2	F_2X
A.............	1	4.03^b	3.73	3.58	3.96	4.69	4.22
	2	4.51	3.70	3.86	4.13	4.18	4.34
	3	4.15	3.78	3.87	4.20^a	3.79^b	3.49^b
	4	4.07	3.78^a	3.52^b	3.28^b	4.92^a	4.67^a
	5	4.82^a	2.28^b	4.53^a	3.69	4.25	4.19
B.............	1	3.95	2.98	3.73^b	3.47	3.77	3.59
	2	4.05^a	2.88^b	4.30^a	3.64	3.73	4.48
	3	3.10^b	3.81	4.16	3.83	4.46^a	3.49^b
	4	3.64	3.24	4.05	4.11^a	3.08^b	3.88
	5	3.84	4.26^a	4.09	2.86^b	4.44	4.53^a
C.............	1	3.78^b	4.48	3.98	4.12^a	4.74^a	3.14^b
	2	3.98	4.64^a	4.37^a	2.95^b	3.82	3.99
	3	4.35	4.15	4.11	4.12	3.71^b	3.43
	4	4.09	3.95^b	3.78	4.10	4.34	4.75^a
	5	4.40^a	4.27	3.68^b	4.10	3.83	3.39
D.............	1	3.39	2.91^b	2.88^b	2.59^b	3.52	3.03
	2	3.75	3.60	3.94^a	3.99	3.72	3.19
	3	3.99^a	3.36	3.21	3.55	3.71	2.87
	4	3.49	4.01	3.21	4.05^a	4.07^a	4.03^a
	5	3.15^b	4.14^a	3.27	3.77	2.71^b	2.11^b

*a = high, b = low F_2 progenies selected for variance tests in Table 7 and Figure 3.

and the differences between the high and low selections by treatment. Each selection was grown in a nursery row in 1958. Twenty plants were harvested individually from each row and tested as F_5 plant progenies in the F_6 generation in blocks of P, PX, F, and FX. The selections from crosses B and C were tested in 1959, those from crosses A and D in 1960. In 1960 the plant progenies of the high and low selections were randomized together by treatment permitting a significance test for high versus low. This was not possible in 1959.

The individual means and variances were estimated for each treatment. The treatment means, together with the indications of significance of differences among progenies in treatments, are shown in Table 7.

The comparisons of s_G by F_2 generation family for the high and low selections from the various treatments are given in Figure 3 for crosses B and C. The cross mean is shown as 100 per cent of all treatments. The bases of the bar graphs represent the F_2 fam-

TABLE 7.—F_2 FAMILY MEAN YIELDS OF F_5 PLANT PROGENIES IN F_6 GENERATION BY TREATMENT IN HIGH AND LOW F_2 FAMILIES OF CROSSES B AND C, 1959; CROSSES A AND D, 1960.

Treatment		B^a	C^a	A		D	
		Yield, lbs.	Yield, lbs.	Yield, lbs.	High vs. low progenies	Yield, lbs.	High vs. low progenies
P1	H	4.13^{NS}	4.08^{NS}	3.85^{NS}	**	4.54^{NS}	NS
	L	4.05^{NS}	3.94^{NS}	3.65^{NS}		4.58^{NS}	
P1X	H	4.21^{NS}	$3.74**$	3.96^{NS}	**	4.61^{NS}	NS
	L	$3.37**$	$3.74**$	$3.33**$		$4.48*$	
P2	H	3.73^{NS}	2.61^{NS}	3.95^{NS}	NS	—^{b}	
	L	3.64^{NS}	2.79^{NS}	$4.01**$		—	
P2X	H	3.73^{NS}	$2.69**$	4.26^{NS}	**	$3.99**$	NS
	L	$3.43**$	$2.78*$	$3.96**$		3.92^{NS}	
F_2	H	3.77^{NS}	$3.64*$	3.90^{NS}	NS	$4.12**$	NS
	L	$3.31**$	$3.52**$	$3.84**$		$3.96*$	
F_2X	H	$3.69*$	$3.98**$	3.66^{NS}	NS	$3.75**$	NS
	L	3.87^{NS}	$3.03**$	$3.55**$		3.85^{NS}	

^{a}Field design did not permit a high vs. low test in crosses B and C.
^{b}Omitted because found to be a hybrid irradiated population selected and tested by accident.
NS = Nonsignificant.
*Significant at the 5 per cent level of probability.
**Significant at the 1 per cent level of probability.
Superscript to the right of each mean refers to significance of differences between F_5 progenies in F_2 families.

ily means. The bars represent s_G among F_5 progenies, given as genotypic coefficients of variability.

Similar results are presented for crosses A and D in Figure 4. Particular attention may be directed toward the comparisons of the treatments in cross A. In cross A, P1 and P2 are high-yielding,

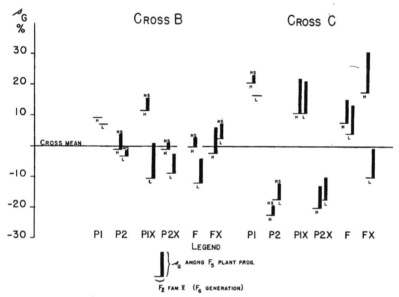

FIGURE 3.—*F_3 variation in yield of P, PX, F, and FX among F_5 plant progenies (in F_6 generation) of high and low F_2 families (selected in F_4) from two crosses (B and C). The variation is shown as in Figure 1.*

X-ray-induced mutants from the same pure line. Each of these parental lines was selected because of its superiority in yield to the pure breeding mother line from which it was derived. The mother line received a dose of 18.5 Kr X-rays in 1949. The two mutants and their hybrid received an additional 15 Kr in the X_5 generation to create what could be called the X_5X_1 and the $X_5F_1X_1$ generations. The results shown in Figure 4 indicate not only a substantial genotypic variance in P1X, P2X, and F_2X but also in the F_2.

The differences between the means of the high and low selections attained significance in 1960 in only 3 of the 11 comparisons made in crosses A and D. However, these tests were weak statistically. It is interesting to observe that although tested in different years and fields, the low selections tended to remain low and the high selections to remain high. There were only 6 reversals of high and low out of a total of the 23 pairs available from the four crosses.

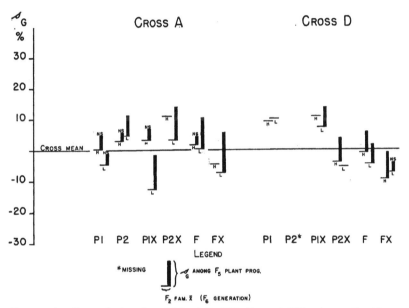

FIGURE 4.—F_3 variation in yield of P, PX, F, and FX among F_5 plant progenies (in F_6 generation) of high and low F_2 families (selected in F_4) from two crosses (A and D). The variation is shown as in Figure 1.

Of the 17 pairs which held their high-low positions, 8 were P1, P2, or F_2 and 9 were P1X, P2X, or F_2X.

Table 8 shows the extent to which the negative regression of s_G for F_2 generation progenies on F_1 generation family mean has disappeared in later generations. In two of the crosses, A and B, s_G was larger in the low selections, while in the other two s_G was larger in the high selections. This difference in the two pairs of crosses appeared to be associated with the crosses irrespective of the treatment. The s_G was larger in the low selections both in P1, P2, and F_2 and in P1X, P2X, and F_2X in crosses A and B; while in crosses C and D, s_G was larger in the high selections in all treatments.

TABLE 8.—COMPARISON OF THE MEANS AND GENOTYPIC STANDARD DEVIATIONS IN THE 17
OUT OF 23 HIGH AND LOW F_2 FAMILIES WHICH MAINTAINED THEIR RELATIVE
HIGH-LOW PERFORMANCE WHEN TESTED AS F_5 PROGENIES.

	Treatment			
Cross	P1,P2,F_2		P1X,P2X,F_2X	
	High	Low	High	Low
	Average s_G			
s_G > in *Low* Sel.				
A....................	0.14	0.27	0.05	0.45
B....................	0.09	0.13	0.12	0.34
s_G > in *High* Sel.				
C....................	0.17	0.16	0.42	0.34
D....	0.28	0.26	0.17	0.12

Discussion

The mutation breeding program with peanuts has been conducted under the hypothesis that coincident with visible mutational change after irradiation numerous nonobservable polygenetic changes occur at loci scattered over the entire genome. This hypothesis grew out of the consideration of the within-X_2-family variability in the expressivity of ordinary morphological mutants. (Robbelen (44) has reported a similar observation for a chlorophyll character in Arabidopsis.) Among the several hypotheses advanced to explain the graded expressivity in morphological mutants of peanuts, the one of genetic variation in the background genotype has proved to be the most plausible. The first test of this hypothesis (19) came with the discovery of a large genotypic variance induced in control-type X_2 sibs of morphological mutants of a pure line.

The next test of the hypothesis was obtained when Loesch (unpublished), in a study of the breeding value of simply inherited deleterious mutants of the same pure line, discovered a remarkable variation in expressivity between and within F_2 progenies of mutant plants derived from intercrosses of different morph-

ological mutants. Variable F_2 individuals of the same mutant phenotypes had arisen from a cross of two highly uniform mutant parents. The most plausible explanation for the new variability observed,
lay in the presumption of the genetic segregation of modifiers
located in the chromosomes of the background genotype. This presumption was further confirmed when Emery, Gregory, and Loesch
(unpublished) made a study of the control type segregates from F_2
generation hybrids of various morphological mutants. They discovered highly significant genotypic variances among nonsegregating control-type F_2 progenies in F_4 generation for several quantitative characters. Finally, Emery (unpublished) showed that even the
quantitative variation of the double recessives of these morphological mutants could be attributed to effects of genetic background.
Furthermore, Emery showed in the double recessives that the background effect was significantly associated with the parental source
of the background, suggesting at least a multi-chromosomal effect
if not a polygenic one.

These observations support the thesis that the mutagenic action
of radiation may be used for the induction of variability in quantitative characters. They also lend support to the hypothesis that
certain species of plants are capable of absorbing relatively large
doses of mutation without crossing the threshold of obvious phenotypic expression and to the further hypothesis that such mutation permits a response to selection sufficient to exceed the reduction in the
mean fitness occasioned by the treatment.

These observations do not lend much support to the general
thesis that high mutation tolerance in the selfbreds has provided
for the function performed by balanced heterozygosity in the crossbreds. Credence here can come only with relatively large accumu-
lations of data on a number of species of different breeding system
and from experiments specifically designed to furnish information
on this subject. Nevertheless it is our working hypothesis.

The data brought forward in the present paper give only a
foretaste of the solution of the problem of the relative efficacy of
conventional and mutational breeding in selfbreds. Much more
must be learned concerning additional environmental agents and
breeding procedures which selectively eliminate undesirable changes
induced by mutagens while permitting the successful exploitation

of changes wanted by the breeder. Likewise it is not known whether there is a saturation effect of mutagenic treatment which, though permitting the initial successes reported in the literature, might prevent repeated success with repeated application of mutagens to chromosomes at short time intervals in recurrent selection cycles. Also still to be learned are the relative efficacies of mutagenic treatments and hybridization when applied in conjunction in species other than peanuts.

References

1. Bell, G. O. H. 1957. Cereal breeding. *Sci. Progr.*, 45 *(178): 193–209.*

2. Borg, G., Fröier, K., and Gustafsson, Å. 1958. Pallas barley, a variety produced by ionizing radiation: Its significance for plant breeding and evolution. *Proc. 2nd Int. Conf. Peaceful Uses Atomic Energy, Geneva,* 27: *341–349.*

3. Brink, R. A. 1956. A genetic change associated with the *R* locus in maize which is directed and potentially reversible. *Genetics,* 41: *872–889.*

4. ———— and Nilan, R. A. 1952. The relation between light variegated and medium variegated pericarp in maize. *Genetics,* 37: *519–544.*

5. ———— and Wood, D. R. 1958. The neutral effect of modulator in maize on plant and pollen-tube growth. *Amer. Jour. Bot.,* 45: *38–44.*

6. Brock, R. D., and Latter, B. D. H. 1961. Radiation-induced quantitative variation in subterranean clover. *Proc. 3rd Aust. Conf. on Radiation Biol.*

7. Burdick, A. B., and Mukai, T. 1958. Experimental consideration of the heterozygous genetic effect of low doses of irradiation on viability in *Drosophila melanogaster. Proc. 10th Int. Cong. Genet.,* 2: *38.*

8. Buzzati-Traverso, A. A. 1954. On the role of mutation rate in evolution. *Proc. 9th Int. Cong. Genet.,* 1: *450–462.*

9. Cooper, W. E., and Gregory, W. C. 1960. Radiation-induced leaf spot resistant mutants in the peanut (*Arachis hypogaea* L.). *Agron. Jour.,* 52: *1–4.*

10. Daly, K. 1960. The induction of quantitative variability by γ radiation in *Arabidopsis thaliana*. Abstract in *Genet. Soc. Amer. 1960 Genetics,* 45: *983.*

11. Darlington, C. D. 1958. Evolution of genetic systems. *Edinburgh: Oliver and Boyd. 2nd Ed.*

12. Dempster, E. R. 1955. Maintenance of genetic heterogeneity. *Cold Spr. Harb. Symp. Quant. Biol.,* **20:** *25–32.*

13. Dobzhansky, Th. 1955. A review of some fundamental concepts and problems of population genetics. *Cold Spr. Harb. Symp. Quant. Biol.,* **20:** *1–15.*

14. Ehrenberg, L., Gustafsson, Å., and Wettstein, D. von. 1956. Studies on the mutation process in plants, regularities and intentional control. *Conf. on Chromosomes, Wageningen, Netherlands.*

15. ———, ———, and Lundquist, U. 1959. The mutagenic effects of ionizing radiations and reactive ethylene derivatives in barley. *Hereditas,* **45:** *351–368.*

16. Fradkin, C-M. W., and Brink, R. A. 1956. Effect of modulator on the frequency of endosperm mosaics in maize. *Amer. Jour. Bot.,* **43:** *267–273.*

17. Gaul, H. 1958. Present aspects of induced mutations in plant breeding. *Euphytica,* **7:** *275–289.*

18. Greenberg, Rayla, and Crow, James F. 1960. A comparison of the effect of lethal and detrimental chromosomes from · Drosophila populations. *Genetics,* **45:** *1153–1167.*

19. Gregory, Walton C. 1955. X-ray breeding of peanuts. *Agron. Jour.,* **47:** *396–399.*

20. ———. 1956. The comparative effects of radiation and hybridization in plant breeding. *Proc. Int. Conf. on Peaceful Uses of Atomic Energy,* **12:** *48–51.*

21. ———. 1956. Induction of useful mutations in the peanut. *Brookhaven Symp. in Biol. (Genetics in Plant Breeding),* **9:** *177–190.*

22. ———. 1957. Progress in establishing the effectiveness of radiation in breeding peanuts. *Proc. 9th Oak Ridge Regional Symp. (Radiation in Plant Breeding),* *36–48.*

23. Gustafsson, Å. 1951. Mutations, environment, and evolution. *Cold Spr. Harb. Symp. Quant. Biol.,* **16:** *263–281.*

24. ———. 1954. Mutations, viability, and population structure. *Acta Agri. Scand.,* **4:** *601–632.*

25. Ham, Wm. T., Jr., and Trout, Dale. 1950. Million-volt beryllium-window X-ray equipment for biophysical and biochemical research. *Radiology,* **55:** *257–270.*

26. Hiraizumi, Yuichiro, and Crow, James F. 1960. Heterozygous effects on viability, fertility, rate of development, and longevity of Drosophila chromosomes that are lethal when homozygous. *Genetics,* **45:** *1071–1083.*

27. Hutchinson, Sir Joseph. 1958. Genetics and the Improvement of Tropical Crops. *Inaugural lecture, Cambridge: Cambridge University Press.*

28. Lerner, I. M. 1954. Genetic Homeostasis. *Edinburgh: Oliver and Boyd.*

29. ————, Dempster, E. R., and Inouye, Nobuo. 1958. Preliminary report on X-ray induction of variability in polygenic traits of chickens. *Proc. 10th Int. Cong. Genet.,* 2: *164.*

30. Lewontin, R. C. 1958. Studies on heterozygosity and homeostasis: II. Loss of heterosis in a constant environment. *Evolution,* 12: *494–503.*

31. MacKey, James. 1956. Mutation breeding in Europe. *Brookhaven Symp. in Biol. (Genetics in Plant Breeding),* 9: *141–156.*

32. Mather, Kenneth. 1955. Response to selection. *Cold Spr. Harb. Symp. Quant. Biol.,* 20: *158–165.*

33. McClintock, Barbara. 1951. Chromosome organization and genic expression. *Cold. Spr. Harb. Symp. Quant. Biol.,* 16: *13–47.*

34. ————. 1956. Controlling elements and the gene. *Cold Spr. Harb. Symp. Quant. Biol.,* 21: *197–216.*

35. ————. 1958. The suppressor-mutator system of control of gene action in maize. *Ann. Rpt. Carnegie Inst. of Washington Year Book,* 57: *415–429.*

36. Mertens, T. R., and Burdick, A. B. 1957. On the X-ray production of "desirable" mutations in quantitative traits. *Amer. Jour. Bot.,* 44: *391–394.*

37. Muller, H. J. 1954. The nature of the genetic effects produced by radiation. *Radiation Biology (Hollaender, editor). New York: McGraw-Hill.*

38. Nybom, Nils. 1957. Vaxtforadling med hjalp au inducerate mutationer. *Sverig. Utsadesforen. Tidskr.,* 67 *(Hafte 1): 51–55.*

39. Oka, Hiko-Ichi, Hayashi, J., and Shiojiri, I. 1958. Induced mutation of polygenes for quantitative characters in rice. *Jour. Hered.,* 49: *11–14.*

40. Prakken, R. 1959. Induced mutation. *Euphytica,* 8: *270–322.*

41. Rawlings, J. O., Hanway, D. G., and Gardner, C. O. 1958. Variation in quantitative characters of soybeans after seed irradiation. *Agron. Jour.,* 50: *524–528.*

42. Rhoades, M. M. 1938. Effect of the *Dt* gene on mutability of the a_1 allele in maize. *Genetics,* 23: *377–395.*

43. ————. 1945. On the genetic control of mutability in maize. *Proc. Nat. Acad. Sci.,* 31: *91–95.*

44. Robbelen, G. 1958. Instabile Genmanifestation nach Rontgen-bestrahlung von *Arabidopsis thaliana* (L.) Heynh. *Zeit. Naturf.,* **13b:** *14–16.*

45. Scossirolli, R. R. 1954. Effectiveness of artificial selection under irradiation of plateaued populations of *Drosophila melanogaster.* *I. U. B. S. Symp. on Genetics of Population Structure, Pavia, 42–66.*

46. Scholz, F. 1959. Strahleninduzierte Mutationen und Pflanzenzuchtung. *Atompraxis,* **5** *(12):* *475–481.*

47. Schuler, J. F. 1954. Natural mutations in inbred lines of maize and their heterotic effect: I. Comparison of parent, mutant, and their F_1 hybrid in a highly inbred background. *Genetics,* **39:** *908–922.*

48. Smith, H. H. 1958. Radiation in the production of useful mutations. *The Bot. Rev.,* **24:** *1–24.*

49. Sprague, G. F., Russell, W. A., and Penny, L. H. 1960. Mutations affecting quantitative traits in the selfed progeny of doubled monoploid maize stocks. *Genetics,* **45:** *855–866.*

50. Stebbins, G. L., Jr. 1950. Variation and evolution in plants. *New York: Columbia University Press.*

51. Stubbe, H. 1958. Advances and problems of research in mutations in the applied field. *Proc. 10th Int. Cong. Genet.,* **1:** *247–260.*

52. Thoday, J. M. 1955. Balance, heterozygosity and developmental stability. *Cold Spr. Harb. Symp. Quant. Biol.,* **20:** *318–326.*

53. ———. 1958. Homeostasis in a selection experiment. *Heredity,* **12:** *401–415.*

54. Wettstein, D. von. 1957. Mutations and the intentional reconstruction of crop plants. *Hereditas,* **43:** *298–302.*

55. ———, Gustafsson, Å., and Ehrenberg, L. 1957. Mutationsforschung und zuchtung Arbeitsgemeinschaft fur Forschung des Landes Nordrhein-Westfalen., **73:** *7–60.*

Comments

MacKey: It is possible that the extremely positive results obtained by Dr. Gregory in his mutation program in peanuts may partly depend on the fact that he is using a polyploid.

Genes interact in a very complicated way, and it is no reason to believe that the dosage for each gene in this interaction should be optimal on the level brought about by a more or less complete reduplication of the somatic chromosome set. A more subtle adjustment may just as well go via loss, inactivation, duplication, and gene diversification.

The better buffered germ plasm able to absorb and tolerate mutation taken in its wide sense makes it also likely that polyploids have a certain advantage in mutation breeding. In addition, they must be younger than the related diploids, and nature has thus had less time to proceed with her own mutation program to a level where improvements will be more and more difficult to achieve. This interpretation will be even more valid for a crop like peanuts which rather recently has been brought from rather restricted to new habitats with new selection pressures.

Discussion of Session V

HERBERT H. KRAMER

Purdue University, Lafayette, Indiana

IN A SHORT TIME Doctor Stephens will attempt to summarize the proceedings of the past few days. In the meantime, however, some speculation on how our knowledge of a gene today might be translated into what we do with it in the future would seem to be in order.

This symposium has emphasized the concept of gene complexity. This concept has long been widely held and in itself represents little advance. The remarkable advance in recent years has been in the elucidation of the nature of the complexity. The nature of the gene has been attacked on almost every conceivable level, molecular, chromosomal, nuclear, cellular, organism, and population. In each case the weapon has been the mutant allele, which for our purpose may be defined as a deviant from an undefined normal form.

Through classical studies of recombination, the one genetic phenomenon unique to, and equally important to, all levels of genetic and plant breeding research, the gene, as a unit of function has been raised to a level of structural sophistication which seems adequate to provide the code for the complex compounds through which it must mediate its effect.

At the cistron level, hope for controlling the direction of change resides in the degree of specificity of reaction to different mutagens. The evidence for such specificity has been thoroughly reviewed by Doctor Smith. Paradoxically, the very studies in bacteriophage which demonstrate so convincingly the extreme specificity of intra-locus reactive sites would argue that similar reactive sites within any cistron would be reactive and lead to inter-locus nonspecificity unless some effect of molecular organization on reactive sites is assumed. This focuses attention on one of the most serious gaps in our knowledge, that of relating the intra-locus or site specificity demonstrated in microorganisms to the spectra of inter-locus mutants observed for different mutagens. Whatever the final explanation, it is certain that induced mutations will play the major role in resolving the problem. It is likely that we will continue to look to

487

the microorganisms for chemical interpretations even though recombination data to date are intelligently interpretable mainly by analogy with behavior demonstrated by higher organisms.

At the chromosome level, the gap between the lower and higher forms needs to be bridged as Doctor Smith has pointed out, and he has suggested fungi as bridging organisms. Other problems left to the future are the relation of the mechanics of replication of the cistron to the duplication of the chromosome and its relation to recombination. Although a great deal is known about recombination, and the intrinsic and extrinsic factors which modify it, the main physico-chemical key is still lacking.

At the nuclear level the relation of cistron structure to the structure of the presumed related enzymes through which effects are mediated needs to be worked out.

At the cellular level, the inter-relations between gene and cytoplasm need clarification. In this important area, practical plant breeding utilization of the phenomenon of cytoplasmic male sterility has far outstripped the explanation of its cause.

At the organism level, the problem of differentiation of tissue is likely to be intriguing for sometime to come. At present it is difficult to define even the nature of the problem. Is it one of progressive gene mutations precisely controlled by special gene-regulating genes? What triggers the reaction of morphogenic factors? Do the morphological mutants represent gaps in the normal gene chain of command, or are they normal genes reacting normally but out of phase? Because they are so common, the chlorophyll mutants have tended to be disregarded except as tools to measure mutation induction efficiency and specificity. Discovery of the cause and the correction of some of their impaired functions might provide powerful tools for controlling physiological states within the plant and its concomitant effects on development.

The organism and population levels are of most concern to plant breeders, to ecologists, to taxonomists, to evolutionists, and to population geneticists. The primary concerns are (a) the relationship of genotype and response to specified environments, and (b) the distribution of genotypes among plant populations. Interlocus specificity with respect to genetic contribution to form and performance has been amply confirmed since Mendel's treatment of it. Evidence of recent origin for such specificity with respect to response to mutagenic agents might have been anticipated and further examples are bound to be forthcoming. This what still encourages the hope and fosters the belief that a degree of control over the type of locus which can be induced to mutate may yet be achieved. This, in turn, should influence the type of variability that can accumulate in populations.

Doctor Gregory has emphasized the idea that the spectrum of diversity of induced mutants which can be immediately recovered from a given organism is a property of the organism itself rather than a property of the mutagen used, and that this spectrum, in turn, is governed by the evolutionary history of the organism. Initial experience with induced mutations with a variety of organisms certainly confirms this view but has led in some instances to questions about the efficacy of mutation induction as a tool in plant breeding.

On the other hand, this view must necessarily be a temporary one if the idea is accepted that the purpose of mutation induction is to accelerate change. Under such accelerated change, accumulated variability might, in turn, be expected to widen the boundaries or raise the ceiling for new variability in an ever-accelerated cycle in which genetic variability must inevitably eventually become divorced from evolutionary history. The tools are available for such an approach.

Herein, it would seem, lies the importance of the work reported by Nilan and Konzak and by Caldecott on increasing the dose tolerances and induced mutation yield, and by Caldecott on the effect of recurrent treatments. It seems reasonable to expect that the coincident quantitative variation so effectively demonstrated by Gregory and his students and coworkers would naturally increase along with the more easily measured yield of qualitative mutants.

Attempts to decide whether "mutation breeding" (perhaps unfortunately coined) is or is not applicable to a given species are too often colored by the immediately foreseeable and sometimes limited objectives of a breeding program. Who can say what breeding objectives for a specific crop will be 50 years hence? Plant breeders are often so restricted by prejudices that hinder acceptance of new products that they must operate within relatively narrow confines of form, taste, and color. Under such confines, plant breeders, often against their better judgment, tend to become variability destroyers to satisfy immediate objectives. The importance of preserving genetic variation in "world collections" to counteract this tendency has been stressed, but must still be regarded as restrictive in that the limits imposed by evolutionary history are still operative.

Technological advances, often unrelated to plant breeding objectives, sometimes dramatically remove the binding restrictions of consumer acceptance. For years we have tailored cultural practices in maize to row widths determined by the convenience of handling a team of horses. We have tailored the form of the corn ear to the esthetic satisfaction of the farmer who contemplated a hand-harvested ear before tossing it against the bang-board of his wagon. I venture to predict that the advent of the picker-sheller for corn harvesting will now permit exploitation of a greater array of corn germ plasm variability than would have been dreamed possible just 3 years ago. The grain combine harvester removed the requirement for smooth awns in barley some years ago.

Whether or not there is an immediate need for attempts to induce specific mutations can be decided only by the interests of each individual breeder in relation to his present objectives. I would be the last to advocate indiscriminate use of mutation induction in every breeding program. On the other hand, I would be among the first to encourage continuation and extension of the use of the powerful tools now available for increasing variability. The use of appropriate mutants and their application to gaining an understanding of, and exploiting heredity and variation at all levels of development by geneticists and breeders, is indispensable to continuing progress.

Comments

SPARROW: Perhaps the variability in crossover frequency may already exist and should be more carefully sought and selected for. In *Trillium erectum,* collected from the wild, there is an extremely wide range of chiasmata frequency. If one assumes this character to be genetically controlled, selection of high chiasmata frequency plants should be considered in plant breeding programs where this character would be useful or desirable.

Ross: There is a development which could have a tremendous impact on mutation genetics and for that reason should be brought to the attention of this symposium. Dr. F. C. Steward of Cornell has been successful in obtaining free-cell cultures and plants from these. This holds the possibility of placing the study of genetics of higher plants on the cellular basis, perhaps, even on the same basis as bacterial genetics.

A second development, also out of this research, is the observation by Dr. J. Mitra and Doctor Steward of what appears to be an excellent exhibition of somatic reduction in cell cultures of Haplopappus. This observation of somatic reduction in Haplopappus suggests that it may also have occurred in sorghum and, thus, explain the homozygous nature of mutants occurring in certain varieties of sorghum as observed by our group at South Dakota State College.

General Résumé of Symposium

Résumé of the Symposium

S. G. STEPHENS

North Carolina State College,
Raleigh, N. C.

A RÉSUMÉ of our proceedings during the past week should properly require a summing-up in organized fashion of all the papers and discussions which have been presented here. But no more than a cursory study of our program is necessary to appreciate that this "Symposium on Mutation and Plant Breeding" includes two rather distinct areas of interest. One is concerned with the nature of genetic material, the other with the use of mutation in plant breeding. One attempts to interpret mutant events in terms of recent advances in genetic theory, the other endeavors to improve methods of producing and using mutants. The goals are different and so are the experimental procedures. They have in common certain "tools of trade" (mutagenic agents) and an interest in refining them for more precise usage. The search for specific mutagens as a source of directed mutations occupies the attention of workers in both areas, though it is by no means clear to this reviewer that the "specificity" they seek is of the same nature or order.

To review adequately two diverse and rapidly growing areas of research requires a level of competence, and first hand experience in both, which I do not possess. The embarrassing fact is that I have never intentionally used a mutagen, induced a mutant, nor applied the same to the betterment of the plant kingdom. Free of first hand experience and the pre-conditioning necessarily associated with it, I can only hope to be objective at the risk of being naive. If I am not mistaken, this is a situation which I share with a number of geneticists and plant breeders, and it is for them that this review is primarily intended.

Mutation and the Nature of Genetic Material

Early in this symposium, Auerbach pointed out that studies on the nature and action of mutagens have not led to new information on the nature of the gene. On the contrary, concepts of the nature of the gene, derived from DNA structure, have led to new

495

interpretations of the actions of mutagens. She lamented the current tendency to assume as a working hypothesis that most mutagenic action involves a direct chemical relation between the mutagen and DNA. Yet in the phages, where the DNA structure is presumably simple and not closely bound with protein, and in the bacteria also where mutant effects can be defined in specific chemical terms, the hypothesis has proved extraordinarily fruitful—as Smith has shown in his excellent review.

The "Freeseian" transitions, involving substitutions of the normal purine and pyrimidine bases by specific analogs, provide satisfying interpretations—even if still theoretical—of the experimental data. The natural extension of the concept is that any agent which can interact with normal cell constituents to produce base analogs is a potential mutagen. It does not follow necessarily that the *in vivo* induction and incorporation of base analogs is a simple process, though it is tempting to assume so in the case of nitrous acid (on the basis of *in vitro* evidence). Among alkylating agents, the evidence that dimethyl sulfate and tri-ethylene melamine react preferentially, with purines and pyrimidines respectively, suggests that *some* of their mutagenic activity *may* involve the production of base analogs *in vivo,* but there is no reason to suppose that this is their sole mode of action.

There are other and cruder ways of disturbing nucleotide sequences in DNA than by stepwise substitution of individual bases followed by miscopying, and these may involve a variety of metabolic pathways. Even when individual base analog substitution is a plausible mechanism, the pathway may be rather indirect. In bacteria, as Haas has reported in this symposium, a number of steps lead from the primary effects of UV irradiation to the phenotypic expression of the mutant. Pre-incubation with the bases normally present in RNA (adenine, guanine, cytosine, and uracil) increased the subsequent mutation rate with UV. Substituting the RNA base uracil by the DNA base thymine in the media reduced the mutation rate considerably. Each step in the pathway leading from mutant induction to phenotypic expression could be blocked by specific changes in the internal or external "environment" of the organism. If his interpretation—or my understanding of it— is correct, the potential mutant or precursor has first to be stabilized by some process requiring the participation of amino-acids. It is then presumably incorporated as a modified component of RNA, since blocking protein and RNA synthesis prevents its "fixation". Finally, it is coded into the DNA system, which, after replication, requiring further protein synthesis, is able to express the new mutant phenotype. Haas has further suggested that some X-ray induced mutants are produced by a similar mechanism; others may be directly incorporated into DNA.

In the higher organisms where the DNA structure is complex (multistranded bundles, possibly linked tandem-wise by divalent cation bonds, and intimately bound with protein) and where mutant effects can be distinguished only as gross morphological deviants, the probability that similar end-effects may be brought about in various ways becomes very great. In view of the complexity it is not surprising that different organisms react very differently to the same mutagen, and that, conversely, a particular mutagen will produce its own characteristic reaction pattern, different from that of other mutagens, in any particular organism. While similar "spectra of reactions" to mutagens may imply common primary effects in microorganisms, it would seem particularly risky to extend the implication to superficially equivalent cases in higher organisms. For instance, from the theoretical point of view, alkylating agents could react with a variety of chemical groups—all available in the components of an intact chromosome, in their possible precursors, and in a number of cellular constituents not obviously associated with nucleic acid synthesis. One cannot assume that where several types of reaction are possible, the same one is operative; either in different organisms or in different mutagens which possess more than one kind of reactive group in common. At this level comparisons become idle until it can be shown that all mutagenic pathways converge at one or other "critical step"—perhaps the incorporation of a base analog, the breakage of a cation bond, or some other plausible event leading to a change in nucleoprotein structure.

Smith has reported his success in incorporating labelled base analogs in the nuclei of *Vicia faba,* but so far it is not known whether this will result in mutagenic effects. If incorporation is followed by miscopying on the Freeseian model, it need not necessarily produce immediate phenotypically recognizable mutants. The multi-stranded structure of the DNA might provide "cover" for miscopying at a specific site in an individual thread; alternatively, matched miscopying in 32 threads simultaneously might allow mutant expression.

Two other kinds of "critical step" in a mutagenic process can be easily visualized as (1) the deletion or rearrangement of groups of nucleotides within an individual DNA sequence and (2) the breakage of cation bonds between neighboring DNA sequences, possibly followed by the deletion or relative rearrangement of whole sequences. In the higher organisms these chemically different "critical steps" may be operationally indistinguishable so long as they can be recognized only by chromosomal breaks or by gross phenotypic changes.

Extra- and Intra-genic Mutation

Stripped to its bare essentials, the DNA-derived concept of genetic structure is a serial coding system in which the determination of specific information depends more on the spatial relationships in the system as a whole than on the properties of its individual components. The latter are only two in number, the base pairs adenine:thymine and guanine:cytosine. Similarly, the number of letters in our alphabet does not exert a serious limitation to the exchange of information; in fact, several letters are superfluous. It is not the number of different kinds of letters which is important but the way in which they are arranged.

Strauss[1] has likened the nucleotide sequence in DNA to a sentence in which the letters are base pairs and the words, genes. What functions as spaces between words is not clearly understood, though it is possible that certain recurring patterns in the nucleotide sequence indicate "stops" between one gene "word" and the next. The ultimate unit is the base pair at a particular site; for example, A-T, T-A, G-C, or C-G. Since the same base pairs occur over and over again throughout the sequence, it follows that they can possess no genetic properties *per se*. They become genetically meaningful only when associated with other base pairs in a local sub-sequence possessing a unique spatial configuration. The sub-

---

[1]Strauss, B. S. 1960. An Outline of Chemical Genetics. *Philadelphia: W. B. Saunders Co.*

sequence is the cistron or functional gene. Within a cistron mutation is due to the deletion, duplication, or re-arrangement of base
pairs. All such changes alter spatial configuration within the cistron, and hence modify its coding system. At this level of organization the concept of "point mutation" is meaningless, e.g., as far as
we know an adenine base cannot mutate into a modified form of
adenine and remain permanently incorporated in the system. *All
mutations within a cistron, i.e., intra-genic mutations, involve a
change in the order of its separable base pairs.* Theoretically, extragenic mutations would involve deletions, duplications, or re-arrangements of whole cistrons without disturbing their internal structures. It is not clear whether intra- and extra-genic mutations would
result from different mechanisms or whether it would be possible
always to discriminate between them. Neither is it clear whether
the cistrons are serially discrete or overlapping.

If we try to extrapolate from the nucleotide level to the chromosomal level in higher organisms, we run into semantic difficulties.
Here the criterion of separability (recombination) is applied to distinguish "extra-genic" from "intra-genic" events, and the latter
are assumed to be true "point mutations". But we have seen at
the nucleotide level that "point mutations" probably do not exist
and that separability is no valid criterion for distinguishing between
extra- and intra-genic mutation. Separability is not only an invalid
criterion, it suffers from the disadvantage that the answer obtained
varies with the sensitivity of the test. This is well illustrated by the
interesting case of the *A* complex in maize whose recent experimental
history has been presented by Laughnan. Earlier work, commenced
by Stadler and developed independently by Laughnan, had shown
that the *A* complex was separable into alpha and beta elements
which could be considered tandem serial repeats. By setting up
test-crosses with appropriately marked chromosomes, it could
be shown that certain mutant events in the *A* complex were always
associated with crossing over, while others were not. Separability, therefore, became the criterion for distinguishing "extragenic" sheep from "intra-genic" goats. Later, a more rigorous examination of the "intra-genic" mutants by Laughnan has shown that
they do not meet the specifications of true "point mutations" but
can best be explained by physical losses of alpha and beta elements.

To account for these losses he has developed an ingenious model of double-loop pairing in the *A* region during meiosis. This allows recombination either normally between homologous strands, or internally between adjacent segments of the same strand (auto-association). Although the proposed model is below the limits of optical verification, it is based on good inferential evidence. (It is known that the losses occur as a result of meiosis and, from independent experiment, that the alpha and beta elements can engage in oblique synapsis). Assuming the model to be correct, the evidence shows that mutations formerly supposed to be "intra-genic" on the basis of standard recombination tests, turn out to be "extra-genic" when the tests are made more rigorous. The goat pen is currently empty and the chances of further screening are limited by the difficulty of obtaining new presumptive goats!

The great technical difficulty of obtaining large enough populations to test rare mutants in this material probably sets a limit to continued analysis along the same lines. However, it is amusing to speculate where its continuation might lead. Because oblique synapsis occurs between the alpha and beta elements it should be possible to synthesize an extended series of repeats. The number of possible recombinants would increase exponentially and models based on triple, quadruple, etc., loop pairing would be permissible. The possible exchanges would be difficult to portray in a 2-dimensional model; conversion to a 3-dimensional model would produce a double helix, resembling a simplified version of the Watson-Crick formula. Diagrammatically, recombinations would involve exchanges between neighboring gyres in the nucleotide sequence and their consequences would not be distinguishable from those resulting from Freeseian transitions. In short, the distinction between "extra-genic" and "intra-genic" mutations would become solely a matter of interpretation.

In the higher organisms it would seem that a criterion other than separability is needed to distinguish between extra- and intra-genic events. An acceptable criterion should be based on function and be capable of critical test. The cistron concept, developed in microorganisms, fulfills these requirements. It is a functional unit in the genetic sense and when the serial order of its base pairs is interrupted, either by rearrangement or physical separation, that function is impaired. In the diploid condition at least one of the homologous cistrons must preserve the normal sequence of base pairs if the normal phenotype is to be produced. If the sequences in both cistrons are interrupted, either at identical or different sites, a mutant phenotype is expressed (the *trans* effect). If the site of interruption is different in the two cistrons, then breakage between the sites followed by recombination may restore the normal sequence in one cistron and produce a double interruption in the other.

The joint action of two homologous cistrons, one with a normal sequence and the other with a double interruption, would produce a normal phenotype (the *cis* effect). According to this concept, the whole lozenge "complex" in Drosophila would be considered a single cistron and its "loci" merely sensitive sites or "hot spots". From the phenotypic point of view, mutant combinations exhibiting different *cis* and *trans* effects should result from rearrangements within a pair of homologous cistrons. On the other hand, if the "mutant" effects are restricted to recombinations in the serial order of adjacent cistrons or cistron "repeats", there should be no phenotypic distinction between *cis* and *trans* arrangements. On this basis, the alpha and beta "elements" in the *A* complex of corn would be considered repeated cistrons—not because they are separable but because *cis* and *trans* combinations of their "mutants" appear to be identical phenotypically.

Although the application of the cistron concept to higher organisms leads to certain interpretative difficulties, e.g., in the case of dominant mutants, it might provide a unifying principle leading to a better understanding of position effects. Breakage *within* a cistron followed by separation of the broken segments through translocation should produce mutant effects; breakage *between* intact cistrons followed by translocation should not. It might be easier chemically for broken cistron ends to be linked to a heterochromatic region than to a euchromatic region. If só, broken cistrons with accompanying mutant effects might be recovered selectively in heterochromatic regions, hence accounting for the association between position effects and heterochromatin. Further, the frequency of position effects in Drosophila and their comparative rarity in other organisms might have nothing to do with relative heterochromatic properties and be only partly explained by the superior screening techniques available in the fruit-fly. A simpler explanation might be that the Drosophila chromosome is composed of relatively longer or more easily broken cistrons.

Today one cannot easily be critical of Goldschmidt's[2] (then controversial) statement of 1955:

> "The important point is that there is no individual locus and no gene which mutates, but a segment of a chromosome which is rather large in chemical terms, which has an orderly, internal structure of a definite sequence, i.e., polarized, and which may even overlap the next one; and that any happening within such a segment which changes this sequence visibly or invisibly appears as a mutant . . . . ."

Specificity

In this Symposium we have heard the word "specificity" used often, in various connections and at several different levels. Dr. Li has told us that a term may be used in any sense we wish, providing that we first define it. I am not so sure that we have a common definition in the present instance; in fact, we have achieved an "unspecific specificity".

Specificity presumably implies a unique property or class of properties. At the chemical level the substitution of a pyrimidine base normally present in DNA by a base analog is highly specific in a chemical sense. But since the natural pyrimidine base occurs over and over again in the nucleotide sequence, the substitution *per se* can hardly be considered genetically specific. Genetic specificity would require that substitution would only take place when the natural base were located in a certain position in the sequence. It does not follow, then, that a base analog needs possess any *genetic* specificity as a mutagen, though it may be "site-specific" within a given locus.

At quite another level of organization, different spectra of reactions resulting from applications of different mutagens have been presented as evidence of "specificity". One feels confident that there is an underlying truth in this viewpoint, but the term has to be employed so loosely that it is not particularly useful. Few of the comparisons seem clear cut. One has a series of more or less

---

[2]Goldschmidt, R. B. 1955. Theoretical Genetics. *Berkeley: University of California Press. Page 161.*

independent variables: (a) between organisms; (b) between types of event, "break' or "mutant"; (c) between heterochromatin and euchromatin; (d) between positions in chromosomes; and (e) between alleles at the same locus. When most of these variables are confounded in a system, the number of possible permutations in effects is quite large, and consequently the chances of obtaining similar effects quite low. The unsophisticated observer may be pardoned for wondering how many of the observed differences are due to "specificity" and how many to the laws of statistical probability. Undoubtedly, there are differences in reaction pattern, but if these shift every time a new variable is added, the concept of specificity does not appear to be helpful.

It has also been suggested that one of the promising lines of research with mutagens is the development of specific mutagens by which mutation can be directed. As a tool in refined genetic and biochemical analysis this is obviously true. But I think it would be difficult for a practising plant breeder to decide what kind of specific mutagen should be given priority and to what end he would direct it. The level of specificity required by the geneticist, as I see it, would be of limited value to the practical breeder. A mutagen which would selectively substitute a particular base pair at a particular site and result in a changed amino acid sequence in a known protein would be invaluable to the geneticist, as Smith has already pointed out. But between a particular protein and the phenotype recognized by the plant breeder lies a long and completely unknown pathway through enzyme syntheses, differentiation, and development. The characteristics with which the plant breeder must work— earliness, disease resistance, stiffness of straw, yield, and the like— are not easily expressed in chemical terms at the *end* of synthetic processes, much less at the *beginning*. Again, to use a directed mutation, it is first necessary to know what direction, and presumably, the nature of the change which would produce it. It is perhaps unlikely that a mutagen which would produce exclusively mutants resistant to a particular disease would be useful against different strains of the disease in the same variety, or against similar strains in different varieties. The more specific the mutant, the more limited its use is likely to be.

On the contrary, I would suggest that the reaction patterns

associated with certain mutagens are already approaching a degree of differentiation which would be of optimum usefulness to the plant breeder. These are:

  (a) Chemical mutagens capable of producing a high yield of visible mutants and a low yield of concomitant breaks. The problems of recurrent screening for fertility and the preservation of unthrifty "carriers" of mutants, which have been described by Nilan, Gaul, and others, should be reduced proportionately with a reduction in chromosome breakage.

  (b) Agents capable of producing chromosomal breakage unaccompanied by visible mutations. These would seem to be of potential use in interspecific breeding where the problem involves the transference of a chromosomal segment from one genome to another, e.g., Sears' transference of rust resistance from *Aegilops umbellulata* to wheat.

It is possible that continued research will lead to the discovery of more efficient mutagens and breakage agents, but, beyond certain limits, the more restricted their action, the less generally useful will they be for plant breeding purposes.

In her review of chemical mutagens, Auerbach concluded that, "while there is good evidence that a common mechanism is responsible for the production of gene mutations, chromosome breaks and minute re-arrangements by chemicals, crossing-over seems to be induced in some different way". From the plant breeding point of view this is an unfortunate situation because methods of controlling chromosomal recombination have at least as great a potential value as methods of inducing mutations. In any heterozygous population of finite size, linkage imposes stringent restrictions on the genetic variation which can be expressed. Its effects only become negligible as the population approaches homozygosity through selection and/or close inbreeding. Following hybridization, linkage favors the recovery of a desirable parental complex, but makes more difficult the insertion of specific improvements. Conversely, the successful recombination of certain selected characteristics can only be gained at the price of disrupting other desirable parental combinations resulting from uncontrolled crossing over. The problem can be alleviated though not removed by the use of recurrent backcross techniques. The latter have usually been limited to the transfer of a few simply inherited characters and in wide crosses are often complicated by close linkages and/or apparent pleiotropic effects. Theoretically, it should be advantageous to minimize crossing over in the early stages of a breeding program so that the genome of one of the parents could be recovered intact save for the particular chromosome or chromosomal segment which carried the property to be transferred. At a later stage enhanced crossing over would permit the "whittling down" of the transferred segment without disrupting the parental combinations in the remainder of the genome. Chemical or other agents which would serve as general crossover enhancers and suppressors might be of considerably practical value to the plant breeder.

"Mutation Breeding" and "Conventional Breeding" Methods

An evaluation of the comparative merits of the so-called "mutation breeding" and "conventional breeding" methods is perhaps premature at this time. A comparison between the novel and relatively untested, on the one hand, and the established and relatively conservative, on the other, could be unfair to both. I would prefer to agree with MacKey's viewpoint that "plant breeding is nothing more than controlled evolution" and that "it seems very likely that mutation work in the future will be more completely interwoven with the so-called conventional methods, with which it is now too often set in opposition."

The degree to which evolution can be controlled by the plant breeder is limited by (1) the amount and kind of genetic variation available in the breeding population, and (2) the efficiency with which the most desirable genetic combinations can be accumulated through selection.

Theoretically, the first limitation can be reduced either by inducing mutations, by outcrossing, or by a combination of both techniques. Because the effects of mutagens can be measured more readily in inbred homozygous populations, we have little critical information on their effects in cross bred populations where the results of mutagenic activity and hybridization are confounded. It would be interesting to know, in a range of crop plants, the relative increases in genetic variance which could be expected from the irradiation of parental lines as compared with those resulting from normal F_2 segregation. The studies reported by Gregory represent pioneer experiments in an otherwise unexplored field. They also raise the interesting question whether the genotypic response to mutagenic action may be a property resident in the breeding system *per se,* and not entirely conditioned by such factors as evolutionary age, level of adaptation, inherent mutagenic resistance, tolerance of aneuploidy, level of ploidy, nuclear volume, etc., which have been considered previously and in more detail by MacKey and Sparrow.

An important limitation to the expression of genetic variation is the effective size of the population which can be studied. In conventional breeding programs population size may be restricted by incompatibility and sterility barriers and by technical difficulties in obtaining F_1 seeds in sufficient quantities. In mutation breeding the effective population size is also limited, not by the number of seeds which can be irradiated, but by the number and kinds of plants which survive after initial injury, induced sterility, and diplontic selection have taken their collective toll. Nevertheless it appears that the judicious use of appropriate mutagens, the manipulation of environmental conditions and the elaboration of screening techniques which have been discussed by Gaul, Nilan and Konzak, Caldecott, and others, provide a greater opportunity for accumulating variation per unit population size than do conventional breeding methods. It may well be, but has yet to be demonstrated, that the amount of *useful* variation is proportional to the *total* amount of variation present.

Many of the conventional plant breeder's efforts are applied to selection in polygenic systems. These sometimes appear to be genetically incomprehensible and statistically exhausting, but they have contributed considerably to long-term plant improvement. The breeder cannot always expect to find in his material economically valuable characters which behave as Mendel thought they should. But when he finds opportunities to obtain disease resistance and other specific qualities through simple gene transference, he accepts them gratefully. Some of the most spectacular advances have been made through the selection of qualitative characters, and some crops like sorghum, as Quinby has shown, are particularly well-endowed with such opportunities. In general, though, it is probably fair to say that spectacular advances through the incorporation of qualitative characters have always been accompanied by the long-term selection of appropriate polygenic complexes which have led to the production of improved, well-adapted varieties. If the genetic variation in an irradiated population is essentially of the same kind as that existing in natural populations, it is a little difficult to understand why the mutation breeder should pay so much attention to the production of "the occasional, spectacular, useful variant" (Laughnan) and so little to the polygenic background. It is possible that some of the potentially most useful variation is discarded when selection is made on the basis of mutant phenotype.

The second limitation is common both to mutation breeding and conventional breeding, and it involves certain problems which are not currently by-passed by either system. Whether the initial pool of genetic variation is enriched by mutation induction or by hybridization, the fact remains that in subsequent generations most of it is expressed as recombinational variance. Every new mutant introduced into a population can, theoretically, generate three new genotypes: with n mutants there are 3^n genotypes. A single mutational event therefore triples the potential number of different genotypes present in the initial population. This means that a relatively low mutation rate, or a relatively small amount of hybridization, can generate a large amount of recombinational variance. In many conventionl programs the time-consuming element is not the inclusion in the breeding population of a new character *per se,* but its stabilization and effective selection from a bewildering variety of possible combinations. While it is possible that this time-consuming task may be reduced in certain mutation breeding systems, I do not think that a reduction need follow as a natural consequence and advantage of mutation breeding in general. Gaul has suggested that if a character is available either in a "primitive or non-adapted form" or in a "mutant collection of an adapted variety", it will be much easier to use the latter in a breeding program. Caldecott has pointed out the potential advantages of a recurrent irradiation program, aimed at accumulating favorable mutants in a common breeding pool. These advantages over conventional methods are readily appreciated if we can assume that an adapted variety will remain adapted in respect of its residual genotypic background after mutagenic action. Otherwise the recovery of adapted forms from a segregating population remains a problem—no less for the mutation breeder than for the conventional breeder. Gregory has shown in irradiated populations of peanuts that the amount of readily visible mutants expressed may be quite unrepresentative of the total (polygenic) variation generated. Moreover, "the remarkable variation in (morphological) mutant expressivity" attributed to polygenic segregation in this material appears to be of a similar order to that described in segregating progenies of inter-specific hybrids in certain other genera, e.g., Lycopersicon and Gos-sypium. There remains the possibility that the amount of poly-genic variation engendered by radiation may be of the same order but differ in kind from that resulting from interspecific hybridiza-tion. If so, it may be easier to recover adapted types from irradiated populations than from segregating populations of interspecific ori-gin. This remains to be demonstrated.

The combination of a long generation cycle and complex heterozygosity, sometimes accompanied by aneuploidy and self-in-compatibility, provides a particularly nasty problem for the breed-er of vegetatively propagated crops. The compensating advantages associated with clonal propagation and abundance of meristems cannot be used readily in a conventional (sexual) breeding system and have led to a supplement of the latter by a form of "histolog-ical engineering". This involves the recognition of somatic mutants and chimeras, and their extraction and isolation by pruning followed by asexual propagation. Under natural conditions the limiting fac-tor is the rate at which somatic mutation occurs. The possibilities of increasing mutation frequency, re-arranging somatic tissues, and breaking self-incompatibility mechanisms through a combination of irradiation and ingenious manipulative techniques have been reviewed by Nybom. In this area, where the opportunities for improving conventional methods seem to be extremely limited, the use of induced "mutation" in a broad sense would seem to offer much promise.

Mutation induction and conventional breeding systems are not necessarily opposed. Certain specific problems indicate rather clearly the advantages of a combined attack. The breakage of self-incompatibility systems, certainly, and the reconstitution of sexual from apomictic systems, possibly, can be achieved through muta-genic treatment. Success in either would open up new potentialities in conventional breeding. Recombination between nonhomologous genomes becomes practically feasible when irradiation is combined with conventional cytogenetic techniques, as Sears has shown. It may be possible to break cross-incompatibility barriers by gamete irradiation as suggested by MacKey. These and other possibilities share the requirement that an experience with mutagenic techniques be combined with a knowledge of the cytogenetics and breeding potential of the material under investigation.

From these, still early, indications it would appear likely that a distinction between mutation breeding and conventional breeding is only a transitory phase in the development of both. In my opinion their delayed incorporation is in large part due to the fact that mutation breeding has developed as a by-product of "mutagenic exercises". These tend to concentrate on the immediate production of novel morphological deviants and to neglect the possibilities of recombination among apparently normal segregates in subsequent generations. Yet the successful manipulation of recombination is a keystone in practical breeding.

Appendix

List of Those Attending
or Participating in the
Symposium

List of Those Attending or Participating in the Symposium

Allison, J. Lewis, The Farm Seed Research Corporation, 3289 Chester Lane, Bakersfield, California

Anderson, Ronald E., Department of Plant Breeding, Cornell University, Ithaca, New York

Ariyanayagam, David V., The Rockefeller Foundation, Department of Genetics, University of California, Davis, California

Atkin, John D., Department of Vegetable Crops, New York State Agricultural Experiment Station, Geneva, New York

Atwood, Kimball C., Department of Microbiology, University of Illinois, Urbana, Illinois

Auerbach, Charlotte, Institute of Animal Genetics, University of Edinburgh, Edinburgh, Scotland

Bai, Daihan, (International Atomic Energy Agency Exchange Visitor, Agricultural Experiment Station, Suwon, Korea), Department of Biology, Brookhaven National Laboratory, Upton, Long Island, New York

Bammi, R. K., Department of Botany, Ontario Agricultural College, Guelph, Canada

Banks, Harlan P., Department of Botany, Cornell University, Ithaca, New York

Bishop, C. J., Department of Agriculture, Research Branch, Central Experimental Farm, Ottawa, Canada

Bollich, Charles N., Crops Research Division, Agricultural Research Service, United States Department of Agriculture, and Louisiana State University, Rice Experiment Station, Crowley, Louisiana

Brawn, Robert I., Department of Agronomy and Genetics, McGill University, Macdonald College, Quebec, Canada

Brewbaker, James L., Department of Biology, Brookhaven National Laboratory, Upton, Long Island, New York

Britton, Donald M., Department of Botany, Ontario Agricultural College, Guelph, Canada

Burton, Glenn W., Crops Research Division, Agricultural Research Service, United States Department of Agriculture, and University of Georgia, Coastal Plain Experiment Station, Tifton, Georgia

Caldecott, Richard S., Division of Biology and Medicine, United States Atomic Energy Commission, Washington, D. C.

Cameron, James W., Department of Horticulture, University of California, Riverside, California

513

Camp, Lewis M., Research Director, Pfister Associated Growers, Inc., Aurora, Illinois

Casas, Eduardo, The Rockefeller Foundation, Department of Field Crops, North Carolina State College, Raleigh, North Carolina

Caspar, Alan L., The Blandy Experimental Farm, University of Virginia, Boyce, Virginia

Chang, Lilli, Department of Plant Breeding, Cornell University, Ithaca, New York

Chase, Sherret S., DeKalb Agricultural Association, Inc., Dekalb, Illinois

Clark, J. Allen, Agricultural Board, National Academy of Science, 2101 Constitution Avenue, Washington, D. C.

Coe, E. H. Jr., Crops Research Division, Agricultural Research Service, United States Department of Agriculture, and the University of Missouri, Columbia, Missouri

Cole, Randall K., Department of Poultry Husbandry, Cornell University, Ithaca, New York

Cushing, R. L., Pineapple Research Institute of Hawaii, Honolulu, Hawaii

Davies, D. Roy, Isotope Division, Wantage Radiation Laboratory, Grove, Wantage, Berkshire, England

Dermen, Haig, Crops Research Division, Agricultural Research Service, United States Department of Agriculture, Beltsville, Maryland

Dewey, Wade G., Department of Agronomy, Utah State University, Logan, Utah

Emery, Donald A., Department of Field Crops, North Carolina State College, Raleigh, North Carolina

Emmerling-Thompson, Margaret, Department of Plant Breeding, Cornell University, Ithaca, New York

Emsweller, S. L., Crops Research Division, Agricultural Research Service, United States Department of Agriculture, Beltsville, Maryland

Evans, H. J., Biology Department, Brookhaven National Laboratory, Upton, Long Island, New York

Evans, Marshall, Green Giant Company, LeSueur, Minnesota

Everett, Herbert L., Department of Plant Breeding, Cornell University, Ithaca, New York

Ewart, Lowell C., Joseph Harris Company, Inc., Moreton Farm, Rochester, New York

Federer, W. T., Biometrics Unit, Department of Plant Breeding, Cornell University, Ithaca, New York

Flor, H. H., Crops Research Division, Agricultural Research Service,

United States Department of Agriculture, and North Dakota Agricultural College, Fargo, North Dakota

Flory, W. S., The Blandy Experimental Farm, University of Virginia, Boyce, Virginia

Frakes, Rod V., Department of Farm Crops, Oregon State College, Corvallis, Oregon

Frey, K. J., Department of Agronomy, Iowa State University, Ames, Iowa

Gableman, W. H., Department of Horticulture, University of Wisconsin, Madison, Wisconsin

Garber, E. D., Department of Botany, University of Chicago, Chicago, Illinois

Gaul, Horst, Max-Planck-Institute für Züchtungsforschung, Köln-Vogelsang, Germany

Gershoy, Alec, Department of Botany, University of Vermont, Burlington, Vermont

Gibson, P. B., Crops Research Division, Agricultural Research Service, United States Department of Agriculture, and Department of Agronomy and Soils, Clemson College, Clemson, South Carolina

Grahn, Douglas, Division of Biology and Medicine, United States Atomic Energy Commission, Washington, D.C.

Grant, W. F., Department of Genetics, McGill University, Macdonald College, Quebec, Canada

Gregory, M. Pfluge, Department of Genetics, North Carolina State College, Raleigh, North Carolina

Gregory, Walton C., Department of Field Crops, North Carolina State College, Raleigh, North Carolina

Grobman, Alexander, Cooperative Program for Maize Research, Universidad Agraria, La Molina, Lima, Peru

Grün, Paul, Department of Botany and Plant Pathology, Pennsylvania State University, University Park, Pennsylvania

Haas, Felix L., Department of Biology, M. D. Anderson Hospital, The University of Texas, Texas Medical Center, Houston, Texas

Hadley, Henry H., Department of Agronomy, University of Illinois, Urbana, Illinois

Hanson, A. A., Crops Research Division, Agricultural Research Service, United States Department of Agriculture, Beltsville, Maryland

Hanson, C. H., Crops Research Division, Agricultural Research Service, United States Department of Agriculture, Beltsville, Maryland

Hanway, D. G., Department of Agronomy, University of Nebraska, Lincoln, Nebraska

Harvey, Paul H., Department of Field Crops, North Carolina State College, Raleigh, North Carolina

Henderson, C. R., Department of Animal Husbandry, Cornell University, Ithaca, New York

Heyne, E. G., Department of Agronomy, Kansas State University, Manhattan, Kansas

Ho, Ti, The Blandy Experimental Farm, University of Virginia, Boyce, Virginia

Hough, L. Fredric, Department of Horticulture, Rutgers University, New Brunswick, New Jersey

Hunter, A. W. S., Department of Agriculture, Research Branch, Genetics and Plant Breeding Research Institute, Ottawa, Canada

Hutt, F. B., Department of Poultry Husbandry, Cornell University, Ithaca, New York

Jensen, N. F., Department of Plant Breeding, Cornell University, Ithaca, New York

John, C. A., Crop Research Department, Research and Quality Control Division, H. J. Heinz Company, Bowling Green, Ohio

Johnson, Elmer C., The Rockefeller Foundation, Calle Londres 40, Mexico, D. F.

Johnson, I. J., Caladino Farm Seeds, Inc., Wheaton, Illinois

Jones, C. M., Department of Agronomy and Soils, Clemson College, Clemson, South Carolina

Jump, Lorin, Funk Brothers Seed Company, Bloomington, Illinois

Kirby-Smith, John S., Biology Division, Oak Ridge National Laboratory, Oak Ridge, Tennessee

Konzak, C. F., Department of Agronomy, Washington State University, Pullman, Washington

Kramer, H. H., Department of Agronomy, Purdue University, Lafayette, Indiana

Krull, Charles, The Rockefeller Foundation, Apartado Aereo 58–13, Apartado Nacional 32–79, Bogota, Colombia

Kwack, Beyoung H., Department of Biology, Brookhaven National Laboratory, Upton, Long Island, New York

Kyle, Wendell H., Agricultural Research Service, United States Department of Agriculture, Pioneering Research Laboratory, Population Genetics Institute, Purdue University, Lafayette, Indiana

Larter, E. N., Department of Field Husbandry, University of Saskatchewan, Saskatoon, Canada

Langham, D. G., Botany Department, Yale University, New Haven, Connecticut

Lantican, Ricardo, The Rockefeller Foundation, Department of Agronomy, Iowa State University, Ames, Iowa

Laughnan, J. R., Department of Botany, University of Illinios, Urbana, Illinois

Lewis, Charles F., Crops Research Division, Agricultural Research Service, United States Department of Agriculture, Beltsville, Maryland

Lewontin, Richard, Department of Biology, University of Rochester, Rochester, New York

Li, C. C., Department of Bio-statistics, Graduate School of Public Health, University of Pittsburgh, Pittsburgh 13, Pennsylvania

Loesch, P. J., Crops Research Division, Agricultural Research Service, United States Department of Agriculture, and the University of Missouri, Columbia, Missouri

Love, H. H., Department of Plant Breeding, Cornell University, Ithaca, New York

Luckett, J. D., 555 West North, Geneva, New York

MacKey, James, The Swedish Seed Association, Svalöf, Sweden

Mahoney, Charles H., National Canners Association, 1133 20th Street, N. W., Washington, D. C.

Marx, G. A., Department of Vegetable Crops, New York State Agricultural Experiment Station, Geneva, New York

Matthews, D. L., Eastern States Farmer's Exchange, Inc., 26 Central Street, West Springfield, Massachusetts

Mehlquist, G. A. L., Plant Science Department, University of Connecticut, Storrs, Connecticut

Mericle, L. W., Department of Biology, Brookhaven National Laboratory, Upton, Long Island, New York

Mericle, R. P., Department of Biology, Brookhaven National Laboratory, Upton, Long Island, New York

Metzger, R. J., Crops Research Division, Agricultural Research Service, United States Department of Agriculture, and Oregon State University, Corvallis, Oregon

Moore, J. N., Department of Horticulture, Rutgers University, New Brunswick, New Jersey

Munger, H. M., Departments of Plant Breeding and Vegetable Crops, Cornell University, Ithaca, New York

Murphy, H. C., Crops Research Division, Agricultural Research Service, United States Department of Agriculture, Beltsville, Maryland

Murphy, R. P., Department of Plant Breeding, Cornell University, Ithaca, New York

Murray, Beatrice, Department of Agriculture, Genetics and Plant Breeding Research Institute, Ottawa, Canada

Myers, W. M., Department of Agronomy and Plant Genetics, University of Minnesota, St. Paul Campus, St. Paul, Minnesota

Nelson, Thomas C., Antibiotics Development and Assay, Eli Lilly and Company, Indianapolis, Indiana

Nilan, R. A., Department of Agronomy, Washington State University, Pullman, Washington

Norris, L. C., Department of Poultry Husbandry, University of California, Davis, California, and Professor Emeritus, Cornell University, Ithaca, New York

North, David T., Department of Agronomy and Plant Genetics, University of Minnesota, St. Paul Campus, St. Paul, Minnesota

Nuffer, M. G., Department of Field Crops, University of Missouri, Columbia, Missouri

Nybom, Nils, Balsgård Fruit Breeding Institute, Fjälkestad, Kristianstad, Sweden

Olmo, Harold P., Department of Viticulture and Enology, University of California, Davis, California

Oberle, George D., Department of Horticulture, Virginia Polytechnic Institute, Blacksburg, Virginia

Ogle, Charles W., Pfister Associated Growers, Aurora, Illinois

Okabe, Shiro, The Rockefeller Foundation, Department of Agronomy, Iowa State University, Ames, Iowa

Osborne, Thomas S., Plant Genetics, UT-AEC Agricultural Research Laboratory, and University of Tennessee, Oak Ridge, Tennessee

Pardee, William D., Department of Plant Breeding, Cornell University, Ithaca, New York

Pate, J. B., Crops Research Division, Agricultural Research Service, United States Department of Agriculture, and The University of Tennessee, United States Cotton Field Station, Knoxville, Tennessee

Patel, K. A., The Rockefeller Foundation, Department of Field Crops, North Carolina State College, Raleigh, North Carolina

Patterson, F. L., Department of Agronomy, Purdue University, Lafayette, Indiana

Peterson, Peter A., Department of Agronomy, Iowa State University, Ames, Iowa

Plaisted, Robert L., Department of Plant Breeding, Cornell University, Ithaca, New York

Poehlman, J. M., Department of Field Crops, University of Missouri, Columbia, Missouri

Pratt, Charlotte, Department of Pomology, New York State Agricultural Experiment Station, Geneva, New York

Quinby, J. Roy, Superintendent and Agronomist in charge of Sorghum Investigations, Texas Agricultural Experiment Station, Chillicothe, Texas

Ramage, R. T., Crops Research Division, Agricultural Research Service, United States Department of Agriculture, and the University of Arizona, Tucson, Arizona

Randolph, L. F., Department of Botany, Cornell University, Ithaca, New York

Reinbergs, E., Department of Field Husbandry, Ontario Agricultural College, Guelph, Canada

Rhoades, M. M., Department of Botany, Indiana University, Bloomington, Indiana

Riedl, William A., Department of Agronomy, University of Wyoming, Laramie, Wyoming

Rinke, E. H., Department of Agronomy and Plant Genetics, University of Minnesota, St. Paul, Minnesota

Robinson, H. F., Department of Genetics, North Carolina State College, Raleigh, North Carolina

Robson, D. S., Biometrics Unit, Department of Plant Breeding, Cornell University, Ithaca, New York

Ross, James G., Department of Agronomy, South Dakota State College, College Station, Brookings, South Dakota

Rudolph, Thomas D., Forest Service, United States Department of Agriculture, Rhinelander, Wisconsin

Sanderson, Kenneth E., Department of Plant Breeding, Cornell University, Ithaca, New York

Sarvella, Patricia, Department of Agronomy, Mississippi State University, State College, Mississippi

Scarascia, G. T., Comitato Nazionale per l'Energia Nucleare, Centro di Studi Nucleari della Casaccia, C. P. 1., S. Maria di Galeria, Roma, Italy

Schaeffer, Gideon W., Department of Biology, Brookhaven National Laboratory, Upton, Long Island, New York

Schaible, Lester W., Campbell Soup Company, Riverton, New Jersey

Schwinghamer, E., Department of Biology, Brookhaven National Laboratory, Upton, Long Island, New York

Sebesta, Emil, E., Department of Agronomy, Oklahoma State University, Stillwater, Oklahoma

Shapiro, Seymour, Department of Biology, Brookhaven National Laboratory, Upton, Long Island, New York

Shifriss, Oved, Department of Horticulture, Rutgers University, New Brunswick, New Jersey

Singleton, W. Ralph, The Blandy Experimental Farm, University of Virginia, Boyce, Virginia

Skirm, George W., Research Department, Asgrow Seed Company, 449 Derby-Milford Road, Orange, Connecticut

Smith, D. C., Department of Agronomy, University of Wisconsin, Madison, Wisconsin

Smith, Harold H., Department of Biology, Brookhaven National Laboratory, Upton, Long Island, New York

Sparrow, Arnold H., Department of Biology, Brookhaven National Laboratory, Upton, Long Island, New York

Sprague, G. F., Crops Research Division, Agricultural Research Service, United States Department of Agriculture, Beltsville, Maryland

Stephens, S. G., Department of Genetics, North Carolina State College, Raleigh, North Carolina

Stewart, Robert N., Crops Research Division, Agricultural Research Service, United States Department of Agriculture, Beltsville, Maryland

Stinson, Harry T., Jr., Department of Genetics, Connecticut Agricultural Experiment Station, New Haven, Connecticut

Strauss, B. S., Department of Microbiology, The University of Chicago, Chicago, Illinois

Streetman, L. J., Cytogeneticist, Texas Research Foundation, Renner, Texas

Suneson, Coit A., Crops Research Division, Agricultural Research Service, United States Department of Agriculture, and The University of California, Davis, California

Teas, Howard J., Puerto Rico Nuclear Center, University of Puerto Rico, College Station, Mayaguez, Puerto Rico

Timothy, David H., The Rockefeller Foundation, Apartado Aereo 58-13, Apartado Nacional 32-79, Bogota, Colombia

Uhl, C. H., Department of Botany, Cornell University, Ithaca, New York

Valentine, Frederick A., Department of Forest Botany and Pathology, State University College of Forestry, Syracuse, New York

Valentyne, J. R., Department of Zoology, Cornell University, Ithaca, New York

Wallace, A. T., Plant Science Department, University of Florida, Gainesville, Florida

Wallace, Donald H., Departments of Plant Breeding and Vegetable Crops, Cornell University, Ithaca, New York

Warmke, H. E., Federal Experiment Station in Puerto Rico, Territorial Experiment Stations Division, Agricultural Research Service, United States Department of Agriculture, Mayaguez, Puerto Rico

Weaver, G. M., Horticulture Section, Department of Agriculture, Research Branch, Harrow, Ontario, Canada

Wiebe, G. A., Crops Research Division, Agricultural Research Service, United States Department of Agriculture, Beltsville, Maryland

Wilcox, A. N., Department of Horticulture, University of Minnesota, St. Paul, Minnesota

Wolff, Sheldon, Biology Division, Oak Ridge National Laboratory, Oak Ridge, Tennessee

Wright, James E., Jr., Department of Botany and Plant Pathology, Pennsylvania State University, University Park, Pennsylvania

Yarnell, S. H., Crops Research Division, Agricultural Research Service, United States Department of Agriculture, United States Vegetable Breeding Laboratory, P. O. 3348, Charleston, South Carolina

Yegian, Hrant M., Department of Agronomy, University of Massachusetts, Amherst, Massachusetts

Young, Donald A., Department of Agriculture, Research Branch, Fredericton, New Brunswick, Canada

Zahler, S. A., Department of Bacteriology, Cornell University, Ithaca, New York

Zoebisch, Oscar, Libby McNeill and Libby, Blue Island, Illinois

THE NATIONAL ACADEMY OF SCIENCES—NATIONAL RESEARCH
COUNCIL is a private, non-profit organization of scientists,
dedicated to the furtherance of science and to its use for the general
welfare. The Academy itself was established in 1863 by the
terms of a Congressional charter under which it is empowered to
provide for all activities appropriate to academies of science
and is required to act as an advisor to the Federal Government
in scientific matters. The National Research Council was established
by the Academy in 1916, at the request of the President
of the United States, to enable scientists generally to associate
their efforts with those of the limited membership of the Academy.
With funds contributed from both public and private
sources, the Academy and its Research Council work to stimulate
research and its applications, to survey the broad possibilities of
science, to promote effective utilization of the scientific and technical
resources of the country, to serve the Government, and to
further the general interests of science.